Natural Computing Series

Series Editors: G. Rozenberg
Th. Bäck A.E. Eiben J.N. Kok H.P. Spaink
Leiden Center for Natural Computing

T0181154

Kenneth V. Price · Rainer M. Storn
Jouni A. Lampinen

Differential Evolution

A Practical Approach to Global Optimization

With 292 Figures, 48 Tables and CD-ROM

 Springer

Authors

Kenneth V. Price
Owl Circle 836
Vacaville, CA 95687
USA

Rainer M. Storn
Rohde & Schwarz GmbH & Co.KG
Mühldorfstraße 15
81671 München
Germany

Jouni A. Lampinen
Lappeenranta University of Technology
Department of Information Technology
P.O.Box 20
53851 Lappeenranta
Finland

Series Editors

G. Rozenberg (Managing Editor)
rozenber@liacs.nl

Th. Bäck, J.N. Kok, H.P. Spaink
Leiden Center for Natural Computing
Leiden University
Niels Bohrweg 1
2333 CA Leiden,
The Netherlands

A.E. Eiben
Vrije Universiteit Amsterdam

ACM Computing Classification (1998): F.1–2, G.1.6, I.2.6, I.2.8, J.6

Additional material to this book can be downloaded from http://extras.springer.com.

ISBN-10 3-642-42416-3 Springer Berlin Heidelberg New York

ISBN-13 978-3-642-42416-8 Springer Berlin Heidelberg New York

Springer is a part of Springer Science+Business Media

springer.com

© Springer-Verlag Berlin Heidelberg 2005
Softcover re-print of the Hardcover 1st edition 2005

Cover Design: KünkelLopka, Werbeagentur, Heidelberg
Typesetting: by the Authors
Production: LE-TₑX Jelonek, Schmidt & Vöckler GbR, Leipzig
Printed on acid-free paper   45/3142/YL – 5 4 3 2 1 0

KP: To my father

RS: To my ever-supportive parents, to my beloved wife, Marion, and to my wonderful children, Maja and Robin

JL: To the memory of my little dog and best friend Tonique, for all the happy countryside and city memories we shared

Preface

Optimization problems are ubiquitous in science and engineering. What shape gives an airfoil maximum lift? Which polynomial best fits the given data? Which configuration of lenses yields the sharpest image? Without question, very many researchers need a robust optimization algorithm for solving the problems that are fundamental to their daily work.

Ideally, solving a difficult optimization problem should not itself be difficult, e.g., a structural engineer with an expert knowledge of mechanical principles should not also have to be an expert in optimization theory just to improve his designs. In addition to being easy to use, a global optimization algorithm should also be powerful enough to reliably converge to the true optimum. Furthermore, the computer time spent searching for a solution should not be excessive. Thus, a genuinely useful global optimization method should be simple to implement, easy to use, reliable and fast.

Differential Evolution (DE) is such a method. Since its inception in 1995, DE has earned a reputation as a very effective global optimizer. While DE is not a panacea, its record of reliable and robust performance demands that it belongs in every scientist and engineer's "bag of tricks".

DE originated with the Genetic Annealing algorithm developed by Kenneth Price and published in the October 1994 issue of *Dr. Dobb's Journal* (DDJ), a popular programmer's magazine. Genetic Annealing is a population-based, combinatorial optimization algorithm that implements an annealing criterion *via* thresholds. After the Genetic Annealing algorithm appeared in DDJ, Ken was contacted by Dr. Rainer Storn, (then with Siemens while at the International Computer Science Institute at the University of California at Berkeley; now at Rohde & Schwarz GmbH, Munich, Germany) about the possibility of using Genetic Annealing to solve the Chebyshev polynomial fitting problem. Determining the coefficients of the Chebyshev polynomials is considered by many to be a difficult task for a general-purpose optimizer.

Ken eventually found the solution to the five-dimensional Chebyshev problem with the Genetic Annealing algorithm, but convergence was very slow and effective control parameters were hard to determine. After this initial find, Ken began modifying the Genetic Annealing algorithm to use floating-point instead of bit-string encoding and arithmetic operations instead of logical ones. He then discovered the differential mutation operator upon which DE is based. Taken together, these alterations effectively transformed what had been a combinatorial algorithm into the numerical optimizer that became the first iteration of DE. To better accommodate parallel machine architectures, Rainer suggested creating separate parent and child populations. Unlike Genetic Annealing, DE has no difficulty determining the coefficients of even the 33-dimensional Chebyshev polynomial.

DE proved effective not only on the Chebyshev polynomials, but also on many other test functions. In 1995, Rainer and Ken presented some early results in the ICSI technical report TR-95-012, "Differential Evolution – A Simple and Efficient Adaptive Scheme for Global Optimization over Continuous Spaces". These successes led Rainer and Ken to enter DE in the First International Contest on Evolutionary Optimization in Nagoya, Japan, that was held during May of 1996 in conjunction with the IEEE International Conference on Evolutionary Computation. DE finished third behind two methods that scored well on the contest functions, but which were not versatile enough to be considered general-purpose optimizers. The first-place method explicitly relied on the fact that the contest functions were separable, while the second-place algorithm was not able to handle a large number of parameters due to its dependence on Latin squares. Buoyed by this respectable showing, Ken and Rainer wrote an article on DE for DDJ that was published in April 1997 (Differential Evolution - A Simple Evolution Strategy for Fast Optimization). This article was very well received and introduced DE to a large international audience.

Many other researchers in optimization became aware of DE's potential after reading, "Differential Evolution – A Simple and Efficient Heuristic for Global Optimization over Continuous Spaces", by Rainer and Ken. Published in the December 1997 issue of *The Journal of Global Optimization*, this paper gave extensive empirical evidence of DE's robust performance on a wide variety of test functions. Also about this time, Rainer established a DE web site (http://www.icsi.berkeley.edu/~storn/code/html) to post code, links to DE applications and updates for the algorithm.

Ken entered DE in the Second International Contest on Evolutionary Optimization that was to be held in Indianapolis, Indiana, USA in April 1997. A lack of valid entries forced the cancellation of the actual contest, although those that qualified were presented. Of these, DE was the best performer. At this conference, Ken met Dr. David Corne who subsequently invited him to write an introduction to DE for the compendium, *New Ideas in Optimization* (1999). Since then, Ken has focused on refining the DE algorithm and on developing a theory to explain its performance. Rainer has concentrated on implementing DE on limited-resource devices and on creating software applications in a variety of programming languages. In addition, Rainer has explored DE's efficacy as a tool for digital filter design, design centering and combinatorial optimization.

Prof. Jouni Lampinen (Lappeenranta University of Technology, Lappeenranta, Finland) began investigating DE in 1998. In addition to contributing to the theory on DE and demonstrating DE's effectiveness as a tool for mechanical engineering, Jouni has also developed an exceptionally simple yet effective method for adapting DE to the particular demands of both constrained and multi-objective optimization. Jouni also maintains a DE bibliography (http://www.lut.fi/~jlampine/debiblio.html).

Like DE, this book is designed to be easy to understand and simple to use. It details how DE works, how to use it and when it is appropriate. Chapter 1, "The Motivation for DE", opens with a statement of the general optimization problem that is followed by a discussion of the strengths and weaknesses of the traditional methods upon which DE builds. Classical methods for optimizing differentiable functions along with conventional direct search methods like those of Hooke–Jeeves and Nelder–Mead are discussed. Chapter 1 concludes with a look at some of the more advanced optimization techniques, like simulated annealing and evolutionary algorithms.

Chapter 2, "The Differential Evolution Algorithm", introduces the DE algorithm itself, first in an overview and then in detail. Chapter 3, "Benchmarking DE", compares DE's performance to that reported for other EAs. Several versions of DE are included in the comparison. Chapter 4, "Problem Domains", extends the basic algorithm to cover a variety of optimization scenarios, including constrained, mixed-variable and multi-objective optimization as well as design centering. All these adaptations are of great practical importance, since many real-world problems belong to these domains.

Chapter 5, "Architectural Aspects", gives explicit advice on how to implement DE on both parallel and sequential machine architectures. In addition, Chapter 5 presents algorithms for auxiliary operations. Chapter 6, "Computer Code", provides instructions for using the software that accompanies this book on CD-ROM. Chapter 7, "Applications", presents a collection of 12 DE applications that have been contributed by experts from many disciplines. Applications include structure determination by X-ray analysis, earthquake relocation, multi-sensor fusion, digital filter design and many other very difficult optimization problems. An appendix contains descriptions of the test functions used throughout this book.

Dr. Storn would like to thank Siemens corporate research, especially Prof. Dr. H. Schwärtzel, Dr. Yeung-Cho Yp and Dr. Jean Schweitzer for supporting DE research. In addition, Prof. Lampinen would like to express his gratitude to members of his DE research group, Jani Rönkkönen, Junhong Liu and Saku Kukkonen, for their help preparing this book. We especially wish to thank the researchers who have contributed their DE applications to Chapter 7.

J.-P. Armspach, Institut de Physique Biologique, Université Louis Pasteur, Strasbourg, UMR CNRS-ULP 7004, Faculté de Médecine, F-67085, Strasbourg Cedex, France ; (Sect. 7.6)

Keith D. Bowen, Bede Scientific Incorporated, 14 Inverness Drive East, Suite H-100, Englewood, CO, USA; (Sect. 7.10)

Nirupam Chakraborti, Department of Metallurgical and Materials Engineering, Indian Institute of Technology, Kharagpur (W.B) 721 302, India; (Sect. 7.1)

David Corcoran, Department of Physics, University of Limerick, Ireland; (Sect. 7.2)

Robert W. Derksen, Department of Mechanical and Industrial Engineering University of Manitoba, Canada; (Sect. 7.3)

Drago Dolinar, University of Maribor, Faculty of Electrical Engineering and Computer Science, Smetanova 17, 2000 Maribor, Slovenia; (Sect. 7.9)

Steven Doyle, Department of Physics, University of Limerick, Ireland; (Sect. 7.2)

Kay Hameyer, Katholieke Universiteit Leuven, Department E.E. (ESAT), Division ELEN, Kaardinal Mercierlaan 94, B-3001 Leuven, Belgium; (Sect. 7.9)

Evan P. Hancox, Department of Mechanical and Industrial Engineering, University of Manitoba, Canada; (Sect. 7.3)

Fabrice Heitz, LSIIT-MIV, Université Louis Pasteur, Strasbourg, UMR CNRS-ULP 7005, Pôle API, Boulevard Sébastien Brant, F-67400 Illkirch, France ; (Sect. 7. 6)

Rajive Joshi, Real-Time Innovations Inc., 155A Moffett Park Dr, Sunnyvale, CA 94089, USA; (Sect. 7.4)

Michal Kvasnička, ERA a.s, Poděbradská 186/56, 180 66 Prague 9, Czech Republic; (Sect. 7.5)

Kevin M. Matney, Bede Scientific Incorporated, 14 Inverness Drive East, Suite H-100, Englewood, CO, USA; (Sect. 7.10)

Lars Nolle, School of Computing and Mathematics, The Nottingham Trent University, Burton Street, Nottingham, NG1 4BU, UK; (Sect. 7.12)

Guy-René Perrin, LSIIT-ICPS, Université Louis Pasteur, Strasbourg, UMR CNRS-ULP 7005, Pôle API, Boulevard Sébastien Brant, F-67400 Illkirch, France ; (Sect. 7. 6)

Bohuslav Růžek, Geophysical Institute, Academy of Sciences of the Czech Republic, Boční II/1401, 141 31 Prague 4, Czech Republic; (Sect. 7.5)

Michel Salomon, LSIIT-ICPS, Université Louis Pasteur, Strasbourg, UMR CNRS-ULP 7005, Pôle API, Boulevard Sébastien Brant, F-67400 Ill-kirch, France ; (Sect. 7. 6)

Arthur C. Sanderson, Rensselaer Polytechnic Institute, 110 8th St, Troy, NY 12180, USA; (Sect. 7.4)

Amin Shokrollahi, Laboratoire d'algorithmique Laboratoire de mathé-matiques algorithmiques, EPFL, I&C-SB, Building PSE-A, 1015 Lausanne, Switzerland; (Sect. 7.7)

Rainer M. Storn, Rohde & Schwarz GmbH & Co. KG, Mühldorfstr. 15, 81671 München, Germany; (Sects. 7.7 and 7.8)

Gorazd Štumberger, University of Maribor, Faculty of Electrical Engineer-ing and Computer Science, Smetanova 17, 2000 Maribor, Slovenia; (Sect. 7.9)

Matthew Wormington, Bede Scientific Incorporated, 14 Inverness Drive East, Suite H-100, Englewood, CO, USA; (Sect. 7.10)

Ivan Zelinka, Institute of Information Technologies, Faculty of Technol-ogy, Tomas Bata University, Mostni 5139, Zlin, Czech Republic; (Sects. 7.11 and 7.12)

We are also indebted to everyone who has contributed the public do-main code that has made DE so accessible. In particular, we wish to thank Eric Brasseur for making plot.h available to the public, Makoto Matsu-moto and Takuji Nishimura for allowing the Mersenne Twister random number generator to be freely used, Lester E. Godwin for writing the C++ version of DE, Feng-Sheng Wang for providing the Fortran90 version of DE, Walter Di Carlo for porting DE to Scilab®, Jim Van Zandt and Arnold Neumaier for helping with the MATLAB® version of DE and Ivan Zelinka and Daniel Lichtblau for providing the MATHEMATICA® version of DE.

A special debt of gratitude is owed to David Corne for his unflagging support and to A. E. Eiben and the editors of Springer-Verlag's Natural Computing Series for their interest in DE. In addition, we want to thank Ingeborg Meyer for her patience and professionalism in bringing our book to print. We are also indebted to Neville Hankins for his exquisitely de-tailed copyediting and to both Ronan Nugent and Ulrike Stricker at Springer-Verlag for helping to resolve the technical issues that arose dur-ing the preparation of this manuscript.

Additionally, this book would not be possible were it not for the many engineers and scientists who have helped DE become so widespread. Although they are too numerous to mention, we wish to thank them all.

Finally, it would have been impossible to write this book without our families' understanding and support, so we especially want to thank them for their forbearance and sacrifice.

Kenneth V. Price
Rainer M. Storn
Jouni A. Lampinen

Table of Contents

1 The Motivation for Differential Evolution

1.1 Introduction to Parameter Optimization

1.1.1 Overview

In simple terms, optimization is the attempt to maximize a system's desirable properties while simultaneously minimizing its undesirable characteristics. What these properties are and how effectively they can be improved depends on the problem at hand. Tuning a radio, for example, is an attempt to minimize the distortion in a radio station's signal. Mathematically, the property to be minimized, distortion, can be defined as a function of the tuning knob angle, x:

$$f(x) = \frac{\text{noise power}}{\text{signal power}}. \tag{1.1}$$

Because their most extreme value represents the optimization goal, functions like Eq. 1.1 are called *objective functions*. When its minimum is sought, the objective function is often referred to as a *cost function*. In the special case where the minimum being sought is zero, the objective function is sometimes known as an *error function*. By contrast, functions that describe properties to be maximized are commonly referred to as *fitness functions*. Since changing the sign of an objective function transforms its maxima into minima, there is no generality lost by restricting the following discussion to function minimization only.

Tuning a radio involves a single variable, but properties of more complex systems typically depend on more than one variable. In general, the objective function, $f(\mathbf{x}) = f(x_0, x_1, \ldots, x_{D-1})$, has D parameters that influence the property being optimized. There is no unique way to classify objective functions, but some of the objective function attributes that affect an optimizer's performance are:

- *Parameter quantization.* Are the objective function's variables continuous, discrete, or do they belong to a finite set? Additionally, are all variables of the same type?

- *Parameter dependence.* Can the objective function's parameters be optimized independently (separable function), or does the minimum of one or more parameters depend on the value of one or more other parameters (parameter dependent function)?
- *Dimensionality, D.* How many variables define the objective function?
- *Modality.* Does the objective function have just one local minimum (uni-modal) or more than one (multi-modal)?
- *Time dependency.* Is the location of optimum stationary (e.g., static), or non-stationary (dynamic)?
- *Noise.* Does evaluating the same vector give the same result every time (no noise), or does it fluctuate (noisy)?
- *Constraints.* Is the function unconstrained, or is it subject to additional equality and/or inequality constraints?
- *Differentiability.* Is the objective function differentiable at all points of interest?

In the radio example, the tuning angle is real-valued and parameters are continuous. Neither mixed-variable types, nor parameter dependence is an issue because the objective function's dimension is one, i.e., it depends on a single parameter. The objective function's modality, however, depends on how the tuning knob angle is constrained. If tuning is restricted to the vicinity of a single radio station, then the objective function is *uni-modal* because it exhibits just one (local) optimum. If, however, the tuning knob scans a wider radio band, then there will probably be several stations. If the goal is to find the station with least distortion, then the problem becomes *multi-modal*. If the radio station frequency does not drift, then the objective function is not time dependent, i.e., the knob position that yields the best reception will be the same no matter when the radio is turned on. In the real world, the objective function itself will have some added noise, but the knob angle will not be noisy unless the radio is placed on some vibrating device like a washing machine. The objective function has no obvious constraints, but the knob-angle parameter is certainly restricted.

Even though distortion's definition (Eq. 1.1) provides a mathematical description of the property being minimized, there is no computable objective function – short of simulating the radio's circuits – to determine the distortion for a given knob angle. The only way to estimate the distortion at a given frequency is to tune in to it and listen. Instead of a well-defined, computable objective function, the radio itself is the "black box" that transforms the input (knob angle) into output (station signal). Without an adequate computer simulation (or a sufficiently refined actuator), the objective function in the radio example is effectively non-differentiable.

Tuning a radio is a trivial exercise primarily because it involves a single parameter. Most real-world problems are characterized by partially non-differentiable, nonlinear, multi-modal objective functions, defined with both continuous and discrete parameters and upon which additional constraints have been placed. Below are three examples of challenging, real-world engineering problems of the type that DE was designed to solve. Chapter 7 explores a wide range of applications in detail.

Optimization of Radial Active Magnetic Bearings

The goal of this electrical/mechanical engineering task is to maximize the bearing force of a radial active magnetic bearing while simultaneously minimizing its mass (Štumberger et al. 2000). As Fig. 1.1 shows, several constraints must be taken into account.

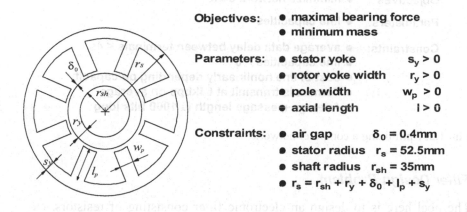

Objectives:	● maximal bearing force	
	● minimum mass	
Parameters:	● stator yoke	$s_y > 0$
	● rotor yoke width	$r_y > 0$
	● pole width	$w_p > 0$
	● axial length	$l > 0$
Constraints:	● air gap	$\delta_0 = 0.4mm$
	● stator radius	$r_s = 52.5mm$
	● shaft radius	$r_{sh} = 35mm$
	● $r_s = r_{sh} + r_y + \delta_0 + l_p + s_y$	

Fig. 1.1. Optimizing a radial active magnetic bearing

Capacity Assignment Problem

Figure 1.2 shows a computer network that connects terminals to concentrators, which in turn connect to a large mainframe computer. The cost of a line depends nonlinearly on the capacity. The goal is to satisfy the data delay constraint of 4 ms while minimizing the cost of the network. A more detailed discussion appears in Schwartz (1977).

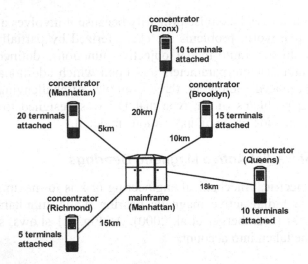

Objectives: ● **minimize network cost**

Parameters: ● **line capacities**

Constraints: ● **average data delay between terminals < 4s**
                   ● **line capacities > 0**
                   ● **cost of line nonlinearly depending on capacity**
                   ● **Terminals transmit at 64kbps on average**
                   ● **Average message length is 1000 bits long**

Fig. 1.2. Optimizing a computer network

Filter Design Problem

The goal here is to design an electronic filter consisting of resistors, capacitors and an operational amplifier so that the magnitude of the ratio of output to input voltages, $|V_2(\omega)/V_1(\omega)|$ (a function of frequency ω), satisfies the tolerance scheme depicted in the lower half of Fig. 1.3.

Classifying Optimizers

Once a task has been transformed into an objective function minimization problem, the next step is to choose an appropriate optimizer. Table 1.1 classifies optimizers based, in part, on the number of points (vectors) that they track through the D-dimensional problem space. This classification does not distinguish between multi-point optimizers that operate on many points in parallel and multi-start algorithms that visit many points in sequence. The second criterion in Table 1.1 classifies algorithms by their reliance on objective function derivatives.

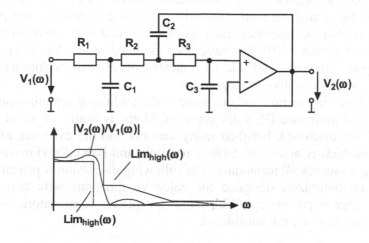

Fig. 1.3. Optimizing an electronic filter

Objectives: • Fit $|V_2(\omega)/V_1(\omega)|$ between $Lim_{high}(\omega)$ and $Lim_{low}(\omega)$

Parameters: • Resistors R_i, Capacitors C_i

Constraints: • $0 < C_i < C_{max}$
• $0 < R_i < R_{max}$
• R_i, C_i from E24 norm series

(discrete set)

Table 1.1. A classification of optimization approaches and some of their representatives

	Single-point	Multi-point
Derivative-based	Steepest descent Conjugate gradient Quasi-Newton	Multi-start and clustering techniques
Derivative-free (direct search)	Random walk Hooke–Jeeves	Nelder–Mead Evolutionary algorithms Differential evolution

Not all optimizers neatly fit into these categories. Simulated annealing (Kirkpartick et al. 1983; Press et al. 1992) does not appear in this classification scheme because it is a meta-strategy that can be applied to any derivative-free search method. Similarly, clustering techniques are general strategies, but because they are usually combined with derivative-based optimizers (Janka 1999) they have been assigned to the derivative-based, multi-point category. As Table 1.1 indicates, differential evolution (DE) is a multi-point, derivative-free optimizer.

The following section outlines some of the traditional optimization algorithms that motivated DE's development. Methods from each class in Table 1.1 are discussed, but their many variants and the existence of other novel methods (Corne et al. 1999; Onwubolu and Babu 2004) make it impossible to survey all techniques. The following discussion is primarily focused on optimizers designed for objective functions with continuous and/or discrete parameters. With a few exceptions, combinatorial optimization problems are not considered.

1.1.2 Single-Point, Derivative-Based Optimization

Derivative-based methods embody the classical approach to optimization. Before elaborating, a few details on notation are in order. First, a D-dimensional parameter vector is defined as:

$$\mathbf{x} = \begin{pmatrix} x_0 \\ x_1 \\ ... \\ x_{D-1} \end{pmatrix} = \begin{pmatrix} x_0 & x_1 & ... & x_{D-1} \end{pmatrix}^T . \tag{1.2}$$

Letters in lowercase italic symbolize individual parameters; bold lowercase letters denote vectors, while bold uppercase letters represent matrices. Introducing several special operator symbols further simplifies formulation of the classical approach. For example, the *nabla* operator is defined as

$$\nabla = \begin{pmatrix} \partial / \partial x_0 \\ \partial / \partial x_1 \\ ... \\ \partial / \partial x_{D-1} \end{pmatrix} \tag{1.3}$$

in order to simplify the expression for the *gradient vector*:

$$g(\mathbf{x}) = \nabla \cdot f(\mathbf{x}) = \begin{pmatrix} \dfrac{\partial f(\mathbf{x})}{\partial x_0} \\[2mm] \dfrac{\partial f(\mathbf{x})}{\partial x_1} \\[2mm] \vdots \\[2mm] \dfrac{\partial f(\mathbf{x})}{\partial x_{D-1}} \end{pmatrix}. \tag{1.4}$$

It is also convenient to define the *Hessian matrix*:

$$\mathbf{G}(\mathbf{x}) = \nabla^2 \cdot f(\mathbf{x}) = \begin{pmatrix} \partial f / \partial x_0 \partial x_0 & \partial f / \partial x_0 \partial x_1 & \dots & \partial f / \partial x_0 \partial x_{D-1} \\ \partial f / \partial x_1 \partial x_0 & & & \\ \dots & & & \\ \partial f / \partial x_{D-1} \partial x_0 & \dots & \dots & \partial f / \partial x_{D-1} \partial x_{D-1} \end{pmatrix}. \tag{1.5}$$

The symbol ∇^2 is meant to imply second-order (partial) differentiation, not that the nabla operator, ∇, is squared.

Using these notational conveniences, the Taylor series for an arbitrary objective function becomes

$$f(\mathbf{x}) = f(\mathbf{x}_0) + \frac{\nabla f(\mathbf{x}_0)}{1!} \cdot (\mathbf{x} - \mathbf{x}_0) + (\mathbf{x} - \mathbf{x}_0)^T \cdot \frac{\nabla^2 f(\mathbf{x}_0)}{2!} \cdot (\mathbf{x} - \mathbf{x}_0) + \dots \tag{1.6}$$

$$= f(\mathbf{x}_0) + g(\mathbf{x}_0) \cdot (\mathbf{x} - \mathbf{x}_0) + (\mathbf{x} - \mathbf{x}_0)^T \cdot \frac{1}{2} \mathbf{G}(\mathbf{x}_0) \cdot (\mathbf{x} - \mathbf{x}_0) + \dots,$$

where $\mathbf{x}_0$ is the point around which the function $f(\mathbf{x})$ is developed. For a point to be a minimum, elementary calculus (Rade and Westergren 1990) demands that

$$g(\mathbf{x}_{\text{extr}}) = \mathbf{0}, \tag{1.7}$$

i.e., *all* partial derivatives at $\mathbf{x} = \mathbf{x}_{\text{extr}}$ must be zero. In the third term on the right-hand side of Eq. 1.6, the difference between $\mathbf{x}$ and $\mathbf{x}_0$ is squared, so in order to avoid a negative contribution from the Hessian matrix, $\mathbf{G}(\mathbf{x}_0)$ must be positive semi-definite (Scales 1985). In the immediate neighborhood about $\mathbf{x}_0$, higher terms of the Taylor series expansion make a negligible contribution and need not be considered.

Applying the chain rule for differentiation to the first three terms of the Taylor expansion in Eq. 1.6 allows the gradient about the arbitrary point $\mathbf{x}_0$ to be expressed as

$$\nabla f(\mathbf{x}_{\text{extr}}) = \mathbf{g}(\mathbf{x}_0) + \mathbf{G}(\mathbf{x}_0) \cdot (\mathbf{x}_{\text{extr}} - \mathbf{x}_0) = \mathbf{0}, \tag{1.8}$$

which reduces to

$$\mathbf{x}_{\text{extr}} = -\mathbf{g}(\mathbf{x}_0) \cdot \mathbf{G}^{-1}(\mathbf{x}_0) + \mathbf{x}_0. \tag{1.9}$$

where $\mathbf{G}^{-1}$ is the inverse of the Hessian matrix.

If the objective function, $f(\mathbf{x})$, is quadratic, then Eq. 1.9 can be applied directly to obtain its true minimum. Figure 1.4 shows how Eq. 1.9 computes the optimum of a (uni-modal) quadratic function independent of where the starting point, $\mathbf{x}_0$, is located.

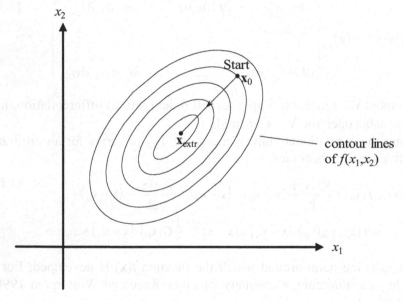

Fig. 1.4. If the objective function is quadratic and differentiable, then Eq. 1.9 can determine its optimum.

Even though there are applications, e.g., acoustical echo cancellation in speakerphones, where the objective function is a simple quadratic (Glentis et al. 1999), the majority of optimization tasks lack this favorable property. Even so, classical derivative-based optimization can be effective as long the objective function fulfills two requirements:

R1   The objective function must be *two-times differentiable*.

R2   The objective function must be *uni-modal*, i.e., have a single minimum.

A simple example of a differentiable and uni-modal objective function is

$$f(x_1, x_2) = 10 - e^{-\left(x_1^2 + 3x_2^2\right)}. \tag{1.10}$$

Figure 1.5 graphs the function defined in Eq. 1.10.

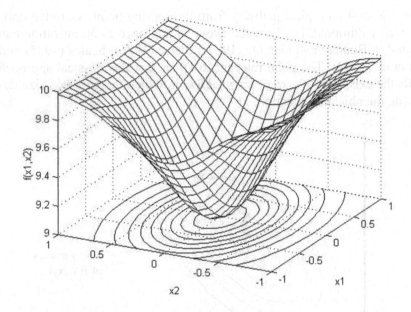

Fig. 1.5. An example of a uni-modal objective function

The method of *steepest descent* is one of the simplest gradient-based techniques for finding the minimum of a uni-modal and differentiable function. Based on Eq. 1.9, this approach assumes that $\mathbf{G}^{-1}(\mathbf{x}_0)$ can be replaced with the identity matrix:

$$\mathbf{I} = \begin{pmatrix} 1 & 0 & \dots & 0 \\ 0 & 1 & \dots & 0 \\ \dots & \dots & \dots & \dots \\ 0 & 0 & \dots & 1 \end{pmatrix}. \tag{1.11}$$

This crude replacement does not lead directly to the minimum, but to the point

$$\mathbf{x}_1 = \mathbf{x}_0 - \mathbf{g}(\mathbf{x}_0). \tag{1.12}$$

Since the negative gradient points downhill, $\mathbf{x}_1$ will be closer to the minimum than $\mathbf{x}_0$ unless the step was too large. Adding a step size, γ, to the general recursion relation that defines the direction of steepest descent provides a measure of control:

$$\mathbf{x}_{n+1} = \mathbf{x}_n - \gamma \cdot \mathbf{g}(\mathbf{x}_n) \qquad (1.13)$$

Figure 1.6 shows a typical pathway from the starting point, $\mathbf{x}_0$, to the optimum $\mathbf{x}_{\text{extr}}$. Additional details of the classical approach to optimization can be found in Bunday and Garside (1987), Pierre (1986), Scales (1985) and Press et al. (1992). The point relevant to DE is that the classical approach reveals the existence of a *step size problem* in which the best step size depends on the objective function.

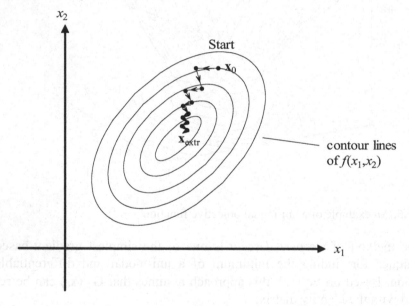

Fig. 1.6. The method of steepest descent first computes the negative gradient, then takes a step in the direction indicated.

Replacing the inverse Hessian, $\mathbf{G}^{-1}(\mathbf{x}_0)$, with the identity matrix introduces its own set of problems and more elaborate techniques like Gauss–Newton, Fletcher–Reeves, Davidon–Fletcher–Powell, Broyden–Fletcher–Goldfarb–Shanno and Levenberg–Marquardt (Scales 1985; Pierre 1986) have been developed in response. These methods roughly fall into two categories. *Quasi-Newton methods* approximate the inverse Hessian by a variety of schemes, most of which require extensive matrix computations.

By contrast, *conjugate gradient* methods dispense with the Hessian matrix altogether, opting instead to use line optimizations in conjugate directions to avoid computing second-order derivatives. In addition to Quasi-Newton and conjugate gradient methods, mixtures of the two approaches also exist. Even so, all these methods require the objective function to be one-time or two-times differentiable. In addition, their fast convergence on quadratic objective functions does not necessarily transfer to non-quadratic functions. Numerical errors are also an issue if the objective function exhibits singularities or large gradients. Methods that do not require the objective function to be differentiable provide greater flexibility.

1.1.3 One-Point, Derivative-Free Optimization and the Step Size Problem

There are many reasons why an objective function might not be differentiable. For example, the "floor" operation in Eq. 1.14 quantizes the function in Eq. 1.10, transforming Fig. 1.5 into the stepped shape seen in Fig. 1.7. At each step's edge, the objective function is non-differentiable:

$$f(x_1, x_2) = \text{floor}\left(10 \cdot \left(10 - \exp\left(-x_1^2 - 3x_2^2\right)\right)\right)/10 \qquad (1.14)$$

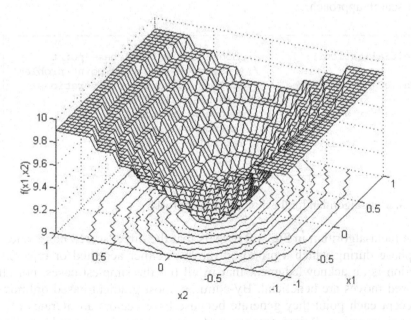

Fig. 1.7. A non-differentiable, quantized, uni-modal function

There are other reasons in addition to function quantization why an objective function might not be differentiable:

- Constraining the objective function may create regions that are non-differentiable or even forbidden altogether.
- If the objective function is a computer program, conditional branches make it non-differentiable, at least for certain points or regions.
- Sometimes the objective function is the result of a physical experiment (Rechenberg 1973) and the unavailability of a sufficiently precise actuator can make computing derivatives impractical.
- If, as is the case in evolutionary art (Bentley and Corne 2002), the objective function is "subjective", an analytic formula is not possible.
- In co-evolutionary environments, individuals are evaluated by how effectively they compete against other individuals. The objective function is not explicit.

When the lack of a computable derivative causes gradient-based optimizers to fail, reliance on derivative-free techniques known as *direct search* algorithms becomes essential. Direct search methods are "generate-and-test" algorithms that rely less on calculus than they do on heuristics and conditional branches. The meta-algorithm in Fig. 1.8 summarizes the direct search approach.

```
Initialization();        //choose the initial base point
                         //(introduces starting-point problem)
while (not converged)    //decide the number of iterations
{                        //(dimensionality problem)
    vector_generation(); //choose a new point
                         //(introduces step size problem)
    selection();         //determine new base point
}
```

Fig. 1.8. Meta-algorithm for the direct search approach

The meta-algorithm in Fig. 1.8 reveals that the direct search has a selection phase during which a proposed move is either accepted or rejected. Selection is an acknowledgment that in all but the simplest cases, not all proposed moves are beneficial. By contrast, most gradient-based optimizers accept each point they generate because base vectors are iterates of a recursive equation. Points are rejected only when, for example, a line search concludes. For direct search methods, however, selection is a central component that can affect the algorithm's next action.

Enumeration or Brute Force Search

As their name implies, one-point, direct search methods are initialized with a single starting point. Perhaps the simplest one-point direct search is the *brute force method*. Also known as *enumeration*, the brute force method visits all grid points in a bounded region while storing the current best point in memory (see Fig. 1.9). Even though generating a sequence of grid points is trivial, the enumerative method still faces a step size problem because if nothing is known about the objective function, it is hard to decide how fine the grid should be. If the grid is too coarse, then the optimum may be missed. If the grid becomes too small, computing time explodes exponentially because a grid with N points in one dimension will have N^D points in D dimensions. Because of this "curse of dimensionality", the brute force method is very rarely used to optimize objective functions with a significant number of continuous parameters. The curse of dimensionality demonstrates that better sampling strategies are needed to keep a search productive.

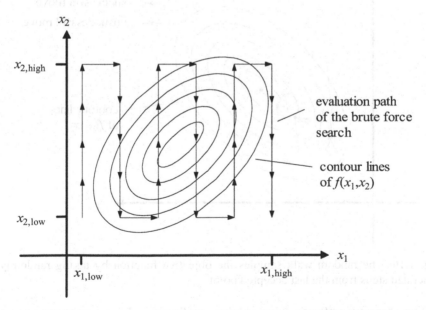

Fig. 1.9. The brute force search tries all grid points in a given region.

Random Walk

The *random walk* (Gross and Harris 1985) circumvents the curse of dimensionality inherent in the brute force method by sampling the objective function value at randomly generated points. New points are generated by adding a random deviation, $\Delta\mathbf{x}$, to a given base point, $\mathbf{x}_0$. In general, each coordinate, Δx_i, of the random deviation follows a Gaussian distribution

$$p(\Delta x_i) = \frac{1}{\sigma_i \cdot \sqrt{2\pi}} \exp\left(-0.5 \cdot \frac{(\Delta x_i - \mu_i)^2}{\sigma_i^2}\right), \qquad (1.15)$$

where σ_i and μ_i are the standard deviation and the mean value, respectively, for coordinate i. The random walk's selection criterion is "greedy" in the sense that a trial point with a lower objective function value than that of the base point is always accepted. In other words, if $f(\mathbf{x}_0 + \Delta\mathbf{x}) \leq f(\mathbf{x}_0)$, then $\mathbf{x}_0 + \Delta\mathbf{x}$ becomes the new base point; otherwise the old point, $\mathbf{x}_0$, is retained and a new deviation is applied to it. Figure 1.10 illustrates how the random walk operates.

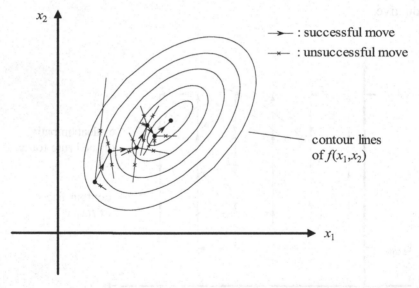

Fig. 1.10. The random walk samples the objective function by taking randomly generated steps from the last accepted point.

The stopping criterion for a random walk might be a preset maximum number of iterations or some other problem-dependent criterion. With luck, a random walk will find the minimum quicker than can be done with a brute force search. Like both the classical and the brute force methods, the random walk suffers from the step size problem because it is very difficult to choose the right standard deviations when the objective function is not sufficiently well known.

Hooke and Jeeves

The Hooke–Jeeves method is a one-point direct search that attacks the step size problem (Hooke and Jeeves 1961; Pierre 1986; Bunday and Garside 1987; Schwefel 1994). Also known as a *direction* or *pattern search*, the Hooke–Jeeves algorithm starts from an initial base point, x_0, and explores each coordinate axis with its own step size. Trial points in all D positive and negative coordinate directions are compared and the best point, x_1, is found. If the best new trial point is better than the base point, then an attempt is made to make another move in the same direction, since the step from x_0 to x_1 was a good one. If, however, none of the trial points improve on x_0, the step is presumed to have been too large, so the procedure repeats with smaller step sizes. The pseudo-code in Fig. 1.11 summarizes the Hooke–Jeeves method. Figure 1.12 plots the resulting search path.

```
...
while (h > h_min)   //as long as step length is still not small enough
{
    x_1 = explore(x_0,h); //explore the parameter space
    if (f(x_1) < f(x_0))  //if improvement could be made
    {
        x_2 = x_1 + (x_1 - x_0);   //make differential pattern move
        if (f(x_2) < f(x_1)) x_0 = x_2;
        else                 x_0 = x_1;
    }
    else h = h*reduction_factor;
}
...

function explore(vector x_0, vector h)
{ //---note that e_i is the unit vector for coordinate i---
    for (i=0; i<D; i++) //for all D dimensions
    {
        if (f(x_0+e_i*h) < f(x_0)) x_0 = x_0 + e_i*h; //check coordinate i
        else if (f(x_0-e_i*h) < f(x_0)) x_0 = x_0 - e_i*h;
    }
    return(x_0);
}
```

Fig. 1.11. Pseudo-code for the Hooke–Jeeves method

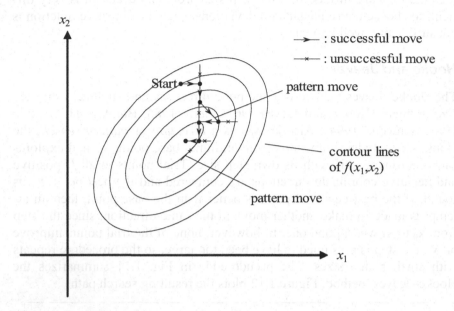

Fig. 1.12. A search guided by the Hooke–Jeeves method. Positive axis directions are always tried first.

On many functions, its adaptive step sizes make the Hooke–Jeeves search much more effective than either the brute force or random walk algorithms, but step sizes that shrink and never increase can be a drawback. For example, if steps are forced to become small because the objective function contains a "valley", then they will be unable to expand to the appropriate magnitude once the valley ends.

1.2 Local Versus Global Optimization

Both the step size problem and objective function non-differentiability can make even uni-modal functions a challenge to optimize. Additional obstacles arise once requirement R2 is dropped and the objective function is allowed to be multi-modal. Equation 1.16 is an example of a multi-modal function. As Fig. 1.13 shows, the "peaks" function in Eq. 1.16 has more than one local minimum:

$$f(x_1, x_2) = 3(1 - x_1)^2 \cdot \exp\left(x_1^2 + (x_2 + 1)^2\right) - 10\left(\frac{x_1}{5} - x_1^3 - x_2^5\right) \cdot \tag{1.16}$$

$$\cdot \exp\left(x_1^2 + x_2^2\right) - \frac{1}{3} \cdot \exp\left((x_1 + 1)^2 + x_2^2\right)$$

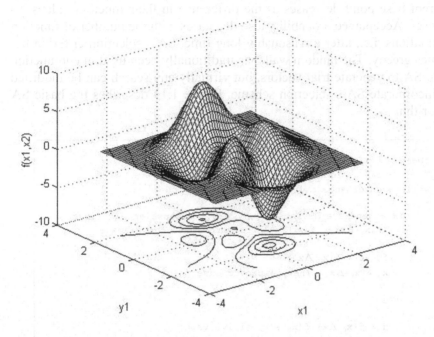

Fig. 1.13. The "peaks" function defined by Eq. 1.16 is multi-modal.

Because they exhibit more than one local minimum, multi-modal functions pose a *starting point problem*. Mentioned briefly in the direct search meta-algorithm (Fig. 1.8), the starting point problem refers to the tendency of an optimizer with a greedy selection criterion to find only the minimum of the basin of attraction in which it was initialized. This minimum need not be the global one, so sampling a multi-modal function in the vicinity of the global optimum, at least eventually, is essential. Because the Gaussian distribution is unbounded, there is a finite probability that the random walk will eventually generate a new and better point in a basin of attraction other than the one containing the current base point. In practice, successful inter-basin jumps tend to be rare. One method that increases the chance that a point will travel to another basin of attraction is simulated annealing.

1.2.1 Simulated Annealing

Simulated annealing (SA) (Kirkpatrick et al. 1983; Press et al. 1992), thoroughly samples the objective function surface by modifying the greedy criterion to accept some uphill moves while continuing to accept all downhill moves. The probability of accepting a trial vector that lies uphill from the current base point decreases as the difference in their function values increases. Acceptance probability also decreases with the number of function evaluations, i.e., after a reasonably long time, SA's selection criterion becomes greedy. The random walk has traditionally been used in conjunction with SA to generate trial vectors, but virtually any search can be modified to incorporate SA's selection scheme. Figure 1.14 describes the basic SA algorithm.

```
. . .
fbest = f(x₀);//start with some base point
T     = T₀;    //and some starting temperature
while (convergence criterion not yet met)
{
    Δx = generate_deviation(); //e.g., a Gaussian distribution
    if (f(x₀+Δx) < f(x₀))   //if improvement can be made
    {
        fbest = f(x₀+Δx);
        x₀ = x₀+Δx; //new, improved base point
    }
    else
    {
        d = f(x₀+Δx)-f(x₀);//positive value
        r = rand(); //generate uniformly distr. variable ex [0,1]
        if (r < exp(-d*beta/T)) //Metropolis algorithm
        {
            x₀ = x₀+Δx; //new base point derived from uphill move
        }
    }
    T = T*reduction_factor;
}
. . .
```

Fig. 1.14. The basic simulated annealing algorithm. In this implementation, the random walk generates trial points.

The term "annealing" refers to the process of slowly cooling a molten substance so that its atoms will have the opportunity to coalesce into a minimum energy configuration. If the substance is kept near equilibrium at temperature T, then atomic energies, E, are distributed according to the Boltzmann equation

$$P(E) \sim \exp\left(\frac{E}{k \cdot T}\right), \qquad (1.17)$$

where k is the Boltzmann constant.

By equating energy with function value, SA attempts to exploit nature's own minimization process *via* the Metropolis algorithm (Metropolis et al. 1953). The Metropolis algorithm implements the Boltzmann equation as a selection probability. While downhill moves are always accepted, uphill moves are accepted only if a uniformly distributed random number from the interval [0,1] is smaller than the exponential term:

$$\Theta = \exp\left(-\frac{d \cdot \beta}{T}\right). \qquad (1.18)$$

The variable, d, is the difference between the uphill objective function value and the function value of the current base point, i.e., their "energy difference". Equation 1.18 shows that the acceptance probability, Θ, decreases as d increases and/or as T decreases. The value, β, is a problem-dependent control variable that must be empirically determined.

One of annealing's drawbacks is that special effort may be required to find an *annealing schedule* that lowers T at the right rate. If T is reduced too quickly, the algorithm will behave like a local optimizer and become trapped in the basin of attraction in which it began. If T is not lowered quickly enough, computations become too time consuming. There have been many improvements to the standard SA algorithm (Ingber 1993) and SA has been used in place of the greedy criterion in direct search algorithms like the method of Nelder–Mead (Press et al. 1992). The step size problem remains, however, and this may be why SA is seldom used for continuous function optimization. By contrast, SA's applicability to virtually any direct search method has made it very popular for combinatorial optimization, a domain where clever, but greedy, heuristics abound (Syslo et al. 1983; Reeves 1993).

1.2.2 Multi-Point, Derivative-Based Methods

Multi-start techniques are another way to extensively sample an objective function landscape. As their name implies, multi-start techniques restart the optimization process from different initial points. Typically, each sample point serves as the initial point for a greedy, local optimization method (Boender and Romeijn 1995). Often, the local search is derivative-based, but this is not mandatory and if the objective function is non-differentiable, any direct search method may be used. Without detailed knowledge of the objective function, it is difficult to know how many different starting points will be enough, especially since many points might lead to the same local minimum because they all initially fell within the perimeter of the same basin of attraction.

Clustering methods (Törn and Zelinkas 1989; Janka 1999) refine the multi-start method by applying a clustering algorithm to identify those sample points that belong to the same basin of attraction, i.e., to the same cluster. Ideally, each cluster yields just one point to serve as the base point for a local optimization routine. *Density clustering* (Boender and Romeijn 1995; Janka 1999) is based on the assumption that clusters are shaped like hyper-ellipsoids and that the objective function is quadratic in the neighborhood of a minimum. Other methods, like the one described in Locatelli and Schoen (1996), use a proximity criterion to decide if a local search is justified. Because this determination often requires that all previously visited points be stored, highly multi-modal functions of high dimension can strain computer memory capacity. As a result, clustering algorithms are typically limited to problems with a relatively small number of parameters.

1.2.3 Multi-Point, Derivative-Free Methods

Evolution Strategies and Genetic Algorithms

Evolution strategies (ESs) were developed by Rechenberg (1973) and Schwefel (1994), while genetic algorithms (GAs) are attributed to Holland (1962) and Goldberg (1989). Both approaches attempt to evolve better solutions through recombination, mutation and survival of the fittest. Because they mimic Darwinian evolution, ESs, GAs, DE and their ilk are often collectively referred to as *evolutionary algorithms*, or EAs. Distinctions, however, do exist. An ES, for example, is an effective continuous function optimizer, in part because it encodes parameters as floating-point numbers and manipulates them with arithmetic operators. By contrast, GAs are often better suited for combinatorial optimization because they encode parameters as bit strings and modify them with logical operators. Modifying a GA to use floating-point formats for continuous parameter optimization typically transforms it into an ES-type algorithm (Mühlenbein and Schlierkamp-Vosen 1993; Salomon 1996). There are many variants to both approaches (Bäck 1996; Michalewicz 1996), but because DE is primarily a numerical optimizer, the following discussion is limited to ESs.

Like a multi-start algorithm, an ES samples the objective function landscape at many different points, but unlike the multi-start approach where each base point evolves in isolation, points in an ES population influence one another by means of recombination. Beginning with a population of μ parent vectors, the ES creates a child population of $\lambda \geq \mu$ vectors by recombining randomly chosen parent vectors. Recombination can be *discrete* (some parameters are from one parent, some are from the other parent) or *intermediate* (e.g., averaging the parameters of both parents) (Bäck et al. 1997; Bäck 1996). Once parents have been recombined, each of their children is "mutated" by the addition of a random deviation, Δx, that is typically a zero mean Gaussian distributed random variable (Eq. 1.15).

After mutating and evaluating all λ children, the (μ, λ)-ES selects the best μ children to become the next generation's parents. Alternatively, the $(\mu + \lambda)$-ES populates the next generation with the best μ vectors from the combined parent and child populations. In both cases, selection is greedy within the prescribed selection pool, but this is not a major drawback because the vector population is distributed. Figure 1.15 summarizes the meta-algorithm for an ES.

```
Initialization();       //choose starting population of μ members
while (not converged)   //decide the number of iterations
{
    for (i=0; i<λ; i++) //child vector generation: λ > μ
    {
        p₁(i) = rand(μ);    //pick a random parent from μ parents
        p₂(i) = rand(μ);    //pick another random parent  p₂(i) != p₁(i)
        c₁(i) = recombine(p₁(i),p₂(i)); //recombine parents
        c₁(i) = mutate(c₁(i));          //mutate child
        save(c₁(i));            //save child in an intermediate population
    }
    selection();        //μ new parents out of either λ, or λ+μ
}
```

Fig. 1.15. Meta-algorithm for evolution strategies (ESs)

While ESs are among the best global optimizers, their simplest implementations still do not solve the step size problem. Schwefel addressed this issue in Schwefel (1981) where he proposed modifying the Gaussian mutation distribution with a matrix of adaptive covariances, an idea that Rechenberg suggested in 1967 (Fogel 1994). Equation 1.19 generalizes the multi-dimensional Gaussian distribution to include a covariance matrix, C (Papoulis 1965):

$$p(\Delta \mathbf{x}) = \frac{1}{\det(\mathbf{C}) \cdot \sqrt{(2\pi)^D}} \exp\left(-0.5(\Delta \mathbf{x} - \mu)^T \cdot \mathbf{C}^{-1} \cdot (\Delta \mathbf{x} - \mu)\right) \qquad (1.19)$$

In Eq. 1.19, μ is the *mean vector* and $\mathbf{C}$ is the *covariance matrix*

$$\mathbf{C} = \begin{pmatrix} \sigma_{0,0}^2 & \sigma_{0,1} & \cdots & \sigma_{0,D-1} \\ \sigma_{1,0} & \sigma_{1,1}^2 & \cdots & \sigma_{1,D-1} \\ \cdots & \cdots & \cdots & \cdots \\ \sigma_{D-1,0} & \sigma_{D-1,1} & \cdots & \sigma_{D-1,D-1}^2 \end{pmatrix} = \mathbf{S} \cdot \mathbf{R} \cdot \mathbf{S} \qquad (1.20)$$

with the *scatter matrix*

$$\mathbf{S} = \begin{pmatrix} \sigma_0^2 & 0 & \cdots & 0 \\ 0 & \sigma_1^2 & \cdots & 0 \\ \cdots & \cdots & \cdots & \cdots \\ 0 & 0 & \cdots & \sigma_{D-1}^2 \end{pmatrix} \qquad (1.21)$$

and the *correlation matrix*

$$\mathbf{R} = \begin{pmatrix} 1 & \rho_{0,1} & \cdots & \rho_{0,D-1} \\ \rho_{1,0} & 1 & \cdots & \rho_{1,D-1} \\ \cdots & \cdots & \cdots & \cdots \\ \rho_{D-1,0} & \rho_{D-1,1} & \cdots & 1 \end{pmatrix} \qquad (1.22)$$

By permitting the otherwise symmetrical Gaussian distribution to become ellipsoidal, the ES can assign a different step size to each dimension. In addition, the covariance matrix allows the Gaussian mutation ellipsoid to rotate in order to adapt better to the topography of non-decomposable objective functions. A *decomposable* function (Salomon 1996) can always be written as

$$f(\mathbf{x}) = \sum_{i=0}^{D-1} f_i(x_i). \qquad (1.23)$$

Because decomposable functions lack cross-terms, their parameters can be optimized independently. Thus, decomposability replaces the task of optimizing one function having D dimensions with the much simpler task of optimizing D one-dimensional functions. The hyper-ellipsoid is a simple example of a decomposable function:

$$f(\mathbf{x}) = \sum_{i=0}^{D-1} \alpha_i x_i^2. \tag{1.24}$$

If, however, the hyper-ellipsoid is rotated in all dimensions, it becomes impossible to optimize one parameter independent of the others. This parameter dependence is often referred to as *epistasis*, an expression from biology (www 01). Salomon (1996) shows that unless an optimizer addresses the issue of parameter dependence, its performance on epistatic objective functions will be seriously degraded. This important issue is discussed extensively in Sect. 2.6.2.

Adapting the components of **C** requires additional "strategy parameters", i.e., the variances and position angles of the D-dimensional hyperellipsoids for which **C** is positive definite (Sprave 1995). Thus, the ES with correlated mutations increases a problem's dimensionality because it characterizes each individual by not only a vector of D objective function parameters, but also an additional vector of up to $D \cdot (D - 1)/2$ strategy parameters. For problems having many variables, the time and memory needed to execute these additional (matrix) calculations may become prohibitive.

Nelder and Mead

The Nelder–Mead polyhedron search (Nelder and Mead 1965; Bunday and Garside 1987; Press et al. 1992; Schwefel 1994), tries to solve the step size problem by allowing the step size to expand or contract as needed. The algorithm begins by forming a $(D + 1)$-dimensional polyhedron, or *simplex*, of $D + 1$ points, $\mathbf{x}_i$, $i = 0, 1, \ldots, D$, that are randomly distributed throughout the problem space. For example, when $D = 2$, the simplex is a triangle. Indices of the points are ordered according to ascending objective function value so that $\mathbf{x}_0$ is the best point and $\mathbf{x}_D$ is the worst point. To obtain a new trial point, $\mathbf{x}_r$, the worst point, $\mathbf{x}_D$, is reflected through the opposite face of the polyhedron using a weighting factor, $F1$:

$$\mathbf{x}_r = \mathbf{x}_D + F1 \cdot (\mathbf{x}_m - \mathbf{x}_D). \tag{1.25}$$

The vector, $\mathbf{x}_m$, is the *centroid* of the face opposite $\mathbf{x}_D$:

$$\mathbf{x}_m = \frac{1}{D} \left(\sum_{i=0}^{D-1} \mathbf{x}_i \right). \tag{1.26}$$

Figure 1.16 illustrates the *reflection operation* defined in Eq. 1.25.

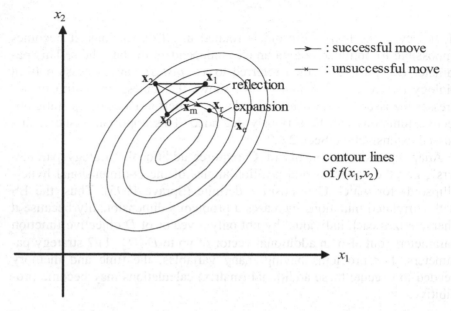

Fig. 1.16. Reflection and expansion in the Nelder–Mead method where $D = 2$

If a reflection through the centroid improves on the best point, x_0, i.e., if $f(x_r) < f(x_0)$, then the Nelder–Mead algorithm takes another step in the same direction based on the assumption that still further improvement may be possible. When weighted by a second scale factor, $F2$, this *expansion step* generates a new trial point, x_e:

$$x_e = x_r + F2 \cdot (x_m - x_D) \qquad (1.27)$$

If this expansion step also improves on x_0, then x_e replaces x_D. This new set of $D + 1$ points becomes the next simplex and the procedure begins again by ordering points based on their objective function value. If, however, x_e did not improve upon x_0, then x_r replaces x_D. If x_r did not improve upon x_0 in the first place, then x_r is compared to the next worst point, x_{D-1}. If x_r is better than x_{D-1}, then x_r replaces x_D. If, however, x_r is worse than x_{D-1}, a third scaling constant, $F3$, *shrinks* the entire simplex. Pseudo-code for the Nelder–Mead algorithm appears in Fig. 1.17. Figures 1.18–1.21 illustrate how the simplex moves in a two-dimensional parameter space.

```
...
while (convergence criterion not yet met)
{
    //---sort all D+1 points of the simplex according to-----
    //---ascending objective function value-----------------
    sort(x_i,D+1);
    //---compute centroid-----------------------------------
    x_m = 0;
    for (i=0; i<D; i++) x_m = x_m + x_i;
    x_m = x_m/D;
    //---start exploration of surface-----------------------
    x_r = x_m + F1(x_m - x_D);//reflection
    if (f(x_r) < f(x_0))    //if best point is improved
    {
        x_e = x_r + F2(x_m - x_D);//expansion
        if (f(x_e) < f(x_0)) x_D = x_e;
        else                 x_D = x_r;
    }
    else if (f(x_r) < f(x_{D-1}))//if next worst point is improved
    {
        x_D = x_r;
    }
    else//if best and next worst point are not improved
    {
        if (f(x_r) < f(x_D))
        {
            x_D = x_r;//replace worst point with reflected point
            x_c = x_m + F3(x_m - x_D);//contract around centroid
        }
        else
        {
            x_c = x_m - F3(x_m - x_D);//contract around centroid
        }
        if (f(x_c) < f(x_D))//if contraction was successful
        {
            x_D = x_c;
        }
        else //contract around the best point
        {
            for (i=1; i<=D; i++) x_i = 0.5*(x_0 + x_i);
        }
    }
}//end while
...
```

Fig. 1.17. Pseudo-code for the Nelder–Mead algorithm

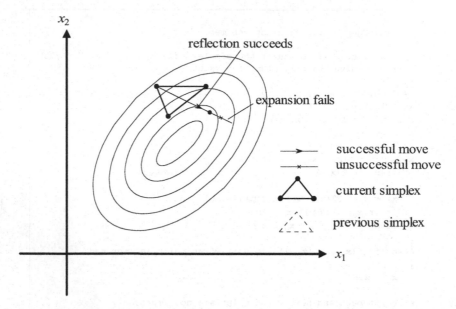

Fig. 1.18. Evolution of the Nelder–Mead simplex: first iteration. The reflection succeeds but the following expansion fails.

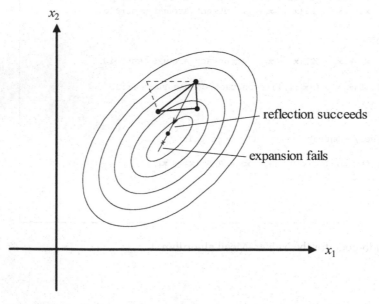

Fig. 1.19. Evolution of the Nelder–Mead simplex: second iteration. Again the reflection succeeds but the expansion fails.

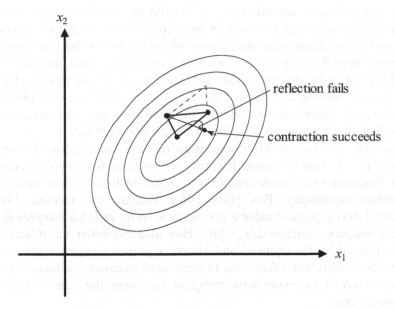

Fig. 1.20. Evolution of the Nelder–Mead simplex: third iteration. This time, even the reflection fails so a contraction must be tried. The contraction is successful.

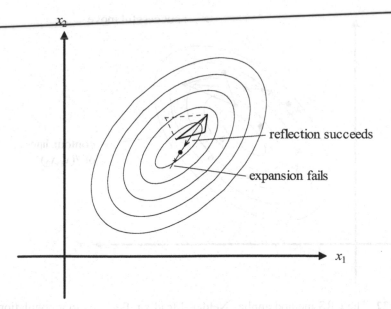

Fig. 1.21. Evolution of the Nelder–Mead simplex: fourth iteration. The reflection succeeds, but the expansion does not.

The Nelder–Mead method is one of the oldest optimization algorithms to heavily rely on difference vectors for exploring the objective function landscape. One advantage of the Nelder–Mead method is that the simplex can shrink as well as expand to adapt to the current objective function surface. This makes the step size a variable that depends on the topography of the objective function landscape. Like the Nelder–Mead method, DE also exploits vector differences but without the positional bias inherent in simplex reflections. Section 2.6.3 explores this distinction in detail.

Unlike DE, the Nelder–Mead algorithm restricts the number of sample points to $D + 1$. This limitation becomes a drawback for complicated objective functions that require many more points to form a clear model of the surface topography. Box (Box 1965; Bunday and Garside 1987; Schwefel 1994) suggested using a geometrical entity called a *complex* that, unlike a simplex, contains $2D$ points. Box also exploited the difference vectors formed by the centroid and all other points except for the worst one, but for multi-modal functions in particular, excessive reliance on the centroid as a reference point is meaningless, or, worse, the cause of premature convergence.

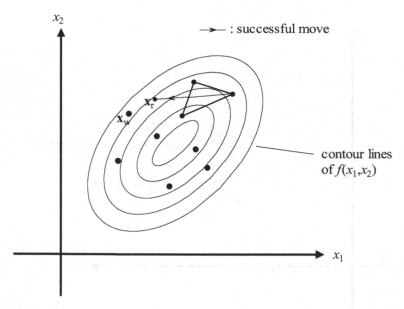

Fig. 1.22. The CRS method applies Nelder–Mead's reflections to a population of points.

```
   ...
   while (convergence criterion not yet met)
   {
       //---sort all N_p points according to--------------------
       //---ascending objective function value------------------
       sort(x_i,N_p);  //x_0 is best, x_D is worst point x_w
       //---compute centroid (Points of centroid should be------
       //---all different. Code to achieve that is not shown.)--
       j = rand(N_p); //pick a random point from the population
                       //as a variant pick j=0 (best point)
       x_m = x_j;
       for (i=1; i<D; i++)
       {
           j = rand(N_p); //pick a random point from the population
           x_m = x_m + x_j;
       }
       x_m = x_m/D;
       //---start exploration of surface----------------------
       x_r = x_m + F1(x_m - x_j);//reflection from last j, usually F1=1
       if (bounds_ok(x_r) == TRUE)//if inside the region of interest
       {
           if (f(x_r) < f(x_D))   //if worst point is improved
           {
               x_D = x_r;
           }
       }
       //---optionally there follows a local search------------
       //---starting from the best points----------------------
       ...
   }//end while
```

Fig. 1.23. Pseudo-code for the CRS-type algorithms

Controlled Random Search

Price's (no relation to the author of this book) *controlled random search* (CRS) also uses difference vectors for reflection operations (Price 1978). CRS employs a Nelder–Mead-like simplex consisting of $D + 1$ points drawn at random from a population of $Np > D + 1$ vectors as shown in Figure 1.22. A reflection through the centroid generates a new point x_r. If this point is better than the current worst point x_w, x_r replaces x_w. Figure 1.23 presents pseudo-code for the CRS.

CRS resembles DE because the population size is a control variable and because vector differences generate new points. Like the Nelder–Mead algorithm, though, CRS's reflection operations are a form of arithmetic recombination (see Sect. 2.6.3), whereas DE's vector operations more closely resemble a mutation operation (see Sect. 2.5).

One drawback of the CRS algorithm is that continually replacing the current worst point exerts high selective pressure that may force the population to prematurely converge. Even though it is a multi-point strategy, this "replace worst" selection strategy also makes it difficult to implement the CRS algorithm in parallel. Conflicts can arise because the current worst point can change after every reflection. There have been several improvements to the CRS algorithm, most notably by Ali et al. (1997) and Ali and Törn (2004).

1.2.4 Differential Evolution – A First Impression

Price and Storn developed DE to be a reliable and versatile function optimizer that is also easy to use. The first written publication on DE appeared as a technical report in 1995 (Storn and Price 1995). Since then, DE has proven itself in competitions like the IEEE's International Contest on Evolutionary Optimization (ICEO) in 1996 and 1997 and in the real world on a broad variety of applications. Recently, Mathematica® added DE to its numerical optimizer package.

Like nearly all EAs, DE is a population-based optimizer that attacks the starting point problem by sampling the objective function at multiple, randomly chosen initial points. Preset parameter bounds define the domain from which the Np vectors in this initial population are chosen (Fig. 1.24). Each vector is indexed with a number from 0 to $Np - 1$. Like other population-based methods, DE generates new points that are perturbations of existing points, but these deviations are neither reflections like those in the CRS and Nelder–Mead methods, nor samples from a predefined probability density function, like those in the ES. Instead, DE perturbs vectors with the scaled difference of two randomly selected population vectors (Fig. 1.25). To produce the trial vector, u_0, DE adds the scaled, random vector difference to a third randomly selected population vector (Fig. 1.26). In the selection stage, the trial vector competes against the population vector of the same index, which in this case is number 0. Figure 1.27 illustrates the select-and-save step in which the vector with the lower objective function value is marked as a member of the next generation. Figures 1.28–1.29 indicate that the procedure repeats until all Np population vectors have competed against a randomly generated trial vector. Once the last trial vector has been tested, the survivors of the Np pairwise competitions become parents for the next generation in the evolutionary cycle.

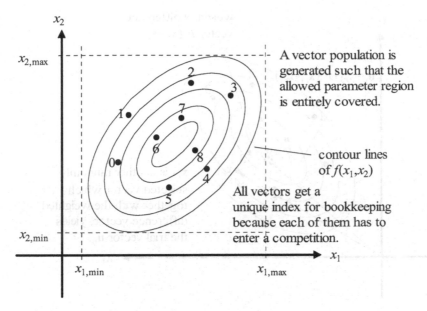

Fig. 1.24. Initializing the DE population

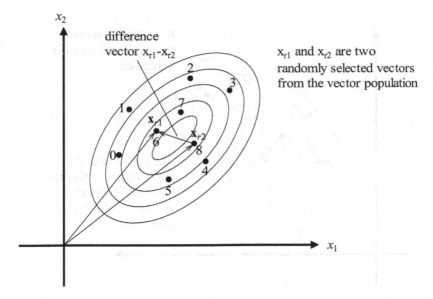

Fig. 1.25. Generating the perturbation: $\mathbf{x}_{r1} - \mathbf{x}_{r2}$

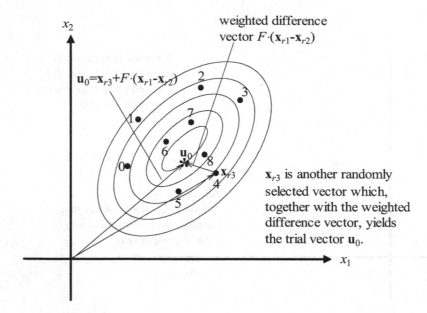

Fig. 1.26. Mutation

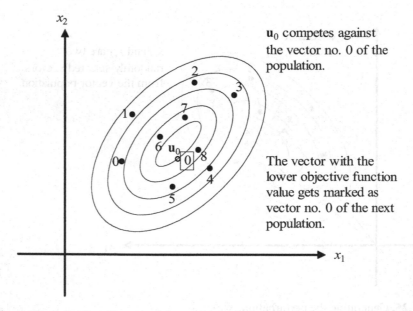

Fig. 1.27. Selection. Because it has a lower function value, $\mathbf{u}_0$ replaces the vector with index 0 in the next generation.

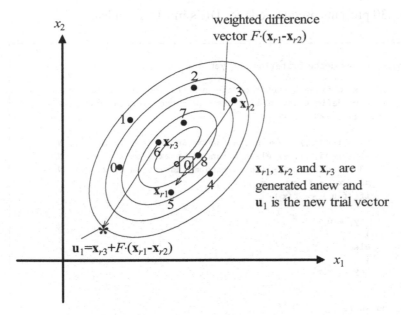

Fig. 1.28. A new population vector is mutated with a randomly generated perturbation.

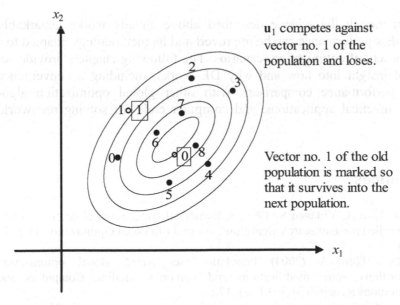

Fig. 1.29. Selection. This time, the trial vector loses.

Figure 1.30 presents pseudo-code for DE's most basic idea.

```
    ...
    while (convergence criterion not yet met)
    {
        //---xᵢ defines a vector of the current vector population-------
        //---yᵢ defines a vector of the new vector population-----------
        for (i=0; i<Nₚ; i++)
        {
            r1 = rand(Nₚ); //select a random index from 1, 2, ..., Nₚ
            r2 = rand(Nₚ); //select a random index from 1, 2, ..., Nₚ
            r3 = rand(Nₚ); //select a random index from 1, 2, ..., Nₚ
            uᵢ = xᵣ₃ + F*(xᵣ₁ - xᵣ₂);
            if (f(uᵢ) <= f(xᵢ))
            {
                yᵢ = uᵢ;
            }
            else
            {
                yᵢ = xᵢ;
            }
        }
    }//end while
    ...
```

Fig. 1.30. Pseudo-code for a simplified form of DE's generate-and-test operations

Even though the scheme described above already works remarkably well, DE's performance can be improved and its methodology adapted to a wide variety of optimization scenarios. The following chapters provide additional insight into how and why DE works, including a convergence proof, performance comparisons with other global optimization algorithms, practical applications, and computer code for solving real-world tasks.

References

Ali MM, Törn A, Viitanen S (1997) A numerical comparison of some modified controlled random search algorithms. Journal of Global Optimization 11:377–385

Ali MM, Törn A (2004) Population set based global optimization algorithms: some modifications and numerical studies. Computers and Operations Research 31(10):1703–1725

Bäck T (1996) Evolutionary algorithms in theory and practice. Oxford University Press

Bäck T, Hammel U, Schwefel H-P (1997) Evolutionary computation: comments on the history and current state. IEEE Transactions on Evolutionary Computation 1(1):3–17

Bentley PJ, Corne DW (2002) Creative evolutionary systems. Morgan Kaufmann, San Francisco

Boender C, Romeijn H (1995) Stochastic methods. In: Horst R, Pardalos P (eds) Handbook of global optimization. Kluwer, Dordrecht

Box MJ (1965) A new method of constrained optimization and a comparison with other methods. Computer Journal 8:42–52

Bunday BD, Garside GR (1987) Optimisation methods in PASCAL. Edward Arnold, London

Corne D, Dorigo M, Glover F (1999) New ideas in optimization. McGraw-Hill, London

Fogel DB (1994) Guest editorial on evolutionary computation. IEEE Transactions on Neural Networks 5(1):1–14

Glentis GO, Berberidis K, Theodoridis S (1999) Efficient least squares adaptive algorithms for FIR transversal filtering. IEEE Signal Processing Magazine July:13–41

Goldberg DE (1989) Genetic algorithms in search optimization and machine learning. Addison-Wesley, Reading, MA

Gross D, Harris CM (1985) Fundamentals of queuing theory. Wiley, New York

Holland JH (1962) Outline for a logical theory of adaptive systems. Journal of the Association for Computing Machinery 3:297–314

Hooke R, Jeeves TA (1961) Direct search solution of numerical and statistical problems. Journal of the Association for Computing Machinery 8:212–229

Ingber L (1993) Simulated annealing: practice versus theory. Journal of Mathematical and Computer Modeling 18(11): 29–57

Janka E (1999) Vergleich stochastischer Verfahren zur globalen Optimierung. Diplomarbeit, University of Wien

Kirkpatrick S, Gelatt CD, Vecchi MP (1983) Optimization by simulated annealing. Science 220:671–680

Locatelli M, Schoen F (1996) Simple linkage: analysis of a threshold-accepting global optimization method. Journal of Global Optimization 9:95–111

Metropolis N, Rosenbluth AE, Rosenbluth NM, Teller AN, Teller E, (1953) Equation of state calculation by fast computing machines, Journal of Chemical Physics 21:1087-1091

Michalewicz Z (1996) Genetic algorithms + data structures = evolution programs, 3rd ed. Springer, Berlin Heidelberg New York

Mühlenbein H, Schlierkamp-Vosen D (1993) Predictive models for the breeder genetic algorithm I. Evolutionary Computation 1(1):25–49

Nelder JA, Mead R (1965) A simplex method for function minimization. Computer Journal 7:308–313

Onwubolu GC, Babu BV (eds) (2004) New optimization techniques in engineering. Studies in Fuzziness and Soft Computing, vol 141. Springer, Berlin Heidelberg New York

Papoulis A (1965) Probability, random variables and stochastic processes. McGraw-Hill, New York

Pierre DA (1986) Optimization theory with applications. Dover, New York

Press WH et al. (1992) Numerical recipes in C. Cambridge University Press

Price WL (1978) A controlled random search procedure for global optimization. In: Dixon LCW, Szegö GP (eds) Towards global optimization 2. North-Holland, Amsterdam, pp 71–84

Rade L, Westergren B (1990) Beta mathematics handbook. CRC Press, Boca Raton, FL

Rechenberg I (1973) Evolutionsstrategie. Frommann-Holzboog, Stuttgart

Reeves CR (1993) Modern heuristic techniques for combinatorial problems. Wiley, New York

Salomon R (1996) Reevaluating genetic algorithm performance under coordinate rotation of benchmark functions. BioSystems 39(3):263–278.

Scales LE (1985) Introduction to non-linear optimization. Macmillan

Schwartz M (1977) Computer communication network design and analysis. Prentice Hall, Englewood Cliffs, NJ

Schwefel H-P (1981) Numerical optimization of computer models. Wiley, New York

Schwefel H-P (1994) Evolution and optimum seeking. Wiley, New York

Sprave J (1995) Evolutionäre Algorithmen zur Parameteroptimierung. Automatisierungstechnik 43(3):110–117

Storn R, Price KV (1995) Differential evolution – a simple and efficient adaptive scheme for global optimization over continuous spaces. Technical Report TR-95-012, ICSI

Štumberger G, Dolinar D, Pahner U, Hameyer K (2000) Optimization of radial active magnetic bearings using the finite element technique and the differential evolution algorithm. IEEE Transactions on Magnetics 36(4):1009–1013

Syslo MM, Deo N, Kowalik JS (1983) Discrete optimization algorithms with Pascal programs. Prentice-Hall

Törn A, Zilinskas A (1989) Global optimization. Lecture Notes in Computer Science, vol 350, 250pp Springer, Berlin Heidelberg New York

www 01 PCB 4673 About Epistasis. [Online]. Available: http://www.bio.fsu.edu/courses/pcb4673/about_epistasis.html

2 The Differential Evolution Algorithm

2.1 Overview

2.1.1 Population Structure

DE's most versatile implementation maintains a pair of vector populations, both of which contain Np D-dimensional vectors of real-valued parameters. The current population, symbolized by P_x, is composed of those vectors, $x_{i,g}$, that have already been found to be acceptable either as initial points, or by comparison with other vectors:

$$P_{x,g} = (x_{i,g}), \quad i = 0,1,...,Np-1, \quad g = 0,1,...,g_{max},$$
$$x_{i,g} = (x_{j,i,g}), \quad j = 0,1,...,D-1. \tag{2.1}$$

Indices start with 0 to simplify working with arrays and modular arithmetic. The index, $g = 0, 1, ..., g_{max}$, indicates the generation to which a vector belongs. In addition, each vector is assigned a population index, i, which runs from 0 to $Np - 1$. Parameters within vectors are indexed with j, which runs from 0 to $D - 1$.

Once initialized, DE mutates randomly chosen vectors to produce an intermediary population, $P_{v,g}$, of Np mutant vectors, $v_{i,g}$:

$$P_{v,g} = (v_{i,g}), \quad i = 0,1,...Np-1, \quad g = 0,1,...,g_{max},$$
$$v_{i,g} = (v_{j,i,g}), \quad j = 0,1,...,D-1. \tag{2.2}$$

Each vector in the current population is then recombined with a mutant to produce a trial population, P_u, of Np trial vectors, $u_{i,g}$:

$$P_{u,g} = (u_{i,g}), \quad i = 0,1,...,Np-1, \quad g = 0,1,...,g_{max},$$
$$u_{i,g} = (u_{j,i,g}), \quad j = 0,1,...,D-1. \tag{2.3}$$

During recombination, trial vectors overwrite the mutant population, so a single array can hold both populations.

2.1.2 Initialization

Before the population can be initialized, both upper and lower bounds for each parameter must be specified. These $2D$ values can be collected into two, D-dimensional initialization vectors, $\mathbf{b}_L$ and $\mathbf{b}_U$, for which subscripts L and U indicate the lower and upper bounds, respectively. Once initialization bounds have been specified, a random number generator assigns each parameter of every vector a value from within the prescribed range. For example, the initial value ($g = 0$) of the jth parameter of the ith vector is

$$x_{j,i,0} = \text{rand}_j(0,1) \cdot \left(b_{j,U} - b_{j,L}\right) + b_{j,L}. \tag{2.4}$$

The random number generator, $\text{rand}_j(0,1)$, returns a uniformly distributed random number from within the range [0,1), i.e., $0 \le \text{rand}_j(0,1) < 1$. The subscript, j, indicates that a new random value is generated *for each parameter*. Even if a variable is discrete or integral, it should be initialized with a real value since DE internally treats all variables as floating-point values regardless of their type.

2.1.3 Mutation

Once initialized, DE mutates and recombines the population to produce a population of Np trial vectors. In particular, *differential mutation* adds a scaled, randomly sampled, vector difference to a third vector. Equation 2.5 shows how to combine three different, randomly chosen vectors to create a mutant vector, $\mathbf{v}_{i,g}$:

$$\mathbf{v}_{i,g} = \mathbf{x}_{r0,g} + F \cdot \left(\mathbf{x}_{r1,g} - \mathbf{x}_{r2,g}\right). \tag{2.5}$$

The scale factor, $F \in (0,1+)$, is a positive real number that controls the rate at which the population evolves. While there is no upper limit on F, effective values are seldom greater than 1.0.

The *base vector* index, $r0$, can be determined in a variety of ways, but for now it is assumed to be a randomly chosen vector index that is different from the *target vector* index, i. Except for being distinct from each other and from both the base and target vector indices, the *difference vector* indices, $r1$ and $r2$, are also randomly selected once per mutant. Figure 2.1 illustrates how to construct the mutant, $\mathbf{v}_{i,g}$, in a two-dimensional parameter space.

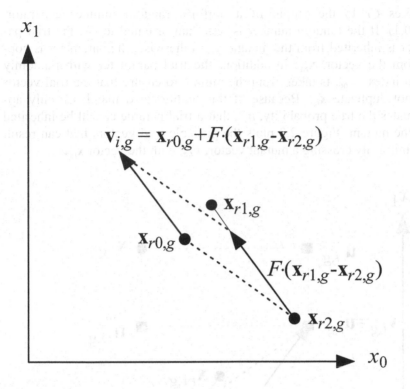

Fig. 2.1. Differential mutation: the weighted differential, $F \cdot (\mathbf{x}_{r1,g} - \mathbf{x}_{r2,g})$, is added to the base vector, $\mathbf{x}_{r0,g}$, to produce a mutant, $\mathbf{v}_{i,g}$.

2.1.4 Crossover

To complement the differential mutation search strategy, DE also employs *uniform crossover*. Sometimes referred to as *discrete recombination*, (dual) crossover builds trial vectors out of parameter values that have been copied from two different vectors. In particular, DE crosses each vector with a mutant vector:

$$\mathbf{u}_{i,g} = u_{j,i,g} = \begin{cases} v_{j,i,g} & \text{if} \left(\text{rand}_j(0,1) \le Cr \text{ or } j = j_{\text{rand}}\right) \\ x_{j,i,g} & \text{otherwise.} \end{cases} \quad (2.6)$$

The crossover probability, $Cr \in [0,1]$, is a user-defined value that controls the fraction of parameter values that are copied from the mutant. To determine which source contributes a given parameter, uniform crossover compares Cr to the output of a uniform random number generator, $\text{rand}_j(0,1)$. If the random number is less than or equal to Cr, the trial parameter is inherited from the mutant, $\mathbf{v}_{i,g}$; otherwise, the parameter is copied from the vector, $\mathbf{x}_{i,g}$. In addition, the trial parameter with randomly chosen index, j_{rand}, is taken from the mutant to ensure that the trial vector does not duplicate $\mathbf{x}_{i,g}$. Because of this additional demand, Cr only approximates the true probability, p_{Cr}, that a trial parameter will be inherited from the mutant. Figure 2.2 plots the possible trial vectors that can result from uniformly crossing a mutant vector, $\mathbf{v}_{i,g}$, with the vector $\mathbf{x}_{i,g}$.

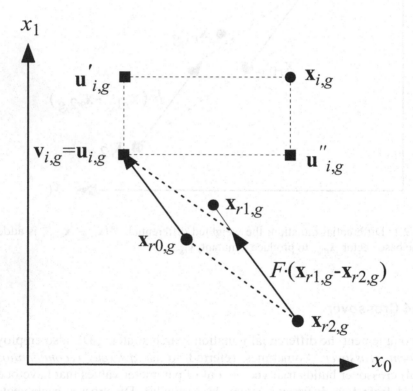

Fig. 2.2. The possible additional trial vectors $\mathbf{u}'_{i,g}$, $\mathbf{u}''_{i,g}$ when $\mathbf{x}_{i,g}$ and $\mathbf{v}_{i,g}$ are uniformly crossed

2.1.5 Selection

If the trial vector, $\mathbf{u}_{i,g}$, has an equal or lower objective function value than that of its target vector, $\mathbf{x}_{i,g}$, it replaces the target vector in the next generation; otherwise, the target retains its place in the population for at least one more generation (Eq. 2.7). By comparing each trial vector with the target vector from which it inherits parameters, DE more tightly integrates recombination and selection than do other EAs:

$$\mathbf{x}_{i,g+1} = \begin{cases} \mathbf{u}_{i,g} & \text{if } f(\mathbf{u}_{i,g}) \leq f(\mathbf{x}_{i,g}) \\ \mathbf{x}_{i,g} & \text{otherwise.} \end{cases} \tag{2.7}$$

Once the new population is installed, the process of mutation, recombination and selection is repeated until the optimum is located, or a pre-specified termination criterion is satisfied, e.g., the number of generations reaches a preset maximum, g_{max}.

2.1.6 DE at a Glance

Here are three different ways to describe the DE algorithm known as "classic DE".

Generate-and-Test

The simplicity of DE's generate-and-test loop becomes apparent once Eqs. 2.5–2.7 are combined:

$$u_{j,i,g} = \begin{cases} x_{j,r0,g} + F \cdot (x_{j,r1,g} - x_{j,r2,g}), & \text{if } (\text{rand}_j(0,1) \leq Cr \text{ or } j = j_{rand}), \\ x_{j,i,g} & \text{otherwise.} \end{cases} \tag{2.8}$$

$$j = 0,1,\ldots,D-1; \quad j_{rand} \in \{0,1,\ldots,D-1\}$$
$$i = 0,1,\ldots,Np-1$$
$$g = 0,1,\ldots,g_{max}$$
$$r0,r1,r2 \in \{0,1,\ldots,Np-1\}, \quad r0 \neq r1 \neq r2 \neq i$$

$$\mathbf{x}_{i,g+1} = \begin{cases} \mathbf{u}_{i,g} & \text{if } f(\mathbf{u}_{i,g}) \leq f(\mathbf{x}_{i,g}), \\ \mathbf{x}_{i,g} & \text{otherwise.} \end{cases}$$

C-Style Pseudo-code

Figure 2.3 presents C-style pseudo-code for classic DE. The vector indices $r0$, $r1$ and $r2$ are all different and distinct from the target index, i. In addition, selection is delayed until the trial population is complete.

```
// initialize...

do // generate a trial population
{
   for (i=0; i<Np; i++)  // r0!=r1!=r2!=i
   {
      do r0=floor(rand(0,1)*Np); while (r0==i);
      do r1=floor(rand(0,1)*Np); while (r1==r0 or r1==i);
      do r2=floor(rand(0,1)*Np); while (r2==r1 or r2==r0 or r2==i);
      jrand=floor(D*rand(0,1));

      for (j=0; j<D; j++)  // generate a trial vector
      {
         if (rand(0,1)<=Cr or j==jrand)
         {
            u_{j,i}=x_{j,r0}+F*(x_{j,r1}-x_{j,r2});  //check for out-of-bounds ?
         }
         else
         {
            u_{j,i}=x_{j,i};
         }
      }
   }

   // select the next generation

   for (i=0; i<Np; i++)
   {
      if ( f(u_i)<=f(x_i) ) x_i=u_i;
   }
} while (termination criterion not met);
```

Fig. 2.3. Classic DE; $0 \le \text{rand}(0,1) < 1$ so that indices never equal Np.

Flow Chart

Figure 2.4 shows a flow chart of DE. That $r0$, $r1$, $r2$ and i are distinct indices is not made explicit in this figure.

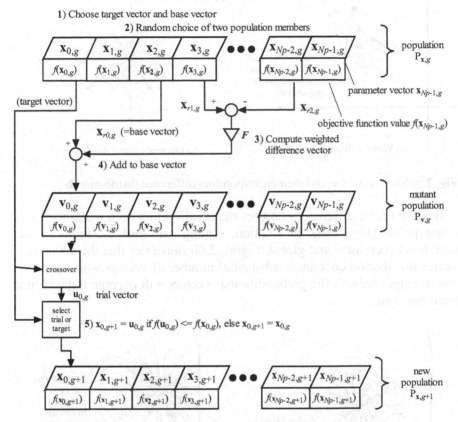

Fig. 2.4. A flow chart of DE's generate-and-test loop

2.1.7 Visualizing DE

The Difference Vector Distribution

Figure 2.5a shows the difference vectors formed by all possible pairings of nine vectors. Transporting the difference vectors to a common origin more clearly shows their distribution (Fig. 2.5b). Because all difference vectors have both a negative counterpart and an equal chance of being chosen, their distribution's mean is zero.

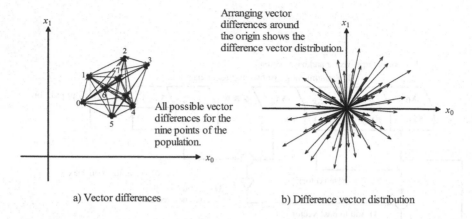

a) Vector differences
b) Difference vector distribution

Fig. 2.5. Nine vectors **a**, and their corresponding difference distribution **b**

Scaling vector differences ensures that trial vectors do not duplicate existing points (Fig. 2.6a). In addition, scaling can shift the focus of the search between local and global. Figure 2.6b illustrates that the difference vector distribution contains a substantial number of vectors whose considerable length reduces the probability that vectors will become trapped in a local minimum.

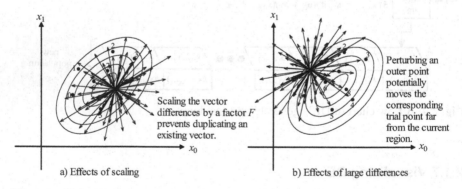

a) Effects of scaling
b) Effects of large differences

Fig. 2.6. The effects of scaling **a**, and large vector differences **b**

Contour Matching

One of the biggest advantages that difference vectors afford is that both a step's size and its orientation automatically adapt to the objective function landscape. The series of plots in Figs. 2.7–2.13 demonstrate this property for the "peaks" function (Eq. 1.16). For clarity, the difference vector distribution plot only shows the difference vector endpoints.

As it evolves, the population coalesces around competing minima (Figs. 2.7–2.10). During this phase, the difference distribution is multi-modal, like the function itself. It contains not only steps adapted to searching within each basin, but also larger steps capable of transporting vectors between basins and beyond. Once the population settles into the optimal basin (Figs. 2.11–2.13), the difference vector distribution becomes uni-modal and steps exhibit both a scale and an orientation that is appropriate for a local search.

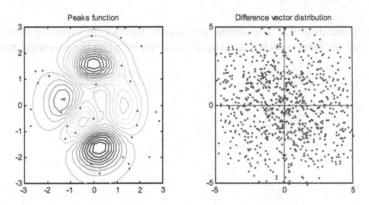

Fig. 2.7. Generation 1: DE's population and difference vector distributions

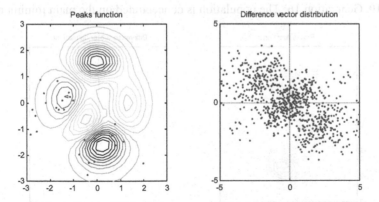

Fig. 2.8. Generation 6: The population coalesces around the two main minima

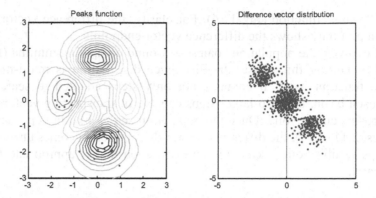

Fig. 2.9. Generation 12: The difference vector distribution contains three main clouds – one for local searches and two for moving between the two main minima.

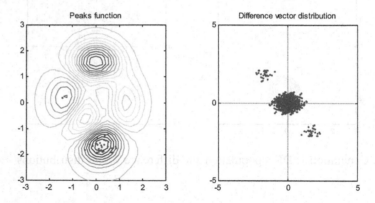

Fig. 2.10. Generation 16: The population is concentrated on the main minimum.

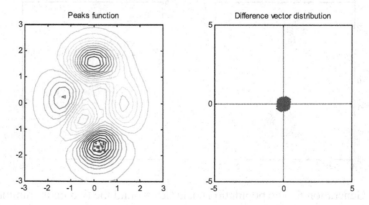

Fig. 2.11. Generation 20: Convergence is imminent. The difference vectors automatically shorten for a fine-grained, local search.

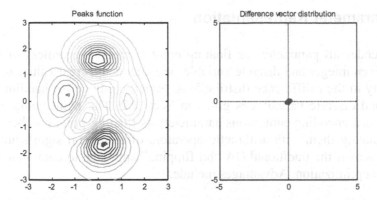

Fig. 2.12. Generation 26: The population has almost converged.

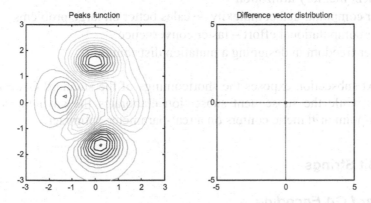

Fig. 2.13. Generation 34: DE finds the global minimum.

2.1.8 Notation

The technical name for the method illustrated in this overview is "DE/rand/1/bin" because the base vector is *rand*omly chosen, *1* vector difference is added to it and because the number of parameters donated by the mutant vector closely follows a *bin*omial distribution. More often, however, this book refers to this method simply as "classic DE". This version will probably suffice for most applications, but a number of variations are possible, each with its own strengths and weaknesses. The most successful of these alternative strategies will be explored later in this chapter, but first the next few sections examine the details missing from this brief overview.

2.2 Parameter Representation

DE encodes all parameters as floating-point numbers, regardless of their type. Even integer and discrete variables are encoded as real values to add diversity to their difference distributions. Specific advice for handling integer and discrete variables is given in Sect. 4.2. The point being made here is that encoding continuous parameters as floating-point numbers and manipulating them with arithmetic operators offer several significant advantages over the traditional GA "bit flipping" approach to continuous parameter optimization. Advantages include:

- ease of use
- efficient memory utilization
- lower computational complexity – scales better on large problems
- lower computational effort – faster convergence
- greater freedom in designing a mutation distribution.

The next subsection exposes the shortcomings of the standard GA coding scheme, while the subsequent subsection elaborates the advantages that floating-point arithmetic confers on a real-parameter optimizer.

2.2.1 Bit Strings

Standard GA Encoding

Typically, GAs encode a continuous parameter, x, as an integer string of q bits, a_k, $k = 0, 1, \ldots q - 1$, each of which is a coefficient for a power of 2:

$$x = b_{\mathrm{L}} + \frac{\left(b_{\mathrm{U}} - b_{\mathrm{L}}\right)}{2^q - 1} \cdot \sum_{k=0}^{q-1} a_k 2^k. \tag{2.9}$$

When decoded, integers are normalized by a factor of $2^q - 1$ and multiplied by $b_{\mathrm{U}} - b_{\mathrm{L}}$ so that values span the range between a parameter's upper and lower bounds, b_{U} and b_{L}, respectively. Assuming that equal resources are devoted to each parameter, a vector of D parameters will require $l = q \cdot D$ bits in all.

For functions with independent parameters, both theory and experiment suggest that the optimal mutation rate, i.e., the probability that a bit should be inverted, or "flipped", is $p_m = 1/l$ (Mühlenbein 1992; Potter and DeJong 1994). The problem with the GA approach is that even on uni-modal objective functions, the computational effort to optimize a parameter is a function of l that depends on a parameter's value. For example, if the initial parameter value is $x = 15$ ($x = 01111$ binary) and the optimal value is $x = 16$ ($x = 10000$ binary), then 5 bits must *simultaneously* be flipped to make the final improving move. When $p_m = 1/l$, the probability of this event is $p = (1/l)^5$. Because this "Hamming cliff" prevents *incremental* improvement, $x = 15$ is one of many *local* optima even if the objective function is uni-modal. In effect, the function that maps bit strings to real-parameter values is itself multi-modal (Bäck 1993).

By contrast, progress does not depend on simultaneously flipping multiple bits if the optimum happens to be $x = 0$. Instead, inverting non-zero bits in any *sequence* produces a series of parameter values each of which is closer to $x = 0$ than the last. If the objective function is separable and uni-modal, these intermediate steps constitute improving moves. In this *very* special case, the computational complexity to optimize a parameter is constant at $O(l \cdot \ln(l))$ (Salomon 1996b). The factor, $\ln(l)$, occurs because the neighborhoods around parameters that are already optimized continue to be re-sampled (Salomon 1997). In the worst case scenario, however, all q bits must be inverted to make an improving move, so the upper bound on the computational complexity for optimizing an independent parameter of a *uni-modal* function becomes $O(l^q \cdot \ln(l))$.

The requirement that all q bits simultaneously be inverted is also a demand when the objective function is separable and *multi-modal*. For example, it may be that two competing local minima are positioned at points whose representations differ at each bit position. Since an improving move from one local minimum to the other must simultaneously change all q bits, the complexity is $O(l^q \cdot \ln(l))$ (Salomon 1996b). That the computational complexity to optimize an independent parameter is the same in the worst case regardless of whether the function is uni- or multi-modal reflects the aforementioned fact that the standard GA coding scheme imposes multi-modality on even uni-modal objective functions (Bäck 1993).

Gray Codes

Gray codes eliminate Hamming cliffs by reassigning bit groupings to integers so that representations for adjacent integers differ by a single bit, i.e., so that the Hamming distance between consecutive integers is 1 (Wright 1991). As long as the objective function is both uni-modal and separable, sequentially flipping single bits in Gray-coded variables can always produce monotonously decreasing objective function values regardless of both the starting point and the optimal parameter value. Since it no longer matters what the optimal parameter value is, the complexity for optimizing a separable, uni-modal function with Gray codes when $p_m = 1/l$ is constant at

$O(l \cdot \ln(l))$ (Salomon 1996b). Because of their constant low complexity, Gray codes are more efficient than the standard GA representation when the objective function is uni-modal (Bäck 1993). If, however, the objective function is multi-modal, then all bits must be inverted simultaneously in the worst case scenario, so the computational complexity again rises to $O(l^q \cdot \ln(l))$ – the same complexity demonstrated for standard GA coding (Salomon 1996b).

2.2.2 Floating-Point

Unlike the standard GA representation in which all bits are potentially significant, the floating-point format retains only a limited number of significant digits. For example, the ANSI C **float** data type encodes a real number with $q = 32$ bits. Twenty-four bits are dedicated to precision, while the remaining eight bits are assigned to an exponent that locates the decimal point. By contrast, a fixed-point integer variable requires 256+ bits to span as many orders of magnitude as the **float** data type. In the final answer, most of the bits in this very long integer format will be either leading zeros or bits of unneeded precision. By contrast, the floating-point format retains only a limited number of significant bits while spanning a vast dynamic range with minimal resources.

The floating-point format is convenient not only because it can efficiently handle parameter values that span a wide dynamic range, but also because most modern programming languages support common floating-point formats. No special routines are needed to define, input, manipulate or output a floating-point value. When representing continuous parameters in floating-point, the encoding process is transparent to the user.

Logical Versus Arithmetic Operators

GAs typically operate on bit strings with *logical* operators like the XOR (exclusive or) which has the effect of inverting specified bits. By contrast, DE and other floating-point optimizers *add* a floating-point deviation to one or more parameters. Compared to bit flipping, arithmetic provides two benefits: it reduces the complexity of the algorithm and it provides greater flexibility in designing a mutation distribution.

The most efficient way for an EA to optimize a function with independent parameters is to change one parameter at a time before evaluating the result (Salomon 1996a). Typically, both standard and Gray-coded GAs implement this strategy by setting $p_m = 1/l$ so that, on average, only one parameter value changes per function evaluation (Potter and DeJong 1994).

For EAs that add a small deviation to a floating-point parameter, the corresponding mutation probability is $p_m = 1/D$ – a value which also, on average, perturbs just one parameter before evaluating the result (Mühlenbein and Schlierkamp-Voosen 1993; Salomon 1996a).

When the objective function is multi-modal, all bits in an *independent* floating-point parameter may have to be set to the correct value to make progress. If there are q bits in the floating-point representation, then the probability of making progress in this worst case scenario is $(1/2)^q$. While this number may be very small, it is constant and independent of D. As a result, the computational complexity for optimizing separable, multi-modal functions with floating-point representations and arithmetic operators is $O(D \cdot \ln(D))$ (Salomon 1996a). Compared to the $O(l^q \cdot \ln(l))$ complexity for optimizing an independent, Gray-coded parameter of a multi-modal function, the floating-point format representation is faster by a factor of up to $q \cdot l^{q-1} \cdot (1 + \ln(q)/\ln(l))$. The rules of complexity mathematics (Beckman 1980), however, replace *leading* constants, like q, with 1 and substitute 0 for terms like $\ln(q)/\ln(l)$ that are negligible for large l. Under these rules, the ratio of Gray to floating-point complexities reduces to l^{q-1} (Salomon 1996b).

Parameter dependence amplifies this disparity between the computational complexity of the Gray and floating-point approaches. For example, if a multi-modal function has two parameters that depend on each other, then progress in the worst case scenario will require flipping all bits in both parameters simultaneously. The probability of this event is $p = (1/l)^{2q}$ and the corresponding computational complexity is $O(l^{2q} \cdot \ln(l))$. Under similar circumstances, all bits in both parameters' floating-point representations also must be changed. When $p_m = 1/D$, this event occurs with a probability of $(1/D)^2$, so the computational complexity for optimizing two, dependent, floating-point parameters of a multi-modal function is $O(D^2 \cdot \ln(D))$. Under the rules of complexity mathematics, the gain over Gray-coded parameters rises to $l^{2(q-1)}$.

Crafting a Mutation Distribution

Arguably the most important advantage that floating-point arithmetic confers on a real-parameter optimizer is the freedom to decide how perturbations are distributed. Because floating-point's computational complexity does not depend on the mutation operator's probability density, distributions can be crafted to implement a particular search strategy (Salomon 1996b). For example, the Breeder Genetic Algorithm perturbs parameters with non-adaptive step sizes that are distributed according to a power law

(Mühlenbein and Schlierkamp-Voosen 1993). Evolution Strategies (Bäck and Schwefel) and Fast Evolution Strategies (Yao and Liu 1997) adaptively modify steps sampled from Gaussian and Cauchy distributions, respectively. The same freedom that these floating-point optimizers enjoy also allows DE to tap the pool of vector differences as its mutation distribution.

2.2.3 Floating-Point Constraints

The number of bits that a floating-point format dedicates to an exponent limits the minimum and maximum values that it can represent. These limits are rarely exceeded in practical applications because physical properties of such extreme magnitude are uncommon. Of greater consequence for DE is the number of significant digits (precision) that a format supports. If the objective function contains terms that differ by many orders of magnitude, contributions from smaller terms will be lost if there are not enough significant bits available. For example, the **float** data type holds about seven decimal digits of precision. If two numbers differ by more than seven decimal orders of magnitude, then the smaller contribution is not taken into account.

$$x = 1.2 \times 10^{10},$$

(2.10)

$$y = 17,$$

$$x + y = 12000000017 \rightarrow 1.200000 \times 10^{10} = x.$$

For the same reason, the lack of precision can be a problem not only when computing the objective function, but also when forming vector differences. Because DE relies on vector differences, the inability to record the effect of small perturbations might cause DE to stagnate (Zimmons n.d.).

In most cases, the **double** format with 15 digits of decimal precision will be enough. Because they evaluate high-order polynomials, however, functions like the high-dimensional versions of the Chebyshev function (see Appendix) require **long doubles**. Except for requiring additional memory and bandwidth, there is little penalty for declaring **long doubles** and their 19 digits of decimal precision because floating-point units compute values to full precision by default.

2.3 Initialization

In order for DE to work, the initial population must be distributed throughout the problem space. One-point optimizers do not require this initial diversity and even the $(1, \lambda)$-ES begins with a single point. If, however, DE is initialized with Np replicas of a single vector, uniform crossover and differential mutation will only clone more replicas. Consequently, DE requires a predefined *probability distribution function*, or PDF, to seed the initial population. When specifying an initial distribution, steps must be taken to ensure that its scale sufficiently broad.

2.3.1 Initial Bounds

As a matter of convenience, test function parameters are often initialized with values that are constrained to lie between a single set of upper and lower bounds. By contrast, bounds for parameters that define real-world objective functions are seldom equal, often because the parameters they delimit correspond to different physical or mathematical entities. In many cases, the existence of natural physical limits or logical constraints makes prescribing bounds for each parameter straightforward. For example, ordinary optical glass can never have an index of refraction less than or equal to 1, nor can a gear have less than one tooth. In cases like these where parameter limits are inviolable, initialization bounds should not only delimit the initial population, but also constrain the subsequent search. Section 4.3.1 discusses several methods for keeping parameters constrained within pre-specified bounds.

Far Initialization

When parameters exhibit no obvious limits, their upper and lower bounds, $b_{j,U}$ and $b_{j,L}$, respectively, should be set so that the initial bounding box they define encompasses the optimum. If the optimum's general location is uncertain, then the possibility exists that it lies outside the initial bounding box. Figure 2.14 shows an example of *far initialization* in which the upper parameter limit has been reduced to the point where the initial bounding box no longer contains the optimum, $\mathbf{x}^*$. In cases of far initialization, bounds on otherwise unconstrained parameters must be ignored once the population has been initialized so that DE can explore beyond the initial bounding box.

Table 2.1 records the effect that far initialization has on DE's ability to discover the optima of ten common test functions (descriptions of test functions can be found in the Appendix). Although each function has a different set of initialization bounds, in each case, these bounds define a D-dimensional box that encloses the function's global optimum.

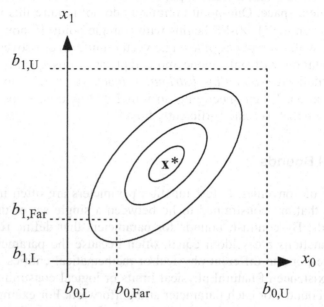

Fig. 2.14. Far initialization shrinks the initial bounding box so that it no longer contains the optimum, $\mathbf{x}^*$.

For the results in Table 2.1, each parameter was initialized with a uniformly distributed random value from within a range that has been reduced by a factor, h, when compared to the originally prescribed bounds:

$$x_{j,i,0} = b_{j,L} + h \cdot \mathrm{rand}_j (0,1) \cdot \left(b_{j,U} - b_{j,L} \right). \qquad (2.11)$$

After far initializing the population with the given value of h, bounds were relaxed to their normal values to constrain the subsequent search.

For each of the functions in Table 2.1, the initial bounding box encloses the optimum when $h = 1$. Setting $h \leq 0.1$ far initializes the population by restricting it to a corner of the original bounding box where it cannot surround the optimum. Table 2.1 reports the average number of function evaluations ("Evals.") taken to find a point whose objective function value differs from the optimum objective function value by less than a preset minimum. Finding such a point within the maximum allowed number of generations constitutes a success; otherwise, the trial is considered to be a failure. (See the Appendix for details on the minimum function value to reach.) Only "successes" contribute to the results in Table 2.1. The fraction of successful trials, P, records the impact of failures. Results are 100-trial averages obtained using classic DE with $F = Cr = 0.9$ and $r0 \neq r1 \neq r2 \neq i$ (distinct indices).

Table 2.1. The effects of far initializing DE with a uniformly random population

Function	D	Np	$h = 1$		$h = 0.1$		$h = 0.01$	
			Evals.	P	Evals.	P	Evals.	P
Sphere	10	30	30,994.5	1	31,514.9	1	31,722.2	1
Ridge	10	30	48,520.2	1	48,825.5	1	48,820.2	1
Rosenbrock	10	30	59,643.4	1	59,721.9	1	60,315.4	1
Chebyshev	9	30	69,522.1	1	72,211.3	1	71,068.5	1
Ackley	10	30	48,385.2	1	49,853.3	0.90	–	0
Rastrigin	5	100	59,840.4	1	60,199.2	1	–	0
Schwefel	5	100	16,245.6	1	22,432.4	0.25	–	0
Griewangk	5	100	19,4202	0.98	19,5551	0.99	19,4921	0.99
Langerman	5	100	38,405.7	0.98	37,196.1	0.54	34,873.2	0.21
Michalewicz	5	100	27,749.5	1	29,291.6	0.95	32,061.6	0.96

As Table 2.1 shows, far initialization's effect on the sphere, ridge, Rosenbrock, Chebyshev, Michalewicz and Griewangk functions is minimal. In most cases, far initialization penalizes these six functions with a very slight increase in the average number of function evaluations and a very slight decrease in the estimated probability of success. For both the sphere and ridge functions, this result is not surprising. Both functions are uni-modal and convex, so neither poses obstacles to the population's expansion toward the minimum. (Pictures of the two-dimensional versions for many of the test functions used in this book appear in the Appendix.) Rosenbrock's function is also uni-modal, but unlike the sphere it is non-convex. At least in the case of Rosenbrock's function, non-convexity does not impede DE's ability to locate the minimum when far initialized.

Unlike the sphere, ridge or Rosenbrock functions, the remaining functions in Table 2.1 are all multi-modal. Optimal parameter values for the Chebyshev function vary greatly in magnitude and restricting initial values to a small range means that some parameter values must inflate many orders of magnitude to be on par with their optimal values. Table 2.1 shows that except for a slight increase in the number of function evaluations, diminishing the value of h did not significantly impact DE's ability to converge on the Chebyshev optimum. Similarly, far initialization did not significantly affect DE's performance on either Michalewicz's or Griewangk's function.

DE became unreliable, however, when far initializing Langerman's function and failed altogether on the Ackley, Rastrigin and Schwefel functions once $h = 0.01$. For these highly multi-modal functions, the entire initial population can land inside a single, non-optimal, local basin of attraction when h becomes too small. If competing basins are sufficiently far apart, then classic DE cannot generate difference vectors large enough to escape the local basin. Thus, it is important to use a bounding box of sufficient size when initializing multi-modal functions with a uniform random distribution.

Initializing with a Constant

Occasionally, it may prove productive to experiment with a design by holding one or more of its parameters constant while optimizing the remaining variables. DE automatically leaves a parameter unchanged during optimization if every vector is initialized with the same value for the given parameter. When all vectors have the same value for a parameter, every differential they combine to create for that parameter will be zero. Furthermore, uniform crossover does not change parameter values, so a parameter initialized with a single constant value will never change.

2.3.2 Initial Distributions

DE can be initialized with either a uniform or a non-uniform distribution. The decision regarding which to use depends on how much is known about the location of the optimum. If the optimum's location is fairly well known, a Gaussian distribution may prove somewhat faster, although it may also increase the probability that the population will converge prematurely. In general, uniform distributions are preferred, since they best reflect the lack of knowledge about the optimum's location. The next section looks at two common uniform distributions.

Uniform Distributions

Distributing initial points with random uniformity is not mandatory, but experience has shown $rand_j(0,1)$ to be very effective in this regard. In general, any distribution that uniformly covers the search domain and contains a degree of irregularity or randomness should serve well for initializing the vector population. For example, Hammersley and Halton point sets are often used in the field of numerical integration (Halton and Weller 1964). Based on prime numbers, these pseudo-random distributions are both uniform and irregular, but lack points in close proximity, i.e., they have a minimum resolution that increases as the number of points in the sample increases. Figure 2.15 gives C-style pseudo-code for computing Halton points in up to ten dimensions. The function, halton(i,j), takes the population and parameter indices as input and returns a (rational) number belonging to the interval [0,1).

```
halton(i,j)
{
    prime[10]=[2,3,5,7,11,13,17,19,23,29];
    p1=prime[j];
    p2=p1;
    sum=0;

    do
    {
        x=i%p1; // "%" is the modulo operator
        sum=sum+x/p2;
        i=floor(i/p1);
        p2=p2*p1;
    }while (i>0);

    return(sum);
}
```

Fig. 2.15. C-style pseudo-code for generating Halton point sets, $D \leq 10$

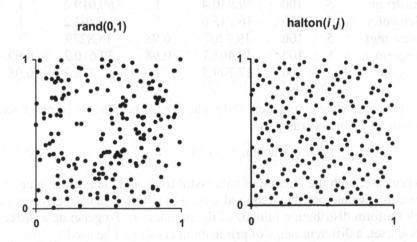

Fig. 2.16. Two hundred points distributed with random uniformity (left) and according to a two-dimensional Halton point set (right).

Figure 2.16 compares the uniform random and Halton distributions in two dimensions. The Halton distribution is more even, but the random distribution displays a wider range of difference vector magnitudes.

Table 2.2 shows how the random and Halton distributions affect DE's performance by reporting the average number of function evaluations ("Evals.") taken to find a point whose objective function value differs from the optimum objective function value by less than a preset minimum. Finding such a point within the maximum allowed number of generations constitutes a success; otherwise, the trial is considered to be a failure. (See Appendix for details on the minimum function value to reach.) Only "successes" contribute to the results in Table 2.2. The fraction of successful trials, P, records the impact of failures. Results are 100-trial averages obtained with classic DE, $F = Cr = 0.9$, distinct indices and with bound constraints imposed. Results for the random uniform distribution have been copied from Table 2.1 ($h = 1$).

Table 2.2. Comparing the effects of uniform initial distributions on performance

Function	D	Np	rand$_j$(0,1)		halton(i,j)	
			Evals.	P	Evals.	P
Sphere	10	30	30,994.5	1	30,971.1	1
Ridge	10	30	48,520.2	1	48,346.8	1
Rosenbrock	10	30	59,643.4	1	59,406.2	1
Chebyshev	9	30	69,522.1	1	72,611.6	1
Ackley	10	30	48,385.2	1	48,354.7	1
Rastrigin	5	100	59,840.4	1	60,019.2	1
Schwefel	5	100	16,245.6	1	16,203.2	1
Griewangk	5	100	19,4202	0.98	18,8279	1
Langerman	5	100	38,405.7	0.98	39,610.2	0.99
Michalewicz	5	100	27,749.5	1	27,130.7	0.98

As Table 2.2 shows, it matters little whether the population is initialized with rand$_j$(0,1), or according to

$$x_{j,i,0} = b_{j,\mathrm{L}} + \mathrm{halton}_j(i,j) \cdot (b_{j,\mathrm{U}} - b_{j,\mathrm{L}}). \tag{2.12}$$

In every case, both the fraction of successful trials and the average number of function evaluations they required were virtually the same regardless of which uniform distribution initialized the population. To generate a different point set, a different range of prime numbers should be used.

Gaussian Distribution

Uniform distributions are reliable, but populations can also be non-uniformly initialized. For example, Fig. 2.17 plots 200 points distributed according to a two-dimensional multi-normal distribution whose mean vector value is $\mu = 0.5$ and whose covariance matrix is $C = \sigma^2 \cdot I$, where $\sigma = 0.5$ and I is the identity matrix, i.e., N(**0.5**, 0.25·**I**). This choice centers the (symmetrical) distribution in the bounding box and places standard deviates (along coordinate axes) on its surface. Unlike the Halton distribution in Eq. 2.12, the multi-normal distribution, whose output is a vector, is sampled only once per initial vector.

N(0.5,0.25·I)

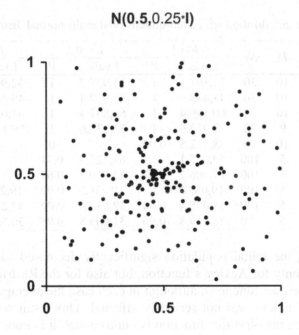

Fig. 2.17. A two-dimensional Gaussian-distributed initial population with mean of 0.5 and a standard deviation of 0.5, i.e., N(**0.5**, 0.25·**I**)

Table 2.3 details how classic DE's performs when the initial population is distributed according to a multi-normal distribution. Unlike Eq. 2.12 in which a new random value is generated for each parameter, the distribution used in both Fig. 2.17 and Table 2.3 generates a single instance of a multi-normally distributed random vector for each initial point. In both cases, the distribution's mean vector is $\mu = (0.5, 0.5, ... 0.5)$ and its covariance matrix is $C = 0.25 \cdot I$. A comparison with Table 2.2 shows that when the population is *not* far initialized ($h = 1$), it makes little difference whether the initial distribution is uniform or Gaussian. The sole exception is Ackley's function. Although initializing Ackley's function with a Gaussian distribution left convergence speed unchanged, it significantly degraded DE's probability of success.

Once the population is far initialized ($h \leq 0.1$), failures become more likely. When compared with Table 2.1, the results in Table 2.3 show that a population far initialized with a Gaussian distribution is less likely to be successful on multi-modal functions than a uniformly distributed one. In every case where uniform distributions failed, the Gaussian-distributed population failed more often.

Table 2.3. Far initialization with a ten-dimensional multi-normal distribution

Function	D	Np	$h = 1$		$h = 0.1$		$h = 0.01$	
			Evals.	P	Evals.	P	Evals.	P
Sphere	10	30	31,801.1	1	31,937.7	1	32,967.6	1
Ridge	10	30	48,483.9	1	48,919.4	1	49,569.9	1
Rosenbrock	10	30	60,198.4	1	60,897.4	1	61,056.4	1
Chebyshev	9	30	72,972.6	1	72,233.6	1	70,129.5	1
Ackley	10	30	48,472.5	0.02	–	0	–	0
Rastrigin	5	100	59,627.1	1	61,125.9	0.71	–	0
Schwefel	5	100	17,406.2	0.97	33,750.1	0.67	–	0
Griewangk	5	100	19,0872	1	19,6362	0.99	19,2462	0.99
Langerman	5	100	34,005.4	0.60	32,630.6	0.09	32,254.3	0.04
Michalewicz	5	100	28,219.8	0.96	31,005.8	0.98	30,828.1	0.73

Clustering the initial population significantly decreased success probabilities not only for Ackley's function, but also for the Rastrigin, Schwefel and Langerman functions, although in each case the average number of function evaluations was not seriously affected. This result reinforces the idea that when the objective function is multi-modal, it is important to disperse the initial population widely enough to contain the optimum. Results also suggest that the penalty for expanding bounds is a small increase in the average number of function evaluations but the reward is often a significantly enhanced probability of success.

DE is based on evolution with vector differences, so it is not surprising that the way in which differences are chosen can have an impact on the optimization process. The following section examines what happens when the base and difference vectors are chosen both with and without restrictions.

2.4 Base Vector Selection

There are four vector indices in classic DE's generating equation (e.g., Eq. 2.8). The target index, i, specifies the vector with which the mutant is recombined and against which the resulting trial vector competes. The remaining three indices, $r0$, $r1$ and $r2$, determine which vectors combine to create the mutant. Typically, both the base index, $r0$, and the difference vector indices, $r1$ and $r2$, are chosen anew for each trial vector from the range $[0, Np - 1]$.

When indices are randomly selected, the possibility exists that some vectors may be chosen repeatedly while others may be omitted altogether. Both omitted and duplicated indices affect DE's performance. Duplicating an index can reduce DE's novel search strategy to a conventional one, while omitting an index may deprive a vector of the opportunity to serve as a base vector. After presenting several alternative schemes for selecting base vectors, this section explores the effects of degenerate vector combinations.

2.4.1 Choosing the Base Vector Index, $r0$

Random Without Restrictions

The base index, $r0$, specifies the vector to which the scaled differential is added. The classic version of DE employs a uniform distribution to randomly select $r0$ anew for each trial vector. To ensure that the index is always less than Np, $rand_i(0,1)$ must return a value that is strictly less than 1.

```
r0=floor(rand_i(0,1)*Np);
```

Fig. 2.18. Base vector selection without restrictions

While base index selection without restrictions (Fig. 2.18) treats all vectors equally in a statistical sense, it may pick some vectors more than once per generation, causing others to be omitted. Stochastic universal sampling provides a more representative population sample.

Stochastic Universal Sampling

Randomly selecting the base vector without restrictions is known in EA parlance as *roulette wheel selection*. Roulette wheel selection chooses Np

vectors by conducting Np separate random trials, much like Np passes at a roulette wheel whose slots are proportional in size to the selection probability of the vector they represent. In many GAs, selection probabilities are biased toward better solutions, meaning that better vectors are assigned proportionally wider slots, but in classic DE, each vector has the same chance of being chosen as a base vector, so all slots are of equal size, just like a real roulette wheel.

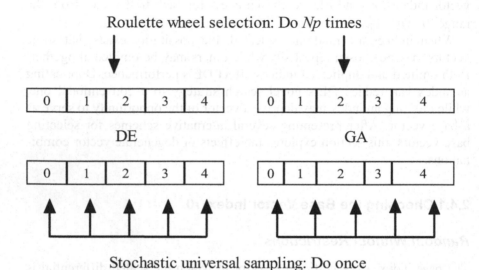

Fig. 2.19. Stochastic universal sampling and roulette wheel selection compared. The fraction of the space allotted to a vector in DE is constant, but in the GA it depends on the vector's objective function value.

Because samples drawn by roulette wheel selection suffer from a large variance, the preferred method for sampling a distribution is *stochastic universal sampling* because it guarantees a minimum spread in the sample (Baker 1987; Eiben and Smith 2003). The relation of stochastic universal sampling to roulette wheel selection is best illustrated if the ball used in real roulette is replaced with a stationary pointer. Once the roulette wheel stops, the vector corresponding to the slot pointed to is selected. Instead of spinning a roulette wheel Np times to select Np vectors with a single pointer, stochastic universal sampling uses Np equally spaced pointers and spins the roulette wheel just once. In the GA, slot sizes are based on a vector objective function value, with better vectors being assigned more space. In DE, each candidate has the same probability of being accepted, so slots are of equal size. Consequently, each of the Np pointers selects one and only one vector regardless of how the roulette wheel is spun (Fig. 2.19)

The following vector selection methods adhere to stochastic universal sampling as it applies to DE since all vectors serve as base vectors once and only once per generation. Both methods described below also establish the one-to-one correspondence needed to pair each target vector with a unique base vector.

2.4.2 One-to-One Base Vector Selection

Permutation Selection

To ensure that each vector serves as a base vector just once per generation, permutation selection draws consecutive base vector indices from an array containing a random permutation of the sequence $[0, 1,..., Np - 1]$. In this scheme, the (target) vector with index i is crossed with is the base vector whose index is the ith element of the permutation. The permutation array can be initialized with consecutive integers and $r0$ can be computed with a single call to a uniform random number generator and one swap of array elements. Another way to permute base vectors assigns to i the vector whose index is the product, modulo Np, of i and an integer that is relatively prime to Np. Details of both methods can be found in Sect. 5.2.

Random Offset Selection

The random offset method is another way to stochastically assign each target vector a unique base vector. Simpler than the permutation method, the random offset method computes $r0$ as the sum, modulo Np, of the target index and a randomly generated offset, r_g. The modulo operator, %, in Fig. 2.20 divides the operand, $(i + r_g)$, by Np and returns the integral remainder.

```
r0=(i+rg)%Np;
```

Fig. 2.20. The base vector is the sum, modulo Np, of the target index, i, and the randomly generated offset, r_g (see Fig. 2.21).

```
rg=floor(randg(0,1)*Np);
```

Fig. 2.21. The random offset, r_g, is chosen anew *at the start of each generation*.

Each of the Np possible values for r_g defines a one-to-one mapping between target and base vectors. These Np rotational mappings are a subset of the set of $Np!$ permutations. The symbol, "!" is the factorial operator. The value of $n!$ is just the product of all of the positive integers less than or equal to n. Figure 2.22 gives examples for each of the aforementioned base vector assignment methods. The target index is the population's running index, i, so each method automatically ensures that each vector serves as a target vector once per generation. Only the last two methods, however, also ensure that each vector serves as a base vector once per generation. permute[i] refers to the ith element of an array containing a randomly generated permutation of the sequence $[0, 1,..., Np - 1]$ ($Np = 7$ in Fig. 2.22).

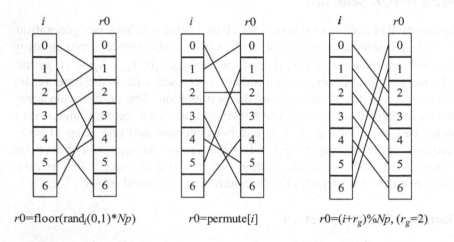

$r0=\mathrm{floor}(\mathrm{rand}_i(0,1)*Np)$ $r0=\mathrm{permute}[i]$ $r0=(i+r_g)\%Np, (r_g=2)$

Fig. 2.22. Three ways to stochastically pair base and target vectors

2.4.3 A Comparison of Random Base Index Selection Methods

Using the ten-dimensional sphere as a test function, Table 2.4 compares the performance of the three stochastic base vector selection methods. (See the Appendix for test function details.) As Table 2.4 shows, all vector selection methods respond similarly when Np is increased. Before convergence becomes regular, increasing Np not only improves the probability of success, but also *decreases* the number of function evaluations needed to reach the optimum. Once convergence becomes regular, however, additional increases in Np only marginally improve the probability of convergence while the number of function evaluations begins to climb. As a result, each method exhibits an optimal population size for which the number of function evaluations is a minimum. In the case of the ten-dimensional sphere, all three stochastic selection methods perform best when $Np = 9$, ($F = Cr = 0.9$ and $i \neq r0 \neq r1 \neq r2$), with each converging reliably in about 6000 function evaluations. Some performance disparities arise, however, once degenerate vector combinations are allowed.

Table 2.4. When best efforts are compared, all the three stochastic selection methods perform similarly. Results are 1000-trial averages of the number of function evaluations needed to reach the optimum to within a pre-specified limit and within the maximum allowed number of generations (see the Appendix for the function value to reach). P is the fraction of trials that were successful. For these results, $F = Cr = 0.9$ and $i \neq r0 \neq r1 \neq r2$.

Np	$r0 = \text{floor}(\text{rand}_i(0,1) \cdot Np)$		$r0 = \text{permute}[i]$		$r0 = (i + r_g)\%Np$	
	Evals.	P	Evals.	P	Evals.	P
5	36,616.0	0.001	17,929.0	0.004	–	0
6	14,215.0	0.074	16,804.2	0.309	17,627.9	0.583
7	12,917.2	0.889	10,017.1	0.961	9047.00	0.977
8	7097.05	0.982	6582.3	0.979	7086.11	0.995
9	6006.70	0.994	5954.05	0.995	5927.24	0.998
10	6039.08	0.996	5969.34	1.0	6669.14	1.0
11	6433.55	0.998	6431.55	0.999	6843.09	1.0
12	71,10.87	0.999	7195.95	1.0	8213.57	1.0
13	79,86.33	0.999	8031.48	1.0	8856.00	1.0
14	90,15.09	1.0	9040.13	1.0	10,509.7	1.0
15	10,095.4	1.0	10,214.1	1.0	11,557.2	1.0

2.4.4 Degenerate Vector Combinations

If indices are chosen without restrictions, there is no guarantee that i, $r0$, $r1$ and $r2$ will be distinct. When these indices are not mutually exclusive, DE's novel trial vector-generating strategy reduces to uniform crossover only, duplication of the base vector, an alternative form of recombination, or mutation only. These possibilities are explored below, first by looking at the three degenerate combinations of indices that comprise the mutant vector, $r0$, $r1$ and $r2$, and then by considering the three interactions of the target index, i, with the mutant indices.

Degenerate Combinations of Mutant Indices: r0, r1, r2

r1 = r2: No Mutation. If $r1 = r2$, then the differential formed by the corresponding vectors will be zero and the base vector, $\mathbf{x}_{r0,g}$, will not be mutated:

$$r1 = r2 (= r0): \quad \mathbf{v}_{i,g} = \mathbf{x}_{r0,g}. \tag{2.13}$$

When indices are chosen without restrictions, $r1$ will equal $r2$ on average once per generation, i.e., with probability $1/Np$. The probability that all three indices will be equal is $(1/Np)^2$, but either way, the result is the same: a randomly chosen base vector that has *not* undergone mutation is recombined with the target vector by means of conventional uniform crossover:

$$\mathbf{u}_{i,g} = u_{j,i,g} = \begin{cases} x_{j,r0,g} & \text{if } \left(\text{rand}_j(0,1) \le Cr \vee j = j_{\text{rand}}\right) \\ x_{j,i,g} & \text{otherwise.} \end{cases} \tag{2.14}$$

Requiring the base vector to contribute a parameter when $j = j_{\text{rand}}$ ensures that the trial vector will not simply reproduce the vector with which it is compared, i.e., the target vector, $\mathbf{x}_{i,g}$. If, however, Cr is greater than 0, the possibility exists that the trial vector will *duplicate* the base vector. When $Cr = 1$, and $r1 = r2$, duplication is a certainty:

$$r1 = r2 (= r0) \wedge Cr = 1: \quad \mathbf{u}_{i,g} = \mathbf{v}_{i,g} = \mathbf{x}_{r0,g}. \tag{2.15}$$

More generally, the probability that the base vector will be duplicated is the product of the probability that $r1 = r2$ and the probability that all parameters are inherited from the mutant, $\mathbf{v}_{i,g}$. Since Cr mediates a random process having just two possible outcomes (mutant or target), the number of parameters inherited from the mutant is governed by a binomial distribution. Thus, the probability of inheriting x mutant parameters in n tries is

$$p(X = x) = \frac{n!}{x!(n-x)!} Cr^x (1 - Cr)^{n-x}, \qquad n! = \prod_{k=1}^{n} k. \tag{2.16}$$

Since one parameter is certain to be taken from the mutant, $n = D - 1$. Thus, the probability, given Cr, that all $D - 1$ of the remaining parameters will also be inherited from the mutant ($x = D - 1$) is

$$p(X = D - 1) = \frac{(D-1)!}{(D-1)!\,0!} Cr^{D-1} (1 - Cr)^0 = Cr^{D-1}, \qquad 0! \equiv 1. \tag{2.17}$$

When difference indices are chosen without restrictions, the probability that the base vector will not be mutated is $1/Np$, making Cr^{D-1}/Np the probability that a base vector will be duplicated.

$r1 = r0$ or $r2 = r0$: Arithmetic Recombination. Another special case occurs when either of the difference indices, $r1$ or $r2$, equals the base index, $r0$. When indices are chosen without restrictions, each coincidence occurs on average once per generation. Equation 2.18 elaborates the two possibilities that result when DE's three-vector mutation formula (Eq. 2.5) reduces to a linear relation between the base vector and a single difference vector:

$$r1 = r0: \quad \mathbf{v}_{i,g} = \mathbf{x}_{r0,g} + F \cdot \left(\mathbf{x}_{r0,g} - \mathbf{x}_{r2,g} \right)$$

$$(2.18)$$

$$r2 = r0: \quad \mathbf{v}_{i,g} = \mathbf{x}_{r0,g} + F \cdot \left(\mathbf{x}_{r1,g} - \mathbf{x}_{r0,g} \right).$$

Each two-vector linear combination defines a line that connects the base vector to one of the two difference vectors (Fig. 2.23). F plays the role of a coefficient of combination that determines which point along the line is targeted. In the parlance of evolutionary computation, this "line search" is usually called either *continuous* or *arithmetic recombination*. This book adopts the term "arithmetic recombination". Section 2.6 explores this process more thoroughly.

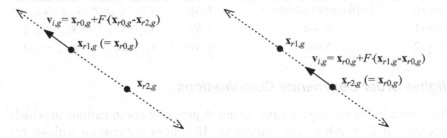

Fig. 2.23. Mutation degenerates into two-vector arithmetic recombination when either $r1 = r0$ (left) or $r2 = r0$ (right).

Degenerate Combinations Involving the Target Index, *i*

$r0 = i$: Mutation Only. If the base index, $r0$, is not different from the target index, i, then crossover reduces to mutation of the target vector. In this scenario, Cr plays the role of a mutation probability:

$$u_{j,i,g} = \begin{cases} x_{j,i,g} + F \cdot \left(x_{j,r1,g} - x_{j,r2,g} \right) & \text{if } \left(\text{rand}_j(0,1) \le Cr \vee j = j_{\text{rand}} \right), \\ x_{j,i,g} & \text{otherwise.} \end{cases} \qquad (2.19)$$

When base vector indices are randomly selected without restrictions, these degenerate vector combinations occur with probability $1/Np$.

$i = r1$ or $i = r2$. Each of the coincidental events, $i = r1$ and $i = r2$, occurs with probability $1/Np$ when indices are chosen without restrictions. Neither coincidence reduces DE's generating process to a conventional one; mutants are still three-vector combinations and crossover recombines distinct base and target vectors (assuming $r0 \ne i$).

Table 2.5 summarizes the possible degenerate vector combinations that can occur when difference indices are chosen without restrictions, i.e., index $= \text{floor}(\text{rand}_i(0,1) \cdot Np)$.

Table 2.5. First-order degenerate combinations

Event	Degenerate process	Prob.	Result
$r1 = r2$	Uniform crossover	$1/Np$	$\mathbf{v}_{i,g} = \mathbf{x}_{r0,g}$
	Duplication of base vector	Cr^{D-1}/Np	$\mathbf{u}_{i,g} = \mathbf{x}_{r0,g}$
$r0 = r1$	Intermediate recombination	$1/Np$	$\mathbf{v}_{i,g} = \mathbf{x}_{r0,g} + F \cdot (\mathbf{x}_{r0,g} - \mathbf{x}_{r2,g})$
$r0 = r2$	Intermediate recombination	$1/Np$	$\mathbf{v}_{i,g} = \mathbf{x}_{r0,g} + F \cdot (\mathbf{x}_{ri,g} - \mathbf{x}_{r0,g})$
$i = r0$	Differential mutation	$1/Np$	$\mathbf{v}_{i,g} = \mathbf{x}_{i,g} + F \cdot (\mathbf{x}_{r1,g} - \mathbf{x}_{r2,g})$
$i = r1$	None	$1/Np$	$\mathbf{v}_{i,g} = \mathbf{x}_{r0,g} + F \cdot (\mathbf{x}_{i,g} - \mathbf{x}_{r2,g})$
$i = r2$	None	$1/Np$	$\mathbf{v}_{i,g} = \mathbf{x}_{r0,g} + F \cdot (\mathbf{x}_{r1,g} - \mathbf{x}_{i,g})$

Higher Order Degenerate Combinations

The above index pairings are first-order degenerate combinations in which only two of four indices are coincident. If indices are chosen without restrictions, the same index may be chosen more than twice. In practice, the effects of higher order degenerate combination are small because their probability is inversely proportional to powers of $Np \ge 2$.

2.4.5 Implementing Mutually Exclusive Indices

Enforcing $i \ne r0$

If base indices are chosen randomly, as they are in classic DE, then $r0 = i$ can be prevented by using a "do–while" loop to reselect $r0$ until it no longer equals the target vector index (Fig. 2.24).

```
do
{
    r0=floor(rand_i(0,1)*Np);
}while(r0==i);
```

Fig. 2.24. To ensure that base and target vectors are different, $r0$ should be reselected.

Similarly, if base vectors are the elements of a permutation, then $r0$ can be redrawn from the remaining list of unused indices, except when i is the last element of the permutation. In the random offset method, choosing r_g from the more restricted range $[1, Np - 1]$, ensures that $r0 = i$ does not occur.

Mutually Exclusive Indices: $i \neq r0 \neq r1 \neq r2$

Once the base vector has been determined, difference indices can be chosen. Perhaps the simplest way to implement mutually exclusive indices is to use a pair of "do–while" loops (Fig. 2.25) to reselect any difference index that happens to equal the target, base or a previously chosen difference index.

```
do
{
    r1=floor(rand_i(0,1)*Np);
}while(r1==i || r1==r0); // "||" is "or"
do
{
    r2=floor(rand_i(0,1)*Np);
}while(r2==i || r2==r0 || r2==r1);
```

Fig. 2.25. Given $r0 \neq i$, distinct indices should be selected with a pair of do–while loops.

Distinct difference indices can be taken from arrays of random permutations of the sequence $[0, 1,..., Np - 1]$. Methods for generating random permutations are presented in Sect. 5.2 and as an option in the Matlab code on this book's companion CD-ROM.

2.4.6 Gauging the Effects of Degenerate Combinations: The Sphere

Table 2.6 calls upon the ten-dimensional sphere to reveal how the presence of degenerate combinations affects both the speed and probability with which each of three stochastic base index selection methods converges. Because it is simple, the sphere provides a good way to interpret the effect of degenerate vector combinations. Performance is measured at the value of Np that minimizes the average number of function evaluations. The first row of results, labeled "All", shows the combined effect of all degenerate combinations (any $r0$, $r1$, $r2$). For the final row of results, labeled "None", indices are mutually exclusive ($i \neq r0 \neq r1 \neq r2$) and degenerate combinations are forbidden. The middle rows record what happens when *only* the designated index coincidence is permitted.

Table 2.6. DE's performance is influenced by the way in which trial vector indices are chosen. Here, the effects of degenerate vector combinations on the three base index selection schemes are compared. Results are 1000-trial averages, with $F = Cr = 0.9$. The value of Np is that which yields the answer in the fewest number of function evaluations, while P is the corresponding probability of success. Data for the last row "None", has been copied from Table 2.4.

Allowed event	$r0 = \text{floor}(\text{rand}_i(0,1) \cdot Np)$			$r0 = \text{permute}[i]$			$r0 = (i + r_g)\% Np$		
	Np	Evals.	P	Np	Evals.	P	Np	Evals.	P
All	14	6479.79	0.992	14	6359.32	0.996	13	6549.55	1.0
$r1 = r2$	13	6585.71	0.797	14	6522.19	0.881	13	6568.55	0.993
$r0 = r1$	9	6388.09	0.967	9	6341.93	0.976	9	6371.36	0.994
$r0 = r1$	13	5832.56	1.0	13	5743.42	1.0	13	5769.72	1.0
$i = r0$	9	6067.18	0.991	9	5940.59	0.993	10	10,829.5	1.0
$i = r1$	9	6009.60	0.992	9	6007.01	0.998	9	6204.13	0.999
$i = r2$	9	5955.17	0.992	9	5847.14	0.998	9	5729.59	1.0
None	9	6006.70	0.994	9	5954.05	0.995	9	5927.24	0.998

All: Any r0, r1 and r2

The first row of data summarizes the combined influence of all degenerate combinations, including higher order degenerate combinations. Although the large optimal population size helps to keep convergence probability competitive, it also slows convergence speed.

r1 = r2

Except for the anomalous behavior of the random offset method when $r0 = i$, all three base index selection methods exhibit their worst performance when equal difference indices are allowed ($r1 = r2$). At $Cr = 0.9$, a significant fraction of these events (about 39%) duplicate base vectors. Reevaluating duplicated vectors wastes time and accepting them reduces the population's effective size. Indeed, Table 2.6 shows that when $r1 = r2$ is allowed, all three base index selection methods require relatively large populations. In this case, increasing Np to compensate for duplicated entries slows convergence without making it reliable.

r0 = r1

Because difference vector $x_{r2,g}$ is preceded by a minus sign, $r0 = r1$ places the recombinant farther away from $x_{r2,g}$ than was $x_{r1,g}$ whenever $F > 0$ (refer back to Fig. 2.23). For the sphere, accepting this recombinant slows convergence and compromises reliability. This form of recombination also tends to slow convergence on multi-modal functions, but its effect on the probability of convergence will not always be detrimental.

r0 = r2

By contrast, all three base index selection methods performed best when they allowed $r0 = r2$ to transform differential mutation into arithmetic recombination. This is because when $0 < F < 2$, $r2 = r0$ produces a recombinant that lies closer to $x_{r1,g}$ than was $x_{r2,g}$. This contractile mapping improves optimization speed even though Np must be increased to compensate for the additional convergence "pressure". Allowing this index combination typically speeds optimizations of multi-modal functions as well, but unlike the case of the sphere, it is less common that convergence probability will also improve.

i = r0

A careful examination of Table 2.6 shows that the random offset method (last column) exhibits the best probability of convergence under all circumstances. In addition, its speed of convergence is competitive except for the case $i = r0$, when the number of function evaluations balloons to nearly twice that of the other two methods. It may seem curious that permitting the combination $r0 = i$ affects the performance of the random offset method so much more than it does the random selection method, even though $r0 = i$ occurs on average once per generation in both cases. The performance disparity arises because when the random offset equals zero ($r_g = 0$), an entire generation of target–base pairings is turned into degenerate combinations, whereas unrestricted random selection spreads them uniformly over the generations. When allowed, the same "identity mapping" of target and base vectors also occurs in the permutation method, but its effect is negligible since it occurs on average only once every $Np!$ generations.

$i = r1$ and $i = r2$

The influence of $i = r1$ and $i = r2$ is more difficult to analyze than that of the corresponding pair of events $r0 = r1$ and $r0 = r2$, but it mirrors their behavior, with one event speeding convergence ($i = r2$) and the other retarding it ($i = r1$). Although its convergence speed distinguishes $i = r1$ from $i = r2$, both events have little effect on either convergence probability or optimal population size when compared to the case of mutually exclusive indices (i.e., "None").

None

Excluding all degenerate target, base and difference vector combinations, i.e., $i \neq r0 \neq r1 \neq r2$, enables DE to achieve both good convergence speed and probability with a relatively small population. Imposing restrictions eliminates the function-dependent effects of degenerate search strategies and ensures that both crossover and differential mutation play a role in the creation of each trial vector.

The effect that degenerate vector combinations have on DE's performance depends in some degree on the objective function. For the hypersphere, however, only $i = r0$ dramatically affected DE's performance. In practice, even these first-order degenerate combinations play only a limited role in the optimization process simply because they become increasingly infrequent as the population grows.

2.4.7 Biased Base Vector Selection Schemes

In GAs, better vectors are more likely to be chosen for recombination (Holland 1973). Similarly, some versions of DE select the base vector based on its objective function value. For example, the algorithm DE/best/1/bin (Storn 1996) always selects the best-so-far vector (*best*) as the base vector, adds a single (*1*) scaled vector difference to it, then creates a trial vector by uniformly crossing (*bin*) the resulting mutant with the target vector. In this algorithm, the base vector always has the lowest objective function value in the current population

$$r0 = \text{best, if } \forall i \in (0,1,...,Np-1), \quad f(\mathbf{x}_{\text{best},g}) \leq f(\mathbf{x}_{i,g}).$$ (2.20)

When compared to random base vector selection *at the same Np*, best-so-far base vector selection usually speeds convergence, reduces the odds of stagnation and lowers the probability of success. Chapter 3 examines this trade-off between speed and reliability when the performance of DE/rand/1/bin and DE/best/1/bin are compared.

Two alternative base vector selection schemes have been proposed that bias solutions toward better vectors without creating the intense selection pressure that the "best" method applies. In Price (1997), a base vector's objective function value must be less than or equal to that of the target vector, $\mathbf{x}_{i,g}$:

$$r0 = \text{better, if } f(\mathbf{x}_{\text{better},g}) \leq f(\mathbf{x}_{i,g}).$$ (2.21)

The other method, DE/target-to-best/1/bin (called "rand-to-best" in (Storn 1996)), uses arithmetic recombination (see Sect. 2.6.3) to generate a base vector that lies on a line between the target vector and the best-so-far vector:

$$\mathbf{x}_{r0,g} = \mathbf{x}_{i,g} + k \cdot (\mathbf{x}_{\text{best},g} - \mathbf{x}_{i,g}), \quad k \in [0,1] = \text{constant}.$$ (2.22)

The constant, k, in Eq. 2.22 controls the bias toward the best-so-far solution.

Compensating for Lost Diversity

Compared to random base vector selection, setting $r0 = \text{best}$ lowers the diversity of the pool of potential trial vectors. Increasing the population size is both a simple and effective way to enhance the diversity of the pool of potential trial vectors, but several other schemes have also been proposed. One idea was to expand the set of vector differences by adding *two difference vectors* together (Price 1996; Storn 1996). Because they are larger than their single difference counterparts, differentials composed of two differences typically require a smaller F to match the convergence rate that one-difference differentials produce. Except for a few early successes on relatively simple functions, this method has not shown much promise, perhaps because adding difference vectors destroys the correlation that the objective function's topography imparts to the one-difference vector differentials (see contour matching in Sect. 2.17).

Making F a random variable is another way to enhance the pool of potential trial vectors. This technique, which is covered extensively in Sect. 2.5.2, has proven useful in cases where stagnation threatens, or when convergence is very slow. In particular, R. Storn has found randomizing the scale factor, F, to be crucial when designing digital filters (Sect. 7.8).

2.5 Differential Mutation

Most dictionaries define mutation as an alteration or change. In the context of genetics and EAs, however, mutation is also seen as change with a random element. Thus, real-valued EAs typically simulate the effects of mutation with additive increments that are randomly generated by a predefined *probability distribution function*, or PDF. DE, however, uses a uniform PDF not to generate increments, but to randomly sample vector differences:

$$\Delta \mathbf{x}_{r1,r2} = (\mathbf{x}_{r1} - \mathbf{x}_{r2}). \qquad (2.23)$$

In a population of Np distinct vectors, there will be $Np \cdot (Np - 1)$ non-zero vector differences and Np null differences having zero magnitude giving a total of Np^2 vector differences. Figure 2.26 pictures an arbitrary population of 5 vectors and the sheaf of 20 non-null difference vectors that they generate.

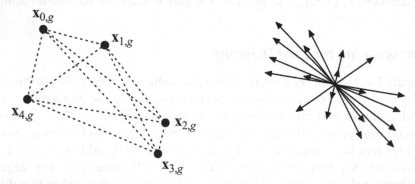

Fig. 2.26. The figure on the right displays the sheaf of 20 vector differences generated by the population of 5 vectors shown on the left. Here, differentials have been scaled by half ($F = 0.5$), and transported to a common origin. Note that the distribution is symmetric about zero.

The distribution of difference vectors will depend on the distribution of vectors and this will be different for each objective function. Each distribution, however, will be symmetric about zero because every pair of vectors gives rise to two opposite but equal difference vectors, since reversing the order of the vectors in the differential reverses the sign of the differential:

$$\Delta\mathbf{x}_{r1,r2,g} = (\mathbf{x}_{r1,g} - \mathbf{x}_{r2,g}) = -(\mathbf{x}_{r2,g} - \mathbf{x}_{r1,g}) = -\Delta\mathbf{x}_{r2,r1,g}. \qquad (2.24)$$

Since each difference vector can be paired with a differential of equal value but opposite sign, and since all vector differences are equally probable, both the sum and average over all Np^2 difference vectors are zero. Equation 2.25 sums the Np^2 vector differences (including the Np cases when $i = k$ and $\Delta\mathbf{x} = 0$) and normalizes the result. The brackets, $\langle\,\rangle$, indicate that $\Delta\mathbf{x}$ is an (ensemble) average taken over all population members, not an expectation or a time average:

$$\langle\Delta\mathbf{x}\rangle_g = \frac{1}{Np^2}\sum_{i,k=0}^{Np-1}(\mathbf{x}_{i,g} - \mathbf{x}_{k,g}) = 0. \qquad (2.25)$$

2.5.1 The Mutation Scale Factor: F

Limits on F

Upper. The stated range for F is $(0,1)$, although 1.0 is an empirically derived upper limit in the sense that no function that has been successfully optimized has required $F > 1$. This is not to say that solutions are not possible when $F > 1$, but only that they tend to be both more time consuming and less reliable than if $F < 1$. When $F = 1$ exactly, otherwise distinct vector combinations become indistinguishable:

$$\mathbf{x}_{r0,g} + \mathbf{x}_{r1,g} - \mathbf{x}_{r2,g} = \begin{cases} \mathbf{x}_{r0,g} + F\cdot(\mathbf{x}_{r1,g} - \mathbf{x}_{r2,g}) \\ \mathbf{x}_{r1,g} + F\cdot(\mathbf{x}_{r0,g} - \mathbf{x}_{r2,g}) \end{cases} \text{when } F = 1. \qquad (2.26)$$

This discontinuity at $F = 1$ reduces the number of mutants by half and can result in erratic convergence unless $Cr < 1$, since $Cr = 1$ further restricts the pool of possible trial vectors by not crossing mutant and target parameters.

Lower. In general, selection tends to reduce the diversity of a population, whereas mutation increases it. To avoid premature convergence, it is crucial that F be of sufficient magnitude to counteract this selection pressure. Zaharie (2002) recently demonstrated the existence of what is effectively a lower limit for F, finding that if F is too small, the population can converge even if selection pressure is absent. In her study, Zaharie measured population diversity as the variance of its parameter values. Because all variables are independent in the absence of selection pressure, population diversity can be measured by tracking the variance of a single parameter of the population. In Eq. 2.27, the subscript, "*x*", in $P_{x,g}$ is set in italics to emphasize that the variance and mean are computed using one *parameter* from each vector in the population (the particular parameter is not specified):

$$\text{Var}(P_{x,g}) = \frac{1}{Np} \sum_{i=0}^{Np-1} \left(x_{i,g} - \langle x \rangle_g\right)^2; \quad \langle x \rangle_g = \frac{1}{Np} \sum_{i=0}^{Np-1} x_{i,g} . \tag{2.27}$$

Using a methodology pioneered by H.-G. Beyer (1999), Zaharie computed the expected variance of DE's mutant and trial populations given the variance of the population. The goal was to determine which combinations of DE control parameters were likely to result in premature convergence due solely to the inability of the algorithm to generate a trial population as diverse as the population. To simplify her analysis, Zaharie dropped DE's usual demand that base and target vectors be different, although the requirement that base and difference vectors be distinct was retained. By dropping the demand that at least one trial parameter be inherited from the mutant, Zaharie also assumed that *Cr* is a true crossover probability, p_{Cr}. In order to compute the expected population variance, Zaharie further modified the standard DE algorithm by multiplying *F* by a Gaussian random variable, ξ_j, that is chosen anew *for each parameter*

$$v_{j,i,g} = x_{j,r0,g} + \tilde{F}_j \cdot \left(x_{j,r1,g} - x_{j,r2,g}\right), \quad \tilde{F}_j = F \cdot \xi_j, \quad \xi_j \approx N(0,1). \tag{2.28}$$

With these caveats, Zaharie determined that the expected variance of the mutant population is related to the variance of the population by the formula:

$$E\left(\text{Var}(P_{v,g})\right) = \left(2F^2 + \frac{Np-1}{Np}\right)\text{Var}(P_{x,g}). \tag{2.29}$$

If this mutant population is then crossed with the original population, the expected trial population variance becomes:

$$E\left(\text{Var}(P_{u,g})\right) = \left(2F^2 p_{Cr} - \frac{2p_{Cr}}{Np} + \frac{p_{Cr}^2}{Np} + 1\right)\text{Var}(P_{x,g}). \tag{2.30}$$

Consequently, DE control parameter combinations that satisfy the equation:

$$2F^2 - \frac{2}{Np} + \frac{p_{Cr}}{Np} = 0 \qquad (2.31)$$

can be considered to be critical since they result in a population whose variance remains constant except for random fluctuations. When selection is "turned off", Eq. 2.31 predicts that F will display a critical value, F_{crit}, such that the population variance decreases when $F < F_{\text{crit}}$ and increases when $F > F_{\text{crit}}$. Solving Eq. 2.31 for F gives F_{crit} as

$$F_{\text{crit}} = \sqrt{\frac{\left(1 - \dfrac{p_{Cr}}{2}\right)}{Np}} . \qquad (2.32)$$

Thus, F_{crit} establishes a lower limit for F in the sense that smaller values will induce convergence even on a level objective function landscape. Figure 2.27 confirms the prediction by Zaharie that $F = 0.1341$ is a critical value when $Np = 50$ and $p_{Cr} = 0.2$.

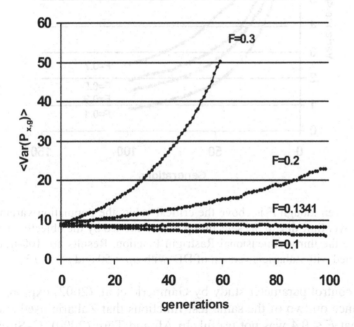

Fig. 2.27. The evolution of the variance of a single parameter is displayed for four different values of F. Note that $F \sim 0.134$ is critical in the sense that the variance is nearly constant. Results are for evolution on a flat surface, i.e., all trial vectors are accepted (no selection pressure). These results are 100-trial averages and were generated using Zaharie's modified version of DE, with $Np = 50$ and $p_{Cr} = 0.2$.

Objective function landscapes are seldom flat. In practice, F must be larger than F_{crit} to counteract the additional reduction in variance that selection induces. For example, Zaharie empirically examined three test functions using $Np = 50$, $p_{Cr} = 0.2$ and found that $F \sim 0.3$ was the smallest reliable scale factor and that $F_{crit} = 0.1341$ was too small to forestall premature convergence.

Figure 2.28 illustrates the effect of this additional selection pressure produced by the 30-dimensional Rastrigin function on the population's variance over time at several different values for F.

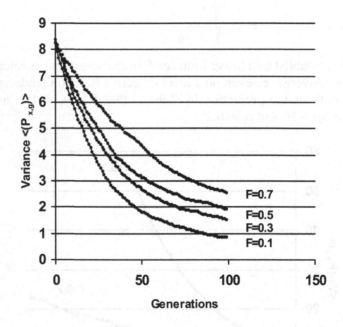

Fig. 2.28. Even though F is above the critical value, the population variance still decreases over time due to the selection pressure exerted by the objective function, in this case the thirty-dimensional Rastrigin function. Results are 100-trial averages obtained with Zaharie's version of DE, with $Np = 50$ and $p_{Cr} = 0.2$.

A DE control parameter study by Gamperle et al. (2002) explored DE's performance on two of the same test functions that Zaharie used and concluded that $F < 0.4$ was not useful. In Ali and Törn (2000), C–Si clusters were optimized with F never falling below $F = 0.4$. On the other hand, Chakraborti et al. (2001; Sect. 7.1) had success minimizing the binding energy of Si–H clusters using values for F ranging from 0.0001 to 0.4, with $F = 0.2$ often proving effective. Such low values for F, however, appear to be atypical. The lower limits suggested by Zaharie and Gamperle et al. more accurately reflect the norm. Zaharie concluded that for the test functions examined, modifying vector differences with a Gaussian distribution did not significantly alter DE's performance compared to when F is held constant. The next section tests this claim and examines several other methods for transforming F into a random variable.

2.5.2 Randomizing the Scale Factor

When compared to the ES, DE shifts the responsibility for adapting step sizes from the mutation distribution's pre-factors to the distribution itself. More specifically, the ES adapts pre-factors (strategy parameters) and multiplies them by the output from a stationary, multi-dimensional PDF, whereas DE multiplies the constant pre-factor, F, by a sample vector difference from an adaptive distribution. Whereas the ES "strategy" parameters adapt to the *absolute* step size, F only affects the *relative* step size since the distribution of vector differences is itself adaptive. Thus, F can be kept constant during optimization without compromising DE's ability to generate steps of the required size. Indeed, keeping F constant has proven effective in the sense that no function that has been solved has required F to be a random variable. Nevertheless, randomizing F offers potential benefits.

Transforming F into a random variable effectively broadens the spectrum of vector differentials beyond the possibilities allowed for by combining vectors. Such an enhanced distribution of differentials might be useful if the population is small and/or symmetrically distributed, since without access to a mutation distribution of sufficient diversity, DE can *stagnate*. When stagnant, DE can no longer find improved solutions because no combination of vector and vector difference leads to a better solution. Instead of coalescing to a single solution, a stagnant population of vectors remains static while still distributed throughout the problem space. The case explored by Lampinen and Zelinka (2000) is hypothetical and subsequent attempts to induce stagnation in test functions with classic DE have been unsuccessful. Nevertheless, randomizing the scale factor is a way to increase the pool of potential trial vectors and minimize the risk of stagnation without increasing the population size.

Transforming F into a random variable also makes the analysis of DE dynamics tractable. By invoking the normal (Gaussian) distribution, Zaharie succeeded not only in predicting critical control parameter combinations, but also in constructing a limited convergence proof (Zaharie 2002). Zaharie based her proof on the general EA convergence criteria set forth by G. Rudolph (1996). Briefly, an evolutionary search algorithm can be proven to converge to within $\varepsilon > 0$ of the global optimum in the long-time limit if its operators fulfill two (sufficient, but not necessary) conditions:

1. The transition probability, through mutation, between any two points in the problem space is strictly positive.

2. Selection is elitist, i.e., that the best-so-far solution is always retained.

DE selection is elitist because the population's current best vector can only be replaced by a better vector. By multiplying F by a normally distributed variable, Zaharie ensured that the unbounded, multi-normal distribution could access any point given enough time. The possibility does exist, however, that all members of a population may have the same value for one or more parameters, in which case no new possibilities for that parameter are generated. Zaharie considers this set to be of zero measure and that it has no impact on the proof that DE is convergent when mutation is augmented by a Gaussian random variable.

Converting F into a random variable, however, involves both selecting a PDF and deciding how often it should be sampled. Zaharie, for example, sampled a zero-mean, normally distributed random variable anew for each parameter, but this is not the only possibility. The next two subsections explore how both the sampling frequency and PDF affect the optimization process.

PDF Sampling Frequency: Dither and Jitter

In Zaharie's version of DE, F_j is a normally distributed random variable that is generated anew for each parameter. For convenience, the practice of generating a new value of F for every *parameter* is called *jitter* and it is signified by subscripting F with the parameter index, j. Alternatively, choosing F anew for each *vector*, or *dithering*, is indicated by subscripting F with the population's running index, i. Dithering scales the length of vector differentials because the same factor, F_i, is applied to all components of a difference vector (Fig. 2.29). As such, dithering does not dramatically depart from traditional DE in which each component of a differential is scaled by the same constant, F. Jitter, however, multiplies each component of the difference vector by a different scale factor, F_j, and this changes not only the scale of the differential, but also its orientation. The rotation that it introduces makes jitter a fundamentally different process than classic DE mutation with F = constant.

When $Cr = 0$, only one trial vector parameter is inherited from the mutated base vector, so it impossible to distinguish jitter from dither, since in both cases only a single instance of F as a random variable occurs per trial vector. In order to compare how jitter and dither affect the optimization process, it is necessary to plot DE's performance *versus Cr*.

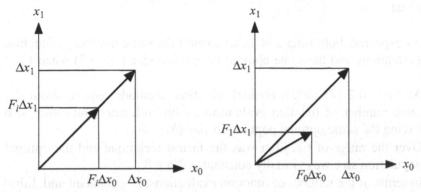

Fig. 2.29. Dithering (left) scales vector differentials, while jitter (right) both scales and rotates them.

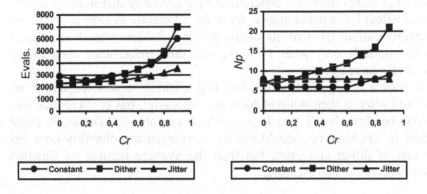

Fig. 2.30. The graph on the left illustrates how implementing jitter and dither with the N(0,1) PDF affects convergence speed compared to holding F constant. Plotted as a function of Cr is the minimum number of function evaluations required to optimize the ten-dimensional hyper-sphere. The graph on the right plots the corresponding optimal Np at which the minimum number of evaluations occurred. For example, at $Cr = 0.8$, the graph on the left shows that jitter took a little more than 3000 evaluations, while the graph on the right shows that the population used to produce this result was $Np = 8$ (at $Cr = 0.8$). Results are 1000-trial averages which, except for the indicated randomization method, were obtained with $F = 0.9$, $r0 = \text{rand}_i(0,1) \cdot Np$, except $r0 \neq r1 \neq r2 \neq i$ (i.e., classic DE).

Using the normal (Gaussian) PDF, N(0,1), to drive dither and jitter, Fig. 2.30 plots both the minimum number of function evaluations and the corresponding optimal population size at which the minimum occurred *versus* *Cr* for each of the three methods when applied to the ten-dimensional hyper-sphere objective function. An inspection of the graphs in Fig. 2.30 reveals that:

- As expected, both jitter and dither exhibit the same number of function evaluations and the same optimal population size (*Np* = 7) when *Cr* = 0.
- At *Cr* = 0.2 (Zaharie's choice), all three methods require about the same number of function evaluations, with both jitter and dither also having the same optimal population size (*Np* = 8).
- Over the range of *Cr*, jitter was the fastest technique and the optimal population size was virtually constant at *Np* = 8.
- In terms of the number of function evaluations, F = constant and dither perform similarly, but dither requires a larger population.

The data in Fig. 2.30 casts suspicion on Zaharie's contention that multiplying each component of a differential by a normally distributed variable does not affect DE's performance. Even for a function as simple as the hyper-sphere, classic DE with its constant F and Zaharie's method of jitter perform similarly only when *Cr* = 0.2 and this performance discrepancy grows as *Cr* increases.

Not shown in Fig. 2.30 is the fact that all trials conducted with both dither and jitter at their optimal *Np* were successful, but convergence was less than perfect when F was kept constant, as Table 2.7 shows. A slight increase in *Np*, however, would put the convergence probability on a par with that of dither and jitter, but then the average number of function evaluations would also increase.

Table 2.7. The fraction of trials that were successful when optimizing the ten-dimensional hyper-sphere using classic DE and F = constant = 0.9. By contrast, all trials with dither and jitter were successful at the specified optimal *Np*.

Cr	0	0.1	0.2	0.3	0.4	0.5	0.6	0.7	0.8	0.9	1.0
P	0.966	0.884	0.95	0.931	0.916	0.881	0.865	0.957	0.971	0.991	1

Also not shown in Fig. 2.30 are the data points associated with *Cr* = 1. These points are not plotted in Fig. 2.30 simply because the large values for dither and F = constant would overwhelm the data for *Cr* ≤ 0.9. Instead, Table 2.8 reports both the average minimum number of function evaluations and the population size at which this minimum occurred for each of the three methods when *Cr* was set equal to 1. When compared to trials using $Cr \leq 0.9$, all three methods required significantly larger populations to offset the loss of diversity that occurs when $Cr = 1$, exactly. In this case, the penalty for enlisting larger populations is slower convergence.

Table 2.8. When $Cr = 1$, both the optimal population size and the number of function evaluations balloon for both dither and F = constant and even jitter takes twice as long to converge as it does when $Cr = 0.9$. Results are 1000-trial averages for the ten-dimensional hyper-sphere using classic DE except for the indicated randomization method using a normal distribution: N(0,1).

Process	Evaluations	Np
F=constant	49,809.5	41
Dither	33,640.1	109
Jitter	6037.11	13

The hyper-ellipsoid (Eq. 2.33) poses a stiffer challenge to optimization algorithms because unlike the symmetrical hyper-sphere, the optimal step size depends on the direction in which the step is taken:

$$f_{\text{ellipsoid}}(\mathbf{x}) = \sum_{j=0}^{D-1} 2^j x_j^2 . \qquad (2.33)$$

Figure 2.31 shows a single contour of constant function value for the two-dimensional version of this function.

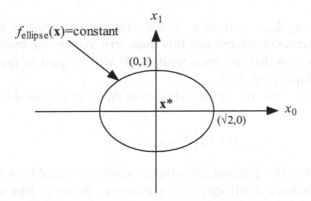

Fig. 2.31. The ellipse is a single contour of the two-dimensional version of the ellipsoidal function described by Eq. 2.33. The optimum, $\mathbf{x}^*$, is located at the origin, $(0,0)$. The principal axes of the ellipse are aligned with the coordinate axes. Taking large steps along x_0 and smaller steps along x_1 efficiently optimizes this function.

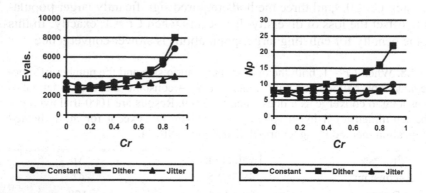

Fig. 2.32. The graph on the left illustrates the effects of jitter, dither and $F = $ constant by plotting, as a function of Cr, the minimum number of function evaluations needed to optimize the ten-dimensional hyper-ellipsoid. The graph on the right plots the corresponding optimal Np at which the minimum number of evaluations occurred. For example, at $Cr = 0.9$, the graph on the left shows that jitter took about 4000 function evaluations when using the population indicated in the graph on the right at $Cr = 0.9$, i.e., $Np = 8$. Results are 1000-trial averages, $F = 0.9$ and classic DE except for randomizing F with a normal distribution $N(0,1)$.

Figure 2.32 profiles how both jitter and dither influence DE's ability to optimize the ten-dimensional hyper-ellipsoid. Except for requiring roughly 15% more function evaluations, the performance profiles for the hyper-ellipsoid are virtually indistinguishable from those generated for the hyper-sphere.

Before taking these profiles to be universal, it is instructive to perform the same experiment, except that this time, trial vectors are evaluated in a coordinate system that has been rotated 45° with respect to the principal axes of the ellipse (Fig. 2.33).

In two dimensions, this rotated version of the ellipse defined by Eq. 2.33 is

$$f_{\text{ellipse}}(\mathbf{x}) = 2(x_0^2 - x_0 x_1 + x_1^2). \tag{2.34}$$

As a result of this rotation, the ellipse, which is separable as defined in Eq. 2.33, becomes nonlinear, i.e., parameters become dependent. The cross-term, $x_0 x_1$, in Eq. 2.34 embodies this parameter dependence (see Sects. 1.2.3 and 2.6.2). Even though rotation does not alter the objective function's topography, the parameter dependence that it induces compromises DE's efficiency in the presence of jitter.

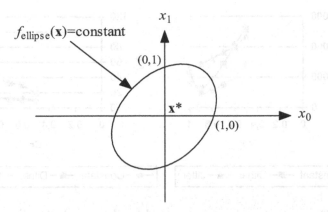

Fig. 2.33. Once rotated, the parameters of the ellipse function become dependent. An efficient search of the long axis along the diagonal now requires that large steps in both coordinate directions occur simultaneously, i.e., that they be *corre-lated*.

As Fig. 2.34 illustrates, transforming the hyper-ellipsoid from a separable function into one with dependent parameters *via* a coordinate system rotation dramatically alters the performance profiles of all three methods. In particular, the data plotted in Fig. 2.34 show that:

- In contrast to results for the separable hyper-ellipsoid, the fastest solutions now occur at high Cr.
- Jitter is now the worst performing method even though population sizes remain relatively small.
- F = constant is the fastest method except when $Cr = 1$, in which case dithering is faster.

At $Cr < 0.5$, the fastest run-times begin to occur at population sizes that are too small to produce reliable convergence. To be fair, the number of function evaluations for different methods must be compared at the same probability of convergence. In previous examples, convergence probabilities were so close to 1 that the small differences between them did not compromise the validity of the performance comparisons. A method for evaluating algorithm performance that gives proper weight to convergence probabilities will be presented in Chap. 3 when several version of DE are tested. For now, higher run-times and worse convergence probabilities make it easy to say that DE's performance on the parameter-dependent (rotated) hyper-ellipsoid deteriorates at low Cr.

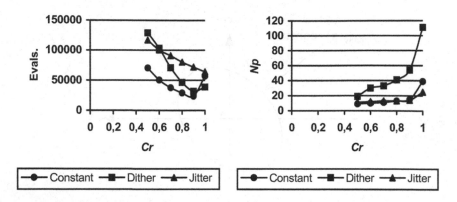

Fig. 2.34. Once rotation induces parameter dependence in the ten-dimensional hyper-ellipsoid, the three techniques become inefficient at low Cr. Population sizes used to produce the graph on the left are plotted in the graph on the right at the corresponding value of Cr. For example, jitter still uses small populations but is slow nonetheless. Despite using large populations, dither is more efficient than jitter when $Cr > 0.6$ and more efficient than F = constant = 0.9 when $Cr = 1$. Keeping F constant, however, uses relatively small populations and gives the overall fastest result at $Cr = 0.9$. All results are 1000-trial averages with classic DE, except for the indicated randomization scheme. For this experiment, the PDF was the normal distribution, N(0,1).

Since it was the fastest method when the hyper-ellipsoid was separable and was competitive with both dither and F = constant on the rotated hyper-ellipsoid at $Cr = 1$, jitter would seem to be a good strategy as long as Cr is chosen wisely. The case of the Chebyshev polynomial, however, suggests differently (see the Appendix for a function description). Like the rotated hyper-ellipsoid, the Chebyshev polynomial fitting problem is a function with dependent parameters that requires correlating step sizes that differ greatly in magnitude from one parameter to the next. Unlike the hyper-ellipsoid, the Chebyshev function is multi-modal. Figure 2.35 compares the number of function evaluations taken by dither to those needed by F = constant to find the coefficients of the nine-dimensional Chebyshev polynomial.

The results in Fig. 2.35 are remarkably similar to those displayed by dither and F = constant for the rotated hyper-ellipsoid in Fig. 2.34, except that now dither gives the overall fastest solution (when $Cr = 1$). Missing from Fig. 2.35 are the results for jitter. Like the case of the rotated hyper-ellipsoid, jitter was most effective when $Cr = 1$, but unlike the case of the rotated hyper-ellipsoid, run-times at this optimal crossover setting were not

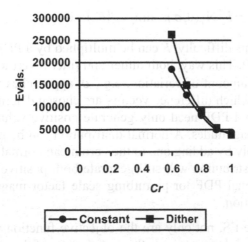

Fig. 2.35. Dither and F = constant perform similarly on the nine-dimensional Chebyshev function, with dither converging in fewer function evaluations than F = constant at $Cr = 1$. Results for jitter are not shown, as run-times were in excess of 6 million function evaluations and convergence was erratic even for large populations. Results are 100-trial averages with $Np = 40$. Both dither and jitter (not shown) used the normal PDF; otherwise, the algorithm was classic DE.

competitive with those turned in by either dither or F = constant. Not only was convergence erratic even with large populations, but the number of function evaluations taken by successful trials never averaged less than 6 *million*, making jitter over 100 times slower than either dither or F = constant. Clearly, these results refute Zaharie's contention that DE's performance is not significantly affected by transforming F into a Gaussian random variable that is sampled anew for each parameter.

Although jitter is effective on separable functions, its poor performance on non-separable, multi-modal functions makes it a questionable strategy for non-linear global optimization with DE *unless the deviations it generates are very small*, e.g., $d = 0.001$ in the case of uniform jitter (see next subsection). The next subsection explores this possibility with some alternatives to the Gaussian PDF.

Other Distributions

The effectiveness of both jitter and dither can be improved by moderating the amount of variation in F_j and F_i, respectively. The problem with Zaharie's formulation in this regard is that as the standard deviation, σ, of the normal (Gaussian) distribution approaches zero, so does F_j (or F_i):

$$F_j = F \cdot \mathrm{N}_j(0,\sigma); \quad \lim_{\sigma \to 0}(F_j) = 0 \tag{2.35}$$

To circumvent this difficulty, F can be multiplied by a PDF whose average value is 1, not 0. This way, both dither and jitter revert to the $F =$ constant model as the amount of variation, e.g., σ, approaches zero. Furthermore, the order in which difference vectors are chosen determines the sign of a differential, so a PDF need only generate positive values in order to scale differential magnitudes. A normal distribution can be given an average value of 1 simply by adding one to the zero mean normal PDF, N(0,1), but the resulting distribution will still generate both positive and negative values. The traditional PDF for perturbing scale factor magnitudes is the *log-normal* distribution.

Log-normal. In the ES, not only are the objective function variables mutated and recombined, but so too are the components of the adaptive correlation matrix. Of the correlation matrix's D^2 components, D are scale factors while the remaining $D \cdot (D - 1)$ are rotation angles. Although the ES perturbs rotation angles with normally distributed random variables, it turns to the *log-normal* PDF to mutate the strategy parameters that regulate step sizes (Bäck 1996). An instance of a log-normal random variable *for DE* can be computed as

$$F_j = F \exp\left(\tau \left(N_j(0,1) - \frac{\tau}{2} \right) \right). \tag{2.36}$$

The factor, τ, controls the spread of the distribution while the term $\tau/2$ is an empirically derived factor that normalizes the expected value of the distribution to 1.0. When $\tau = 0$, the average value of the log-normal PDF is the constant value 1, so all $F_j = F$. In this model, the distribution's variance can be controlled independently of F. Figure 2.36 shows how the spread of the log-normal distribution affects DE's ability to optimize both the rotated hyper-ellipsoid and the Chebyshev polynomial fitting problem.

In both plots, jitter requires an increasing number of function evaluations as τ increases. For the Chebyshev polynomial fitting problem, this increase is explosive. By contrast, dither actually shows a slight *decrease* in the number of function evaluations when compared to $F =$ constant, with the best performance occurring near $\tau = 0.4$. The improvement amounts to roughly 10% for the rotated hyper-ellipsoid and just over 40% for the Chebyshev problem.

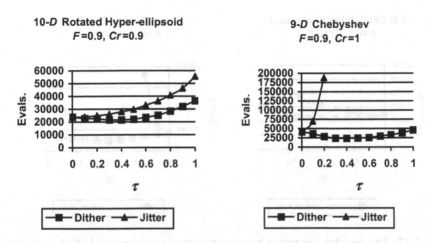

Fig. 2.36. Jitter performs worse as the variance of the log-normal distribution is increased from zero. By contrast, dither is faster than F = constant (τ = 0) on both the rotated hyper-ellipsoid (while τ > 0.6) and the Chebyshev problem (while $\tau \leq$ 0.9). In both cases, the fastest convergence occurs near τ = 0.4. Data points are 1000-trial averages for the rotated hyper-ellipsoid and 100-trial averages for the Chebyshev problem. Results were obtained using classic DE except for the indicated randomization method with a log-normal PDF. $Np = 40$.

Uniform. The uniform distribution can also be transformed into a PDF whose average value is F and whose spread is an independent variable. Equation 2.37 illustrates one possibility:

$$F_j = F + d \cdot \left(\mathrm{rand}_j(0,1) - 0.5\right), \quad d < 2F. \qquad (2.37)$$

To keep F_i positive, d must be less $2F$. Like τ in the log-normal PDF, d controls the amount of variation in the uniform PDF. The log-normal PDF, however, occasionally generates both very large and very small perturbations, both of which can degrade DE's performance because they tend to slow progress toward the optimum. The uniform distribution with $d \sim F$ effectively eliminates these extremes. Figure 2.37 compares DE's performance on both the rotated hyper-ellipsoid and Chebyshev polynomial fitting problem as a function of the spread, d.

Figure 2.37 shows that as long as $d < 0.1$, jitter remains competitive, although once $d > 0.2$, its performance quickly deteriorates. It should be emphasized, however, that a very small amount of jitter can prove useful, sometimes providing solutions that would otherwise be impossible with F = constant.

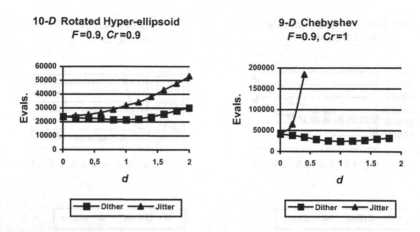

Fig. 2.37. The profiles generated by the uniform and log-normal PDFs are very similar. Jitter's performance worsens as the variation increases and dither converges faster than $F = $ constant ($d = 0$) when $0 < d < 1.4$. Dither's best performance in this case occurs when $d = 0.9$ (not plotted). Results are 1000-trial averages for the rotated hyper-ellipsoid and 100-trial averages for the Chebyshev problem. In both cases, all trials were successful. The algorithm was classic DE except for the specified randomization technique with the uniform PDF. $Np = 40$.

In particular, experiments with the digital filter design program FIWIZ (Sect. 7.8). have shown that uniform jitter on the order of $d = 0.001$ is often indispensable. In addition, jitter can reduce the size of the population that DE needs to solve a given problem.

Dither's performance changes little when log-normal noise replaces the uniform PDF. The slightly larger optimal population size posted by the log-normal PDF suggests that the small steps present in the log-normal PDF but excluded by the uniform PDF only marginally inflate the optimal population size. The similarly of the two results also suggests that the very large steps generated when $\tau = 0.4$ are too infrequent to have much impact on convergence speed.

Power Law. Just as choosing a PDF complicates the optimization task, so too does having to decide what level of variability is suitable for the normal, log-normal and uniform models. One PDF that avoids this difficulty is based on a power law. An instance of a power law variable can be generated by raising a uniformly distributed random value, rand(0,1), to the power, q, where $q = (1/F) - 1$:

$$F_j = \text{pow}\left(\text{rand}_j(0,1), q\right) = \text{rand}_j(0,1)^q, \quad q = \frac{1}{F} - 1. \tag{2.38}$$

This distribution has F as its average value and when F is between 0 and 1, F_j will also lie in this interval. For example, when $F = 0.5$ and $q = 1$, the distribution is uniform between (0,1). As F approaches either 1 or 0, the amount of variation decreases so that when $F = 1$ all $F_j = 1$ and when $F = 0$ all $F_j = 0$. When $F > 1$, q is negative and all F_j are greater than 1. Table 2.9 reports DE's performance on both the rotated hyper-ellipsoid and the Chebyshev polynomial fitting problems when F_j is a random variable distributed according to the power law in Eq. 2.38.

Table 2.9. At $F = 0.9$, the power law distribution has a small variance, so results for jitter and dither on the ten-dimensional rotated hyper-ellipsoid are close to those for $F_i = $ constant. Nevertheless, the variation is large enough to inflate jitter's function evaluations to twice that of dither in the case of the Chebyshev polynomial fitting problem. Results are 1000-trial averages for the rotated hyper-ellipsoid and 100-trial averages for the Chebyshev function. The algorithm was classic DE except for the stated randomization method using a power law PDF. All trials were successful (P = 1.0).

Function	Rotated hyper-ellipsoid $Cr = 0.9$			Chebyshev $Cr = 1$		
Process	Np	Evals.	P	Np	Evals.	P
F = constant = 0.9	16	23,208.2	1.0	36	43,608.8	1.0
Dither	16	23,060.5	1.0	34	35,966.8	1.0
Jitter	15	25,212.4	1.0	22	70,358.4	1.0

In both cases, dither's fast convergence did not require a compensating increase in population size. Jitter, although competitive on the rotated hyper-ellipsoid, took twice as many function evaluations to solve the Chebyshev problem as did dither, even though it operated with a smaller population. Still, this is much faster than the 6 million function evaluations that jitter took when driven by the normal PDF, N(0,1). In this model, the amount of jitter cannot be chosen independently of F. For example, using a very small amount of jitter will require F to be very close to 1.

2.6 Recombination

Recombination randomly exchanges or merges parameters from two or more vectors to create one or more trial vectors. *Discrete recombination*, also known as *crossover*, is an operation in which trial vector parameters are copied from randomly selected vectors. Since it only copies information, crossover can be applied to binary, real-valued or even symbolic data. By contrast, *continuous* or *arithmetic recombination* expresses trial vectors as linear combinations of vectors, so it is inapplicable to symbolic data and inappropriate for binary variables. Both crossover and arithmetic recombination have a variety of implementations. Those with particular relevance to DE are described below.

2.6.1 Crossover

It was originally thought that crossover could exponentially increase the probability of above-average parameter groupings (alleles) while exponentially decreasing the likelihood of less than average groupings (Holland 1973). More recent analysis shows that growth is not exponential because the selective advantage of a parameter grouping decreases as it becomes more prevalent (Macready and Wolpert 1998). Empirical evidence also exists suggesting that (uniform) crossover does not decrease the time complexity of an EA but merely speeds convergence by a constant factor (Mühlenbein and Schlierkamp-Voosen 1993). Nevertheless, crossover plays a significant role in most EAs.

Global discrete recombination refers to the case where both vectors are chosen anew for each trial *parameter* (Bäck and Schwefel 1993). The ES globally recombines its strategy variables, but like DE and most GAs, it crosses objective function parameters from just two vectors (*dual crossover*). Both DE and ES also use crossover to create a single trial vector, whereas most GAs cross two vectors to produce two trial vectors, often by one-point crossover.

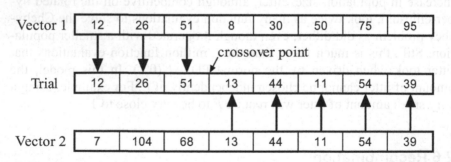

Fig. 2.38. One-point crossover. Each string represents a vector of parameters. In this figure, $D = 8$ and values are integral, although real-valued or symbolic data could also have been used. Each vector contributes a contiguous series of parameter values to the trial vector. The crossover point is randomly chosen. In this case, it occurs between the third and fourth parameters.

One-Point Crossover

There are several ways to assign donors to trial parameters. For example, *one-point crossover* randomly selects a single *crossover point* such that all parameters to the left of the crossover point are inherited from vector 1, while those to the right are copied from the vector 2 (Fig. 2.38) (Holland 1995). GAs often construct a second trial vector by reversing the roles of the vectors, with vector 2 contributing the parameters to the left of the crossover point and vector 1 supplying all trial parameters to the right of the crossover point.

N-Point Crossover

N-point crossover randomly subdivides the trial vector into $n + 1$ partitions such that parameters in adjacent partitions are inherited from different vectors. If n is odd (e.g., one-point crossover), parameters near opposite ends of a trial vector are less likely to be taken from the same vector than when n is even (e.g., $n = 2$) (Eshelman et al. 1989). This dependence on parameter separation is known as *representational* or *positional bias*, since the particular way in which parameters are ordered within a vector affects algorithm performance. Studies of n-point crossover have shown that recombination with an even number of crossover points reduces the representational bias at the expense of increasing the disruption of parameters that are closely grouped (Spears and DeJong 1991). To reduce the effect of their individual biases, DE's exponential crossover employs both one- and two-point crossover.

Exponential Crossover

DE's exponential crossover achieves a similar result to that of one- and two-point crossover, albeit by a different mechanism. One parameter is initially chosen at random and copied from the mutant to the corresponding trial parameter so that the trial vector will be different from the vector with which it will be compared (i.e., the target vector, $x_{i,g}$). The source of subsequent trial parameters is determined by comparing Cr to a uniformly distributed random number between 1 and 0 that is generated anew for each parameter, i.e., $\text{rand}_j(0,1)$. As long as $\text{rand}_j(0,1) \le Cr$, parameters continue to be taken from the mutant vector, but the *first time* that $\text{rand}_j(0,1) > Cr$, the current *and all remaining* parameters are taken from the target vector. The example in Fig. 2.39 illustrates a case in which the exponential crossover model produced two crossover points.

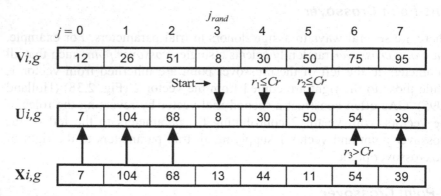

Fig. 2.39. Exponential crossover. Starting at the randomly chosen parameter index, j_{rand} (= 3), trial parameters are inherited from the mutant, $\mathbf{v}_{i,g}$, as long as $rand_j(0,1) \leq Cr$ (e.g., $j = 4, 5$). The first time that $rand_j(0,1) > Cr$, *all* the remaining trial parameters (e.g., $j = 6, 7, 0, 1, 2$) are inherited from the target vector, $\mathbf{x}_{i,g}$. Indices are computed modulo $D = 8$.

Figure 2.40 describes the process in C-style pseudo-code. Parameter indices are computer modulo D. The exponential method's name reflects the fact that the number of inherited mutant parameters is an exponentially distributed random variable. For example, the probability that the initial, randomly chosen parameter is the trial vector's only mutant parameter is equal to the chance that the first comparison of $rand_j(0,1)$ and Cr results in a failure, i.e., that $rand_j(0,1) > Cr$. Thus, the odds of crossover resulting in exactly one mutant parameter are

$$p(x = 1) = 1 - Cr. \tag{2.39}$$

```
jr=floor(rand(0,1)*D); // 0<=jr<D
j=jr;
do
{
    uj,i=vj,i;  // Child inherits a mutant parameter
    j=(j+1)%D;// Increment j, modulo D
}while(rand(0,1)<Cr && j!=jr); // Take another mutant parameter?
while(j!=jr) //Take the rest, if any, from the target
{
    uj,i=vj,i;
    j=(j+1)%D;
}
```

Fig. 2.40. C-style pseudo-code for DE's exponential crossover scheme

Similarly, the probability that two mutant parameters are inherited is the same as the chance that there will be one success before the first failure:

$$p(x=2) = (1-Cr)\cdot Cr. \tag{2.40}$$

In general, the probability that the trial vector will inherit exactly n mutant parameters is

$$p(x=n) = (1-Cr)\cdot Cr^{n-1} = Cr^{n-1} - Cr^{n}. \tag{2.41}$$

Summing these terms gives the cumulative distribution function. Once summed, only the first and last terms remain, since consecutive contributions contain identical terms of opposite sign that cancel. As a result, the probability that n or fewer parameters are inherited from the mutant is

$$p(x \le n) = \sum_{k=1}^{n} Cr^{k-1} - Cr^{k} = 1 - Cr^{n}. \tag{2.42}$$

One way to eliminate any representational bias associated with the crossover process is to shuffle the vector indices, perform crossover and then un-shuffle the trial vector indices (Caruana et al. 1989). Alternatively, the representational bias inherent in n-point crossover can be eliminated if donors are determined by D independent random trials. This alternative, known as *uniform* crossover, is the discrete recombination method that DE employs most often.

Uniform (Binomial) Crossover

G. Syswerda defined uniform crossover as a process in which independent random trials determine the source for each trial parameter (Syswerda 1989). Crossover is uniform in the sense that each parameter, regardless of its location in the trial vector, has the same probability, p_{Cr}, of inheriting its value from a given vector. For this reason, uniform crossover does not exhibit a representational bias. Syswerda's original definition also allows for the possibility that donors are chosen with different probabilities, but $p_{Cr} = 0.5$ is the most commonly cited value (both donors are equally probable).

When the vectors being crossed are randomly chosen from the same population, p_{Cr} and $1 - p_{Cr}$ create the same pool of trial vectors. For example, both $p_{Cr} = 0.3$ and $p_{Cr} = 0.7$ produce a vector that on average inherits 30% of its parameters from one vector and 70% from another. In particular, when two vectors, A and B, are crossed with $p_{Cr} = 0.3$, trial vectors will inherit, on average, 30% of their parameters from A and 70% from B. It is equally probable, however, that B will be drawn first and A second, in which case trial vectors inherit, on average, 30% of their parameters from

B and 70% from A. These trial vectors could also have been generated by taking A first, B second and $p_{Cr} = 0.7$. Reversing the roles of the donor vectors has the same effect as using $1 - p_{Cr}$ instead of p_{Cr}. Since the order in which vectors are chosen is random, p_{Cr} potentially generates the same population as does $1 - p_{Cr}$. DE on the other hand crosses vectors from different populations and their order of crossover is not random. In DE, each value of $Cr \sim p_{Cr}$ generates a different trial population.

As with exponential crossover, DE's version of uniform crossover begins by taking a randomly chosen parameter from the mutant so that the trial vector will not simply replicate the target vector. Comparing Cr to $\text{rand}_j(0,1)$ determines the source for each remaining trial parameter. If $\text{rand}_j(0,1) \leq Cr$, then the parameter comes from the mutant; otherwise, the target is the source. Figure 2.41 illustrates the process.

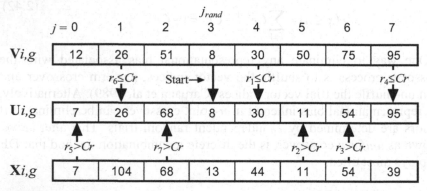

Fig. 2.41. Uniform crossover. Once an initial, randomly chosen parameter is inherited from the mutant (e.g., $j_{rand} = 3$), $D - 1$ independent trials are conducted to determine the source of the remaining parameters. If $\text{rand}_j(0,1) \leq Cr$, the mutant donates a parameter value; otherwise, parameters are copied from the target.

The number of inherited mutant parameters follows a binomial distribution, since parameter origins are determined by a finite number of independent trials having two outcomes with constant probabilities. In particular, the odds of successfully inheriting only one parameter from the mutant is the probability that there will be $D - 1$ "failures" occurring with probability $1 - Cr$

$$p(x = 1) = (1 - Cr)^{D-1}. \tag{2.43}$$

More generally, the probability, given D, that exactly n parameters are inherited from the mutant is

$$p(x = n; D) = {}_{D-1}C_{n-1} \cdot Cr^{n-1} \cdot (1 - Cr)^{D-n}, \qquad (2.44)$$

$$\text{where} \quad {}_{D-1}C_{n-1} \equiv \frac{(D-1)!}{(n-1)! \cdot (D-n)!}.$$

The term ${}_{D-1}C_{n-1}$ represents the number of combinations of $D - 1$ items taken $n - 1$ at a time. Summing the first n terms of Eq. 2.44 gives the probability that the trial vector will inherit at least n mutant parameters. Unlike exponential crossover, the cumulative binomial distribution does not reduce to a simple expression. Because the distribution of inherited mutant parameters is binomial, most DE literature refers to this method as "binomial crossover" to distinguish it from exponential crossover.

Lampinen and Zelinka (2000) have shown that the number of possible trial vectors, n_{trial}, that can be created with DE's uniform (binomial) crossover is

$$n_{\text{trial}} = \begin{cases} Np^3 - 3Np^2 + 2Np & \text{if } Cr = 1 \\ D \cdot Np \cdot (Np^3 - 3Np^2 + 2Np) & \text{if } Cr = 0 \qquad (2.45) \\ 2^D \cdot Np \cdot (Np^3 - 3Np^2 + 2Np) & \text{otherwise.} \end{cases}$$

Although the number of possible trial vectors is constant when $0 < Cr < 1$, uniform crossover suffers from a *distribution bias* because not all configurations are equally likely (Spears and DeJong 1991). DE does not eliminate distribution bias but relies on Cr to provide the means for controlling it. At one extreme, $Cr \sim 0$ minimizes disruption by incrementally altering just a few parameters of a vector at a time, while at the other extreme, $Cr \sim 1$ favors exploration by drawing most trial vectors directly from the mutant population. The next section examines the conditions under which reinforcement and incremental change are useful and in what contexts exploration becomes crucial.

2.6.2 The Role of *Cr* in Optimization

Despite mediating a crossover process, Cr can also be thought of as a *mutation rate*, i.e., the (approximate) probability that a parameter will be inherited from a mutant. In DE, the average number of parameters mutated for a given Cr depends on the crossover model (e.g., exponential or binomial) but in each, a low Cr corresponds to a low mutation rate. Many GAs recommend a mutation rate of $1/D$, meaning that, on average, only one trial parameter is mutated (Potter and DeJong 1994). Indeed, Zaharie's results for Rastrigin, Griewangk and the sphere, as well as those for the simple hyper-ellipsoid in Fig. 2.33, consistently found low Cr to be the most effective values. Similarly, optimizing the extensive test beds in Storn and Price (1997) showed that all functions could be solved with either $0 \le Cr \le 0.2$ or $0.9 \le Cr \le 1$. The reason for the bifurcation of the crossover control space was not at first appreciated until it was realized that functions solvable with low Cr were inevitably decomposable, while those requiring $Cr \sim 1$ were not.

Limitations of a Low Mutation Rate

As Sect. 1.2.3 mentioned, a decomposable function can be written as a sum of D one-dimensional functions (not necessarily all the same)

$$f(\mathbf{x}) = \sum_{j=0}^{D-1} f_j(x_j) \tag{2.46}$$

Decomposability simplifies the task of optimization because each parameter can be optimized *independently*, allowing the task of optimizing a single D-dimensional function to be broken up into D one-dimensional problems. Once the optima of the D one-dimensional functions have been located, they can be combined to specify the optimum of the original D-dimensional function

$$f(\mathbf{x}^*) = f(x_0^*, x_1^*, ..., x_{D-1}^*), \quad \min(f_j(x_j)) = f_j(x_j^*), \quad j = 0,1,...,D-1. \tag{2.47}$$

For such functions, changing just one parameter (e.g., $Cr = 0$) before each evaluation can be viewed as a single step in an independent, one-dimensional optimization. If the parameter being modified is randomly selected, then the D one-dimensional optimizations proceed as arbitrarily sequenced tasks (Salomon 1996a).

Any decomposable uni- or multi-modal function can be optimized in linear time, $O(D)$, but randomly interleaving the order in which these one-dimensional optimization tasks are executed causes EAs to incur an additional penalty of $\ln(D)$, raising their total computational complexity for decomposable functions to $O(D \cdot \ln(D))$ (Salomon 1996a). Thus, DE and other GAs with low mutation rates should not be expected to compete with dedicated decomposable function solvers. Such was the case at the First International Contest of Evolutionary Optimizers, held in Kyoto, Japan, where DE finished behind a method that exploited the fact that the contest functions were decomposable (Storn and Price 1996). Even so, the $\ln(D)$ penalty incurred by EAs when using low mutation rates on decomposable functions is not prohibitive. Once parameters become dependent, however, the penalty incurred by algorithms using low mutation rates does become prohibitive.

Salomon provides two reasons why a low mutation rate is an ill-advised strategy when optimizing parameter-dependent functions (Salomon 1996). The first reason, mentioned briefly in conjunction with the rotated ellipse of Sect. 2.5.2, is illustrated by Fig. 2.42. The picture on the left shows contours of an elliptical objective function whose principal axes are parallel to the coordinate axes. Any trial vector that is interior to the contour on which x_i resides constitutes an improving move. If only one parameter is changed per evaluation, then x_i can move at most Δx_0 in the x_0 direction *or* Δx_1 in the x_1 direction before it produces an unacceptable result. For this ellipse, these intervals are large enough to permit the optimum to be located in just two moves, first to either $x_i + 0.5 \cdot \Delta x_0$ or to $x_i + 0.5 \cdot \Delta x_1$, and then to x^* on the next move.

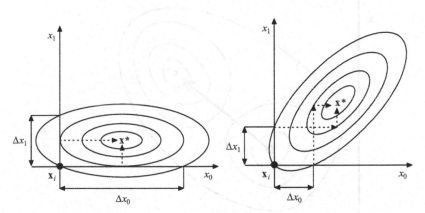

Fig. 2.42. When the principal axes of the ellipse are aligned along coordinate axes, improvement intervals are large compared to those available once the coordinate axes have been rotated by 45°. In the figure on the left, a single pair of moves executed in either order would be able to reach the minimum, but in the figure on the right it takes at least three moves parallel to the coordinate axes to reach the optimum.

By contrast, rotation shortens the improvement intervals to the point where the optimum can no longer be reached in just two consecutive moves *if each step is taken parallel to a coordinate axis*. These additional steps slow convergence and raise the algorithm's time complexity above $O(D \cdot \ln D)$. Both the dimension and eccentricity of the hyper-ellipsoid exacerbate this performance loss. Indeed, the experiments in Sect. 2.5.2 confirmed Salomon's predictions that low Cr, though efficient on the decomposable ellipse, is inefficient on its rotated, non-decomposable counterpart.

As the example of the rotated ellipse in Fig. 2.42 demonstrates, a low-Cr DE strategy can suffer a loss of performance even if the function is unimodal. Salomon's second reason for not using low mutation rates applies only to multi-modal functions whose local minima are not aligned with the coordinate axes. Figure 2.43 shows the contours of a hypothetical multi-modal function having two local optima located on a diagonal. The only way to reach the optimum at $\mathbf{x}^*$ from inside the penultimate basin of attraction is by moving in both the (positive) x_0 and x_1 directions *simultaneously*. Since the current vector is in a local optimum, no single move parallel to a coordinate axis will be acceptable and improving moves into a basin of equal or lower function value will have components in both axes.

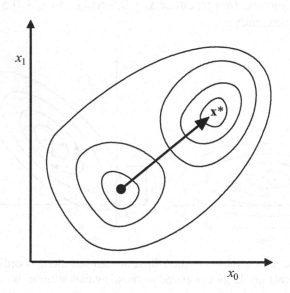

Fig. 2.43. Multi-modal functions with dependent parameters pose additional challenges to low-Cr strategies. The only improving move out of the penultimate basin of attraction requires making changes in both coordinates simultaneously.

Salomon has shown that at $O(D^D) = O(\exp(D \cdot \ln(D)))$, a low mutation rate can actually take longer than a random search to optimize a parameter-dependent, multi-modal function (Salomon 1997). Time complexity of this order is prohibitive in all but the most trivial cases.

In summary, the role of Cr is to provide the means to exploit decomposability, if it exists, and to provide extra diversity to the pool of possible trial vectors, especially near $Cr = 1$. In the general case of parameter-dependent functions, Cr should be close to 1 so that the performance losses associated with using a low mutation rate are minimized.

Rotational Invariance

An algorithm whose performance depends on the objective function being aligned with a privileged coordinate system is a poor choice in general because it is unlikely that the optimal orientation will be known in advance. What is needed instead is a search algorithm that is *rotationally invariant* – one whose performance does not depend on the orientation of the coordinate system in which the objective function is evaluated. For classic DE, this means that $Cr = 1$, i.e., mutation only and no crossover.

That crossover is not rotationally invariant can be seen in Fig. 2.44, which plots the trial vectors generated by a pair of vectors both before and after a coordinate rotation. Although rotation leaves the position of the vectors with respect to one another unaltered, trial vector placement relative to the vector population depends on the angle of rotation. Since each angle samples different regions of the objective function, performance is rotation dependent.

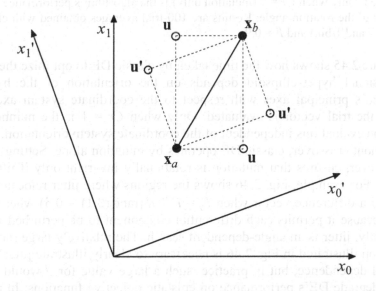

Fig. 2.44. Crossover is not a rotationally invariant process. The trial vectors derived by crossover from vectors $\mathbf{x}_a$ and $\mathbf{x}_b$ change from $\mathbf{u}$ to $\mathbf{u}'$ as the coordinate system is reoriented.

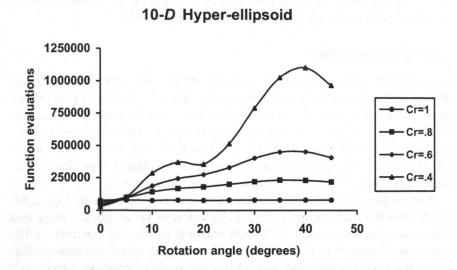

Fig. 2.45. The average number of function evaluations to solve the ten-dimensional hyper-ellipsoid is a function of the angle between the hyper-ellipsoid's principal axes and the axes of the coordinate system in which it is evaluated. Only when $Cr = 1$ (mutation only) is the algorithm's performance independent of the rotation angle. Results are 100-trial averages obtained with classic DE (DE/rand/1/bin) and $F = 0.9$.

Figure 2.45 shows how the time taken by classic DE to optimize the ten-dimensional hyper-ellipsoid depends on the orientation of the hyper-ellipsoid's principal axes with respect to the coordinate system axes in which the trial vector is evaluated. Only when $Cr = 1$ is the number of function evaluations independent of the coordinate system orientation.

Without crossover, classic DE operates by mutation alone. Setting $Cr = 1$, however, ensures that mutation is rotationally invariant only if jitter is absent. For example, Fig. 2.46 shows the regions where jitter relocates the head of a difference vector when $F_j = F + d \cdot (\text{rand}_j(0,1) - 0.5)$ where $d = 0.5$. Because it permits each differential component to be perturbed independently, jitter is an angle-dependent search. The relatively large random deviation illustrated in Fig. 2.46 is necessary to clearly illustrate jitter's rotational dependence, but in practice, such a large value for d would seriously degrade DE's performance on epistatic objective functions. In practice d should be much smaller, e.g., $d = 0.001$.

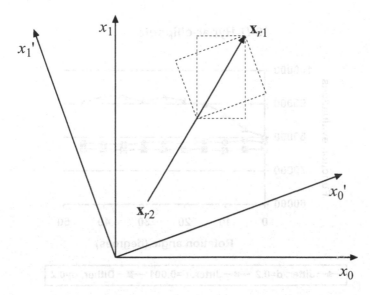

Fig. 2.46. Jitter is not a rotationally invariant process because components of the differential are altered independently. Dashed boxes outline the areas in which jitter with $F_j = 0.5 + 0.5 \cdot \mathrm{rand}_j(0,1)$ can place the head of the difference vector, $\mathbf{x}_{r1} - \mathbf{x}_{r2}$. As the coordinate axes are reoriented, the range of possibilities changes.

Figure 2.47 shows that even with a mutation-only strategy, DE's performance is rotationally dependent if jitter is present (top line). The magnitude of the dependence increases as the magnitude of jitter's deviation increases. On the other hand, dither, like the F = constant model profiled in Fig. 2.45, is rotationally invariant as the lower line in Fig. 2.47 shows. The middle line shows that when jitter is very small (e.g., $d = 0.001$), the penalty for rotational invariance is also small.

Salomon's warnings notwithstanding, DE performs well on parameter-dependent multi-modal functions in practice as long as rotationally invariant processes are the dominant strategies, e.g., when Cr is "close" to 1, say, $Cr = 0.98$, and when jitter's PDF has a "small" variance, e.g., $d = 0.001$ in Eq. 2.37.

The value of such a small value for jitter appears to be that the diversity it adds to the pool of trial vectors lowers the odds that DE will stagnate, particularly when Np is relatively small. This added diversity seems to be of particular benefit to the algorithm DE/best/1/bin, for which reliance on the best-so-far vector as a base vector lowers diversity in the pool of possible trial vectors. In addition, jitter with a suitable PDF makes DE provably convergent. It should be emphasized, however, that jitter's practical value is still a matter of debate.

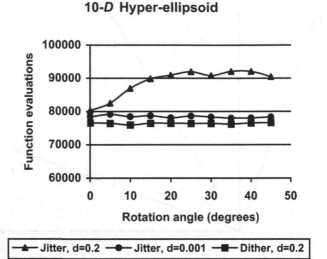

Fig. 2.47. When using jitter, DE's performance on the ten-dimensional hyper-ellipsoid depends on the orientation of the coordinate system relative to the principal axes of the hyper-ellipsoid. Plotted are the number of function evaluations that DE needed to optimize the ten-dimensional hyper-ellipsoid using both jitter and dither in a mutation-only strategy ($Cr = 1$). Unlike jitter, dither is rotationally invariant, but when the level of variation in jitter is very small ($d = 0.001$), rotation does not significantly affect run-times. Results were obtained using $Np = 50$, $Cr = 1$ and classic DE except that $F_i = 0.9 + d \cdot (\text{rand}_i(0,1) - 0.5)$ with $d = 0.2$ for dither and $F_j = 0.9 + d \cdot (\text{rand}_j(0,1) - 0.5)$ with both $d = 0.2$ and $d = 0.001$ for jitter.

If a strictly rotationally invariant scheme is demanded, then $Cr = 1$ and the pool of potential trial vectors is limited to the mutant population. Without crossover or jitter, the only rotationally invariant way to increase the pool of potential trial vectors is by increasing Np or by using dither. If, however, dither's PDF has a high proportion of small perturbations, then optimal population sizes may be larger than if no dithering is used at all. Alternatively, certain forms of *arithmetic recombination* – unlike discrete recombination – can add diversity and complement the mutation search strategy without becoming rotationally dependent.

2.6.3 Arithmetic Recombination

Although crossover creates new combinations of parameters, it leaves the parameter values themselves unchanged. *Continuous* or *arithmetic recombination*, however, operates on individual trial parameter values by expressing them as linear combinations of parameters. Arithmetic recombination's global variant selects both vectors anew for each parameter of a recombinant vector, $\mathbf{w}_{i,g}$ (Bäck and Schwefel 1993), but most EAs select just one set of vectors for all parameters of $\mathbf{w}_{i,g}$:

$$\mathbf{w}_{i,g} = \mathbf{x}_{r0,g} + k_i \left(\mathbf{x}_{r1,g} - \mathbf{x}_{r0,g} \right). \tag{2.48}$$

The *coefficient of combination*, k_i, can be a constant (e.g., $k_i = 0.5$ is *uniform arithmetic recombination* (Eiben and Smith 2003)), or a random variable (e.g., rand(0,1)). More generally, if k_i is either constant or a random variable that is sampled anew for each vector, then the resulting process is called *line recombination* (Eq. 2.48) (Mühlenbein and Schlierkamp-Voosen 1993). If, however, the coefficient of combination is sampled anew for each parameter, then the process is known as *intermediate recombination* (Mühlenbein and Schlierkamp-Voosen 1993):

$$w_{j,i,g} = x_{j,r1,g} + k_j \left(x_{j,r2,g} - x_{j,r1,g} \right). \tag{2.49}$$

Not all sources agree on this terminology. For example, in ES terminology, the coefficient of combination is chosen anew for each parameter only in the global version, i.e., when vectors are also chosen anew for each parameter (Bäck and Schwefel 1993). This book equates intermediate recombination with the two-vector linear combination in Eq. 2.49, where k_j is a random variable that is sampled anew for each parameter, but vectors are chosen once per trial vector. If k_j is allowed to assume values outside the range (0,1), then the process is called *extended intermediate recombination* (Mühlenbein and Schlierkamp-Voosen 1993).

Figure 2.48 compares the regions searched by discrete, line and intermediate recombination when the coefficient of combination is distributed with random uniformity between 0 and 1. The two vectors occupy opposing corners of a hypercube whose remaining corners are the trial vectors created by discrete recombination. Line recombination, as its name suggests, searches along the axis connecting vectors, while intermediate recombination explores the entire D-dimensional volume contained within the hypercube.

Since the hypercube's corners are the possible outcomes of discretely recombining two vectors, intermediate recombination, like both jitter and crossover, is not a rotationally invariant process. Rotation relocates the hypercube's corners, which in turn redefine the area that intermediate recombination searches. On the other hand, line recombination is rotationally invariant. Given that both differential mutation and line recombination are rotationally invariant schemes for adding a weighted vector difference to an existing vector, the question arises: what real difference is there between the two operations?

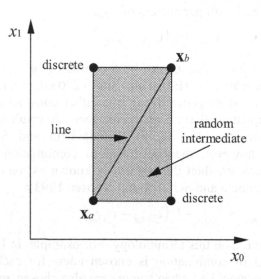

Fig. 2.48. Domains of the possible recombinant vectors generated using discrete, line and intermediate recombination. The coefficient of combination is drawn from the interval [0,1].

Distinguishing Line Recombination from Differential Mutation

Why should some vector differences be associated with recombination and the others not? The reason is that the presence of the base vector in recombination differentials constitutes a bias that makes recombination's dynamics different from those of differential mutation. For example, shifting the base vector's position with respect to the population does not influence its mutation differentials, but it does alter the size and orientation of its recombination differentials. Figure 2.49 shows that if the base vector moves from the population's outer boundary to a more central position, its recombination differentials will become shorter and more symmetrically distributed, whereas mutation differentials – defined by the remaining vectors whose positions are unchanged – are unaffected.

Recombination's positional dependence allows trial vectors to be deliberately placed into the population in locations that mutation can reach only by chance. For example, $k_i = 0.5$ (Eq. 2.48) places the trial vector midway between the base vector and the vector x_{r1}. Moreover, $k_i = 1$ reduces recombination to a replacement operation by placing the trial vector at $\mathbf{x}_{r1}$. By contrast, non-zero mutation differentials place trial vectors on, between or in relation to other vectors only by chance, not by intention.

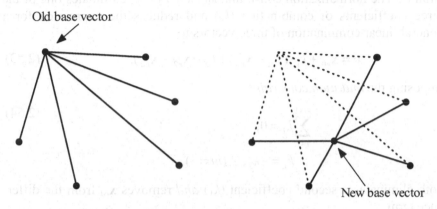

Fig. 2.49. Recombination differentials change in response to a shift in the base vector's position relative to the population.

When a trial vector is a linear combination of only two vectors, the differential's dependence on the base vector is inevitable. For example, let $\mathbf{u}$ be a trial vector that is a linear combination of two, randomly chosen vectors

$$\mathbf{u} = k_0 \cdot \mathbf{x}_{r0} + k_1 \cdot \mathbf{x}_{r1}. \tag{2.50}$$

To prevent trial vectors from expanding ($k_0 + k_1 > 1$) or contracting ($k_0 + k_1 < 1$) over the course of many generations due only to the generating process itself, the coefficients k_0 and k_1 are subject to a *normalization constraint* that requires their sum to equal 1. For a linear combination of m vectors,

$$\sum_{i=0}^{m-1} k_i = 1, \tag{2.51}$$

$$k_0 + k_1 = 1, \quad (m = 2).$$

Substituting $1 - k_1$ for k_0 in Eq. 2.50 yields the familiar formula for line recombination in which the base vector, $\mathbf{x}_{r0}$, *also appears in the difference term*:

$$\mathbf{u} = (1 - k_1) \cdot \mathbf{x}_{r0} + k_1 \cdot \mathbf{x}_{r1}, \tag{2.52}$$

$$= \mathbf{x}_{r0} + k_1 \cdot (\mathbf{x}_{r1} - \mathbf{x}_{r0}).$$

Once three vectors are linearly combined, however, the positional bias inherent in two-vector combinations can be eliminated. For example, a mutant is a three-vector linear combination that is subject to two constraints. The normalization constraint, $k_0 + k_1 + k_2 = 1$, eliminates one of the three coefficients of combination (k_0) and reduces the expression for a general linear combination of three vectors to

$$\mathbf{u} = \mathbf{x}_{r0} + k_1 \cdot (\mathbf{x}_{r1} - \mathbf{x}_{r0}) + k_2 \cdot (\mathbf{x}_{r2} - \mathbf{x}_{r0}). \tag{2.53}$$

Imposing the *mutation constraint*

$$\sum_{i=1}^{m-1} k_i = 0, \tag{2.54}$$

$$k_1 = -k_2, \quad (m = 3),$$

both eliminates a second coefficient (k_1) *and* removes $\mathbf{x}_{r0}$ from the difference term

$$\mathbf{u} = \mathbf{x}_{r0} + k_2 \cdot (\mathbf{x}_{r1} - \mathbf{x}_{r2}) = \mathbf{v}. \tag{2.55}$$

Satisfying Eq. 2.54 cancels out the base vector's contribution to the $m - 1$ differential terms. The one remaining coefficient of combination, k_2, is the mutation scale factor, F. Like the increments generated by a PDF, the mutation differentials contain no reference to the vector they modify.

Two-vector line recombination's positional dependence complements a mutation-driven search, but the existence of only $Np - 1$ possible recombination axes limits its explorative power. More than two vectors can be recombined and elevating line recombination to a three-vector process places it on an equal footing with differential mutation as both consist of a linear combination of three vectors.

Three-Vector Recombination

Equation 2.51 appears to be missing the differential mutation operator because it expresses a trial vector as the sum of the base vector and two recombination differentials that contain the base vector. The reciprocal roles played by recombination and mutation in three-vector linear combinations become clearer once Eq. 2.53 is rewritten with a change of variables that decomposes any normalized, three-vector linear combination into separate recombination and mutation components, K and F, respectively. First, let

$$k_1 \equiv \frac{(K+F)}{\sqrt{2}}, \tag{2.56}$$

$$k_2 \equiv \frac{(K-F)}{\sqrt{2}}.$$

Replacing k_1 and k_2 in Eq. 2.53 with the expressions in 2.56 yields

$$\mathbf{u} = \mathbf{x}_{r0} + \frac{(K+F)}{\sqrt{2}} \cdot (\mathbf{x}_{r1} - \mathbf{x}_{r0}) + \frac{(K-F)}{\sqrt{2}} \cdot (\mathbf{x}_{r2} - \mathbf{x}_{r0}). \tag{2.57}$$

Multiplying out the expressions in Eq. 2.57 and collecting terms reorganizes Eq. 2.53 into a recombination term that contains the base vector and a mutation term from which the base vector is absent:

$$\mathbf{u} = \mathbf{x}_{r0} + \frac{K}{\sqrt{2}} \cdot (\mathbf{x}_{r1} + \mathbf{x}_{r2} - 2\mathbf{x}_{r0}) + \frac{F}{\sqrt{2}} \cdot (\mathbf{x}_{r1} - \mathbf{x}_{r2}). \tag{2.58}$$

The change of variables laid out in Eq. 2.56 defines a 45° rotation of the K–F plane with respect to the k_1–k_2 plane (Fig. 2.50). The mutation constraint, $k_1 = -k_2$, defines a mutation axis, F, that passes through the origin and has a slope of −1, while the *recombination constraint*, $k_1 = k_2$, defines a recombination axis, K, that also passes through the origin but has a slope of +1. The advantage of the K–F decomposition is that it permits two search processes with different dynamics to be controlled independently.

The coordinates, (k_1, k_2), locate the trial vector, $\mathbf{u}$, *relative to* the base vector $\mathbf{x}_{r0}$ using two-vector recombination differentials as basis vectors. Coordinate k_1 measures the distance of the trial vector from the base vector in the direction of the differential $(\mathbf{x}_{r1} - \mathbf{x}_{r0})$, while k_2 measures the distance from $\mathbf{x}_{r0}$ in the direction of the differential $(\mathbf{x}_{r2} - \mathbf{x}_{r0})$. Similarly, K and F measure the distance of the trial vector from the base vector along the direction of the three-vector recombination and mutation axes, respectively.

The medial line (the K–axis in Figs. 2.50 and 2.51) plays an important role in the two-dimensional version of the Nelder–Mead algorithm. As Sect. 1.2.3 explained, the Nelder–Mead strategy tests a point located on the axis defined by the vector being modified (the worst vector in Nelder–Mead, but $\mathbf{x}_{r0}$ in this case) and the centroid of a simplex consisting of D additional vectors. When $D = 2$, this axis is a medial line that passes through not only the centroid, but also the average position of $\mathbf{x}_{r1}$ and $\mathbf{x}_{r2}$.

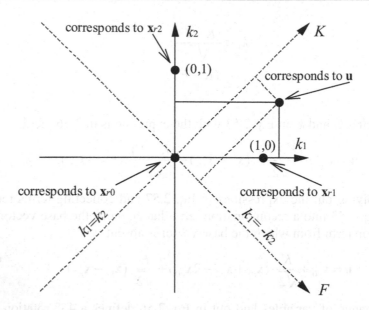

Fig. 2.50. Decomposing the position of a trial vector into separate mutation and recombination components in the K–F plane (refer to Eqs. 2.51 and 2.56). The rotation angle between the k_1–k_2 and K–F coordinate systems is 45°.

Figure 2.51 illustrates some of the important features of the K–F plane. Coordinates are given as (K,F) where K and F are a vector's coordinates along the recombination and mutation axes, respectively. The base vector, $\mathbf{x}_{r0}$, corresponds to the origin, $(K,F) = (0,0)$. The remaining two vectors correspond to (0.5,0.5) and (0.5, −0.5). Together, the three vectors form an inverted triangle whose sides and their extensions constitute the three axes along which three-vector combinations reduce to two-vector line recombination. This triangle of vectors is inscribed inside a larger triangle whose vertices are the three *mutation points* (0,1), (0, −1) and (1,0) corresponding to the vectors $\mathbf{x}_{r0} + \mathbf{x}_{r1} - \mathbf{x}_{r2}$, $\mathbf{x}_{r0} + \mathbf{x}_{r2} - \mathbf{x}_{r1}$ and $\mathbf{x}_{r1} + \mathbf{x}_{r2} - \mathbf{x}_{r0}$, respectively.

Only the order in which vectors are combined distinguishes these three strategies and as long as vectors are randomly selected, the three mutation points are dynamically indistinguishable, i.e., the three strategies cannot be distinguished based on their performance. Similarly, the sides of this larger triangle represent the three possible mutation axes and its three medial lines represent the three possible recombination axes. The figure is bilaterally symmetric (left and right sides are mirror images, with the mirror aligned on the K axis) about the vertical recombination axis because $\mathbf{x}_{r1} - \mathbf{x}_{r2} = - (\mathbf{x}_{r2} - \mathbf{x}_{r1})$. The centroid of both the large and small triangles lies at (1/3,0).

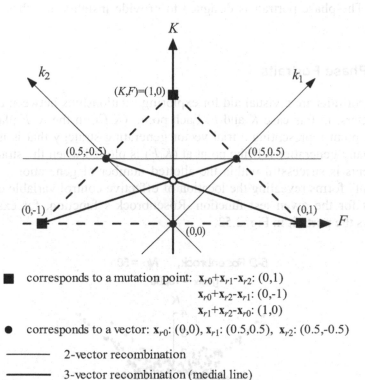

$\blacksquare$ corresponds to a mutation point: $\mathbf{x}_{r0}+\mathbf{x}_{r1}-\mathbf{x}_{r2}$: $(0,1)$
$\qquad\qquad\qquad\qquad\qquad\quad$ $\mathbf{x}_{r0}+\mathbf{x}_{r2}-\mathbf{x}_{r1}$: $(0,-1)$
$\qquad\qquad\qquad\qquad\qquad\quad$ $\mathbf{x}_{r1}+\mathbf{x}_{r2}-\mathbf{x}_{r0}$: $(1,0)$

$\bullet$ corresponds to a vector: $\mathbf{x}_{r0}$: $(0,0)$, $\mathbf{x}_{r1}$: $(0.5,0.5)$, $\mathbf{x}_{r2}$: $(0.5,-0.5)$

$\rule{2cm}{0.4pt}$ 2-vector recombination

$\rule{2cm}{0.4pt}$ 3-vector recombination (medial line)

$-\ -\ -$ mutation axis

Fig. 2.51. The K–F plane exhibits three axes along which two-vector recombination produces trial vectors. Squares plot the three dynamically equivalent mutation points. The vertical axis measures the component along the medial axis while the horizontal mutation axis measures the component in the direction of the difference vector $\mathbf{x}_{r1}-\mathbf{x}_{r2}$. Note that the coordinate values are expressed in the K–F coordinate system. Note also that the vectors that correspond to these points are also mentioned. As an example, the point $(1,0)$ corresponds to the vector $\mathbf{x}_{r1}+\mathbf{x}_{r2}-\mathbf{x}_{r0}$.

Because they represent varying fractions of recombination and mutation, points in the K–F plane also represent different search strategies. For example, classic DE with $Cr = 1$ includes all the points on the mutation axis where trial vectors are pure mutants, whereas those that lie along the medial axes are pure three-vector recombinants similar to those produced by the two-dimensional Nelder–Mead algorithm. Off-axis points possess attributes of vectors that have been subjected to both differential mutation and three-vector line recombination. For optimization, the most important questions regarding K and F are whether they are correlated and whether successful strategies consistently cluster around landmarks in the K–F

plane. The phase portrait is designed to provide insights into these questions.

2.6.4 Phase Portraits

Phase portraits are a visual aid for exploring relationships between control parameters, in this case K and F. Each point, (K,F), in the K–F plane locates a point representing a trial vector generating strategy that is iterated over many generations. If the point at (K,F) is plotted when the strategy it represents is successful within the allotted number of generations, then a "portrait" forms revealing the location of effective control variable combinations for the given test function. Rosenbrock's function, for example, displays the portrait in Fig. 2.52.

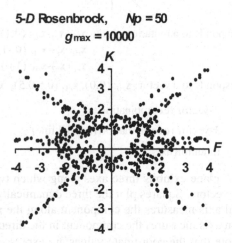

$$5\text{-}D \text{ Rosenbrock,} \quad Np = 50$$
$$g_{max} = 10000$$

Fig. 2.52. The phase portrait for the five-dimensional generalized Rosenbrock function. Points were sampled with random uniformity, i.e., $F = 8\cdot(\text{rand}(0,1) - 0.5)$ and $K = 8\cdot(\text{rand}(0,1) - 0.5)$. The function rand$(0,1)$ lacks a subscript to indicate that a single value is generated anew *for each optimization run*. One optimization was run for each point. If the optimization was successful within the allotted number of generations and with the chosen population size, the point was plotted. Results were obtained with DE/rand/1/bin, $Cr = 1$ and $i \neq r0 \neq r1 \neq r2$.

Regions that are most densely populated correspond to strategies that have the highest probability of convergence. Points in the central triangular void are strategies that converged prematurely for the given Np, while those in the vacant space surrounding the portrait did not converge within the allowed number of generations.

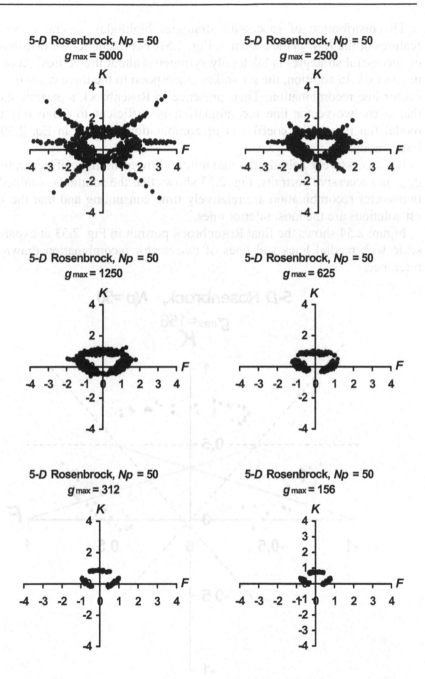

Fig. 2.53. Reducing the maximum allowed number of generations reveals that the fastest solutions are the most interior ones. The three clusters represent symmetric solutions.

The distribution of successful strategies highlights several important features of the K–F plane shown in Fig. 2.51. For example, the distribution of successful strategies is bilaterally symmetric about the vertical recombination axis. In addition, the six spikes correspond to the three cases of two-vector line recombination. Their presence in Rosenbrock's portrait shows that even two-vector line recombination is sufficient to solve this uni-modal function if the coefficient of combination (e.g., k_1 in Eq. 2.50) is large enough.

By successively halving the maximum allowed number of generations, g_{max}, in successive portraits, Fig. 2.53 shows that the solutions obtained by two-vector recombination are relatively time consuming and that the fastest solutions are the most interior ones.

Figure 2.54 shows the final Rosenbrock portrait in Fig. 2.53 at expanded scale with medial lines and lines of two-vector recombination drawn for reference.

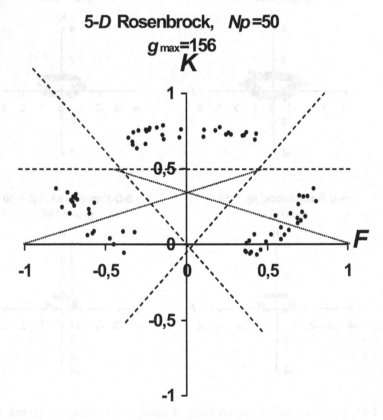

Fig. 2.54. Clusters for Rosenbrock's functions are bisected by a medial line and constrained by the lines of two-vector recombination.

Even though they possess very different topography, many other functions display portraits similar to Rosenbrock's. As Fig. 2.55 shows, the phase portraits for the hyper-ellipsoid, Ackley, Whitley and Lennard-Jones functions all look remarkably similar to Rosenbrock's portrait when g_{max}= 5000.

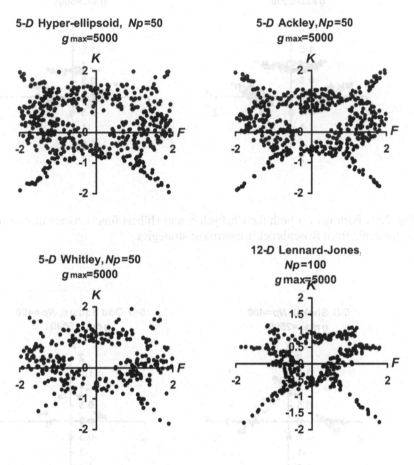

Fig. 2.55. Despite having radically different topographies, these functions produce portraits similar to Rosenbrock's.

Other portraits, such as those of the Chebyshev and Hilbert functions in Fig. 2.56, produce images similar to plots of Rosenbrock's fastest strategies. In each case there are three clusters centered on a medial line that are constrained by the lines representing two-point recombination.

Not all functions conform to this pattern and some have portraits with clusters that lie predominantly along *either* the mutation axis *or* the three-vector recombination axis. Figure 2.57 shows that most of the successful strategies for the Shekel and odd square functions lie on the mutation axis.

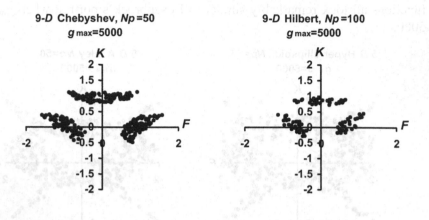

Fig. 2.56. Portraits for both the Chebyshev and Hilbert functions are almost indistinguishable from Rosenbrock's innermost strategies.

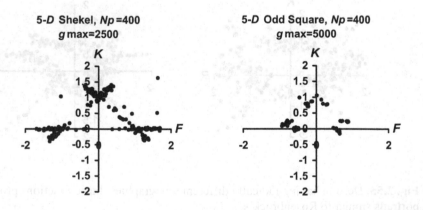

Fig. 2.57. Solutions for both the Shekel and odd square functions lie almost entirely on the mutation axes. Recombination is effective on the Shekel function as long as K is close to 1, but it is an ineffective strategy when applied to the odd square.

At the other extreme, Griewangk's function shows a distribution of points centered on the medial lines that has only a few outlying points approaching the mutation axis, most notably near $F = 1$ and $F = 0.5$.

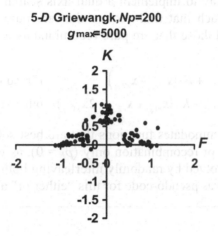

Fig. 2.58. Although a few cluster points intersect the mutation axes, the most robust strategies lie on the medial axes.

These phase portraits show that mutation and recombination differentials do indeed have different effects on the optimization dynamic. For functions like the odd square, mutation is the only viable option, while for those like Griewangk, recombination is a better strategy. Reliance on the wrong operation is likely to result in poor performance for a significant number of functions, but many functions are *generic*, meaning that either mutation or recombination makes an effective strategy. Given the range of behaviors displayed in the phase portraits, what is the best strategy in general?

2.6.5 The Either/Or Algorithm

All portraits in the previous section displayed clusters of successful strategies that were bisected by either a recombination or a mutation axis. In the generic case, both axes intersected clusters. Furthermore, there was no case in which a cluster *only* occupied the spaces between axes. Because these isolated, off-axis clusters are not observed, the best strategy for locating a central cluster point is to look along the mutation axis, the recombination axis, or both, but not between them. Compared to searching the entire two-dimensional K–F plane, a dual-axis search reduces the effort to find a successful strategy because it restricts the search to a pair of one-dimensional axes.

The simplest way to implement a dual-axis search is to define a mutation probability such that trial vectors that are pure mutants occur with probability p_F and those that are pure recombinants occur with probability $1 - p_F$:

$$\mathbf{u}_{i,g} = \begin{cases} \mathbf{v}_{i,g} = \mathbf{x}_{r0,g} + F \cdot \left(\mathbf{x}_{r1,g} - \mathbf{x}_{r2,g} \right) & \text{if } \mathrm{rand}_i(0,1) < p_F, \quad (2.59) \\ \mathbf{w}_{i,g} = \mathbf{x}_{r0,g} + K \cdot \left(\mathbf{x}_{r1,g} + \mathbf{x}_{r2,g} - 2\mathbf{x}_{r0,g} \right) & \text{otherwise.} \end{cases}$$

This scheme accommodates functions that are best solved by either mutation only ($p_F = 1$) or recombination only ($p_F = 0$), as well as generic functions that can be solved by randomly interleaving both operations ($0 < p_F < 1$). Figure 2.59 gives pseudo-code for this "either/or" algorithm.

```
. . .
if (rand_i(0,1)<P_F)              // mutate or recombine ?
{
    u_i=x_r0+F*(x_r1-x_r2);       // mutate
}
else
{
    u_i=x_r0+K*(x_r1+x_r2-2*x_r0); // recombine
}
. . .
```

Fig. 2.59. Pseudo-code for creating a trial vector with the "either/or" algorithm. From experience $K = 0.5 \cdot (F + 1)$ can be recommended as a good first choice for K given F.

2.7 Selection

There are primarily two stages in the evolutionary process where selection can be applied to a population. Some GAs (Goldberg 1989) employ *parent selection* to decide which vectors will undergo recombination. Typically, vectors with the best function values are assigned the highest *selection probability*, making them the most likely to be chosen for mating. This strategy mimics the one employed by breeders and botanists who try to

improve traits by selectively breeding individuals with superior characteristics. In practice, methods for assigning selection probabilities involve additional assumptions about how to map objective function values to a set
of probabilities. Instead of selecting mates based on objective function
value, both ES and classic DE select mates with equal probability. In the
ES, each vector has the same chance to be chosen for mutation and/or recombination. Similarly, classic DE randomly selects base vectors without
regard for their objective function values (see Sect. 2.4).

In contrast to parent selection, *survivor selection*, also called *replacement*, chooses the next generation of vectors from the current generation of
vectors and trial vectors. Most EAs apply selection pressure either when
choosing vectors to recombine or when choosing survivors. GAs typically
bias selection in favor of better vectors, whereas DE, ES and other EAs,
however, combine randomly chosen vectors and apply selection pressure
only when picking survivors. Using both parent (base vector) and survivor
selection can cause premature convergence to a local optimum.

The remainder of this section is primarily concerned with survivor selection and it will be convenient for the following discussion to assume
that the current and trial populations can have different sizes. In keeping
with the naming traditions established by the ES community, μ will denote
the size of the current population and λ will represent the size of the trial
population.

2.7.1 Survival Criteria

In some algorithms, age alone determines which individuals survive. Here,
age distinguishes vectors in the current population from those in the
(younger) trial population. More often, however, both a vector's objective
function value and the luck of the draw are also factors. The simple GA,
however, determines survivors by their age alone.

Age Only

The simple GA replaces μ vectors with $\lambda = \mu$ trial vectors without regard
to whether the trial vectors actually have lower function values than those
in the current generation (Goldberg 1989). This *age-based replacement*
scheme only works if parent selection is driven by an objective function-
based criterion. Without the feedback that an objective function-based parent selection rule provides, there is no bias to drive the population toward
better solutions. For example, the (1,1)-ES with its age-based selection is
nothing more than a random walk in which each trial vector replaces the

current vector regardless of its objective function value (Bäck et al. 1993). Similarly, age-based replacement is unsuitable for classic DE because its parent selection scheme, i.e., random base vector selection, does not choose vectors based on objective function value.

Objective Function Value Only

When only trial vectors are allowed to advance, there is no guarantee that the best-so-far solution will not be lost. Retaining the best-so-far solution is known as *elitism* and part of the task of proving that an algorithm will converge to the global optimum in the long-time limit is proving that it is elitist (Rudolph 1996). For this reason, and because of the speed improvement that it offers, most EAs, including DE, evolutionary programming (EP) and some versions of ES and genetic programming (GP) (Koza 1992), include the current population when determining the membership of the next generation. For example, the $(\mu + \lambda)$-selection scheme (see Sect. 1.2.3) ranks all vectors in both the current and trial populations from best to worst and then populates the next generation with the best μ individuals. Similarly, EP tournament selection (see subsection 2.7.2) compares the objective function value of vectors randomly chosen from the current and trial populations. In both cases, a vector's age is irrelevant and the best-so-far result is always retained.

Age and Objective Function Value

In ES (μ, λ)-selection (see Sect. 1.2.3), age dictates that only trial vectors can survive, while objective function values determine which trial vectors are among the μ best. Using objective function values to pick the μ best survivors from a pool of λ trial vectors biases evolution toward better solutions, unlike the simple GA in which $\mu = \lambda$ trial vectors survive regardless of their objective function values. Since surviving trial vectors can overwrite better current vectors, (μ, λ)-selection is not elitist. Forgetting prior results, however, allows the population both to escape local optima and to track dynamic ones. In addition, (μ, λ)-selection lessens the chance that ES "strategy" parameters will prematurely adapt to a good but sub-optimal solution (Bäck and Schwefel 1995).

As the next section shows, both age and objective function value also play a role in DE selection. Age is a factor because trial vectors can only compete against members of the current population, while their objective function values determine which vector survives. The DE scheme is elitist since the best vector in the current and trial populations always survives.

2.7.2 Tournament Selection

In general, any parent selection scheme can also be adapted for survivor selection, but in practice nearly all EAs, including DE, determine survivors by some form of tournament selection or ranking, which is a special case of tournament selection. The next subsection explores DE selection in the context of the tournament survivor selection method employed by the EP algorithm (Fogel et al. 1966; Fogel 1991).

In EP-style tournament selection, each vector competes against T opponents drawn at random from a selection pool of Ns vectors (Saravanan and Fogel 1997). In deterministic tournaments, vectors are assigned a "win" for each pair wise competition in which they have the lower objective function value (in non-deterministic tournaments, the best vector wins with a user-defined probability). The μ vectors that accumulate the most wins populate the next generation.

The main control variable in tournament selection is the tournament size, T, where $2 \leq T \leq Ns$. A typical tournament size for the EP algorithm is $T = 10$. DE, however, conducts Np, *binary* tournaments ($T = 2$) in which only two individuals compete. In general, the selection pressure increases as T increases, i.e., increasing T speeds convergence, so compared to EP tournament selection ($T = 10$), DE selection is gentler. DE's lower selection pressure helps avoid premature convergence without the introduction of variation operators to enhance the diversity of the pool of potential trial vectors.

Ranking (e.g., ($\mu + \lambda$)-selection in which both the current and trial populations are sorted based on objective function value) is a special case of EP tournament selection for which $T = Ns$. For example, if one vector is better than another, the better vector will win all the same tournaments that the inferior vector wins plus the tournament with the inferior vector itself. Since better vectors always have more wins than inferior vectors, conducting $T = Ns$ tournaments for each vector ensures that ranking vectors by the number of wins also ranks them by objective function value. In practice, ranking is accomplished more efficiently by sorting the population from best to worst based on objective function value and then taking the top μ individuals. Efficient sorting reduces the computational complexity of the each-against-all tournament process from $O(Ns^2)$ to $O(Ns \cdot \log(Ns))$ (Blahut 1984).

Tournament selection is very versatile because it only depends on knowing which of two solutions that have been paired for competition is better. Because it only depends on the difference between objective function values, tournament selection is unaffected when a constant is added to every vector's objective function value (transposition) (Eiben and Smith 2003). By contrast, *fitness proportional selection* selects an individual with a probability based on its function value (Holland 1992)

$$P_i = \frac{f(\mathbf{x}_i)}{\sum_{i=0}^{\mu-1} f(\mathbf{x}_i)} \qquad (2.60)$$

and adding a constant to each vector's objective function value will change its fitness proportional selection probability.

Tournament selection is also well suited for co-evolutionary optimization tasks in which the quality of a given solution is defined only in the context of its performance with respect to the rest of the population. For example, it is difficult to rate an arbitrary checkers strategy, but it is a simple matter to determine which of two strategies is better by actually using them to play one or more games against each other. Similarly, tournament selection is the most effective way to evolve solutions to "subjective" objective functions, like those used in evolutionary art. In such environments, it is easier to decide, for example, which of two pictures is more pleasing than it is to decide how pleasing a picture is in an absolute sense. In addition, tournament selection permits the concept of Pareto-dominance to be implemented for both constraint functions and for multi-objective optimizations (see Sects. 4.3 and 4.6).

A single competition in an EP tournament might select two current population vectors, a current and a trial vector, or two trial vectors to compete against one another. DE, however, restricts tournament selection to this last possibility in which each competition pits a trial vector against a vector in the current population (the target) with the additional proviso that the target and trial vectors are also related by crossover. The next subsection explores this special class of deterministic, binary tournaments, known as one-to-one selection for the way in which population and trial vectors are paired for competition.

2.7.3 One-to-One Survivor Selection

Besides pairing competitors based on age, DE's one-to-one replacement scheme differs from EP tournament selection in other ways. For example, an EP-style binary tournament conducts $2Np$ competitions by pairing each vector in the current population and each trial vector with a randomly selected competitor. Each vector competes once in its own tournament and possibly one or more times as a competitor in another vector's tournament. Consequently, not every vector that wins advances and not every vector that loses fails to advance. For example, a very good vector will lose if it is chosen as a competitor in the best vector's tournament. If, however, an average competitor is chosen to compete in the very good vector's tournament, then the very good vector will win. Even though it loses in competition with the best vector, the very good vector still wins its own tournament, giving it the same chance to enter the next generation as the best vector because both vectors won their tournaments and scored one win apiece. Furthermore, it is possible for more than Np vectors to win their tournaments depending on how competitors are chosen, in which case all winning vectors cannot advance. For example, although improbable, every member of the current population and every member of the trial population might randomly pick the very worst vector in the combined populations as a competitor. In that case, there would be $2Np - 1$ vectors with one win and one vector with no wins. In such cases, the best vector can be lost unless steps are taken to preserve it.

By contrast, DE's one-to-one selection holds only Np "knock-out" competitions. Any vector that loses the single competition in which it competes is eliminated and vectors that win are assured of a spot in the next generation. This form of binary, deterministic, one-to-one tournament selection in which competitors are chosen from different populations is not unique to DE. Like DE, the Particle Swarm Optimization (PSO) algorithm also conducts Np competitions that compare the trial vector with population index i to the best performing vector at population index i (Kennedy and Eberhart 1995). In DE, the best performing vector at the ith position is just the ith vector in the current population, i.e., the target vector $\mathbf{x}_{i,g}$. In both DE and PSO, the trial vector replaces the best-so-far vector with the same index only if it has an equal or lower objective function value.

Comparing each trial vector to the best performing vector at the same index ensures that both DE and PSO retain not only the best vector at each index, but also the very best-so-far solution at any index. Even so, a trial vector that is better than most of the current population will be rejected if its target is even better. Trial vectors that are worse than the worst vector in the current population, however, are never accepted.

2.7.4 Local Versus Global Selection

Local Selection

When an objective function is known to exhibit multiple global optima, some algorithms subdivide the selection pool into subpopulations. Each subpopulation evolves in isolation to prevent the entire population from coalescing about a single optimum. Selection is local because survivors can only replace members of the same subpopulation. For example, in the simple GA with a (μ, λ) survivor selection scheme, age determines the interacting subpopulation, or *selection neighborhood*, because only trial vectors are allowed to compete. In general, the smaller the selection neighborhood is, the lower the selection pressure will be. Just as increasing λ or T increases selection pressure, increasing the size of the population from which the base vector is drawn speeds convergence.

If DE's base and target vectors are the same, vectors evolve in isolation as though there were Np subpopulations. Selection will be local because each population vector will be compared to a mutated version of itself. Although the mutation differential is still drawn from the population at large, there is no interaction with other population members – no comparisons to solutions evolving in other parts of the solution space.

Global Selection

When seeking a single, global optimum, care must be taken to ensure that information about the best solutions can reach all members of the population. If base vectors are randomly chosen, then each vector in the current population is compared to and possibly crossed with the mutated version of another vector. Compared to local selection, global selection speeds convergence and minimizes the risk of stagnation.

2.7.5 Permutation Selection Invariance

When base vectors are the elements of a random permutation of the sequence $(0, 1, \ldots, Np - 1)$, the roles played by the base vector and survivor selection become interchangeable. If a permutation of the sequence $(0, 1, \ldots, Np - 1)$ indexes base vectors, then each vector in the current population serves as a base vector once per generation (Sect. 2.4.2). Each vector in the current population also serves as a target vector once per generation. As such, it makes no difference whether the random permutation indexes either base vectors or target vectors. Either way, each vector in the current population vector is mutated, then matched by permutation to a vector with which it is both crossed and compared and which it potentially replaces. For example, the first expression in Eq. 2.61 shows the traditional DE approach in which permuted indices, permute[i], select the base vector, while the running index, i, points to the target vector. The second expression shows the situation reversed, in which the running index specifies the base vector and the ith permutation entry locates the target vector. For clarity, Eq. 2.61 expresses this symmetry as a vector relationship ($Cr = 1$):

$$ \mathbf{x}_i \text{ vs. } \mathbf{x}_{\text{permute[i]}} + F(\mathbf{x}_{r1} - \mathbf{x}_{r2}) \quad \Leftrightarrow \quad \mathbf{x}_{\text{permute[i]}} \text{ vs. } \mathbf{x}_i + F(\mathbf{x}_{r1} - \mathbf{x}_{r2}). \quad (2.61) $$

These two approaches based on the permutation selection method give identical results, i.e., optimizer performance is the same regardless of which method is employed. In both cases, each vector in the current population is mutated and then crossed with and compared to another vector in the current population not assigned to any other mutant. As such, random assignments derived from permutations can be performed either during parent (base vector) selection (left side of Eq. 2.61), or when selecting a target vector with which to cross and compete (right side of Eq. 2.61).

The "urn" permutation algorithm (see Sect. 5.2) helps illustrate the symmetry between these two selection options. For example, let urn 1 hold Np marbles, each of which is numbered with a unique vector index, $i \in [0, Np - 1]$. Urn 2 also contains Np marbles numbered 0 through $Np - 1$, but this time, numbers indicate a vector in the current population that has been mutated. Permutation selection matches vectors in the current population with mutants for crossing and competing by drawing a marble (at random and without replacement) from each urn. Once all marbles have been randomly paired this way, the final mapping between population and mutant indices will define a permutation. It does not matter to which urn the role of the permutation is assigned, just as it does not matter whether targets or mutants have their indices permuted.

2.7.6 Crossover-Dependent Selection Pressure

Because DE selection compares each trial vector to the vector in the current population with which it is crossed, replacing a vector in the current population can change the population's composition by as little as one parameter ($Cr = 0$), or by as many as D ($Cr = 1$). If, unlike DE, each vector in the current population is compared to and replaced by a trial vector with whom it shares no common parameters, then the composition of (one member of) the vector in the current population changes by D parameters. Similarly, when $(\mu + \lambda)$-selection replaces a vector in the current population with a trial vector, the two vectors usually share no parameter values in common. By contrast, the number of parameters changed when classic DE accepts a trial vector is a function of Cr.

Figure 2.60 compares the selection pressure exerted by classic DE and two other selection schemes, both of which change D parameters each time they replace a vector in the current population. Classic DE (DE/rand/1/bin) generated the trial vectors in each case and algorithms differed only in how they selected survivors. Data points were only plotted if all 20 trials were successful. The top line shows that classic DE selection is the slowest of the three schemes when $Np = 60$, although it is the only method whose selection pressure is gentle enough to prevent premature convergence at small values of Cr. The middle line corresponds to a selection scheme that pairs vectors in the current population and trial vectors with a random permutation. This use of random permutation to pair vectors and trial vectors is distinctly different from the permutation selection method described in Sect. 2.7.5. Instead of drawing vectors from the first urn and mutants from the second, this selection method draws completed trial vectors that have already been crossed with another vector from the second urn. As such, vectors in the current population and the trial vectors that compete against them share no parameters through crossover. Its greater rate of convergence in Fig. 2.60 shows that in the case of the hyper-ellipsoid, accepting D new parameters per trial vector creates more selection pressure than does classic DE selection, except at $Cr = 1$ where both algorithms change the current population by D parameter values for each trial vector accepted. The $(\mu + \lambda)$-selection scheme (bottom line) generates the highest selection pressure because it not only changes D parameters per accepted trial vector, it also uses $T = Np$ tournaments instead of just $T = 2$.

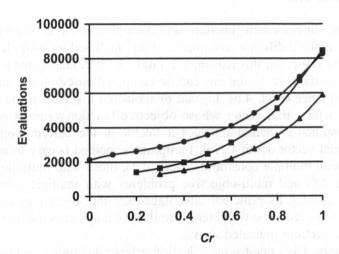

Fig. 2.60. Classic DE selection (top line) is weaker than $(\mu + \lambda)$-selection (bottom line) but both share a similar profile. Pairing target and trial vector (not mutant) adversaries with a random permutation provides an intermediate level of selection pressure (middle line). DE/rand/1/bin ($Np = 60$, $F = 0.9$) generated trial vectors, but survivors were selected by the indicated selection method. Data points are 20-trial averages.

2.7.7 Parallel Performance

Not all survivor selection methods are equally well suited to parallel implementations. For example, $(\mu + \lambda)$-selection is time consuming when implemented as tournament selection without replacement. If instead, $(\mu + \lambda)$-selection is done by sorting, it becomes difficult to implement efficiently in parallel because some comparisons must be performed before others. DE, however, is ideally suited for parallel computing, primarily because each vector in the current population competes in a single tournament against a trial vector that belongs to an intermediate population. Section 5.1 describes several schemes for distributing DE across a network of processors. In addition, Sect. 7.6 describes how DE was implemented in parallel to perform image registration. In that application, performance scaled quasi-linearly.

2.7.8 Extensions

The presence of constraint functions and multiple objectives in an optimization task make it difficult to compare solutions based on a single objective function value. For this reason, J. Lampinen (2002) has extended DE's selection criteria so that solutions can be compared based on the notion of Pareto-dominance (Sect. 4.6). Instead of replacing a vector in the current population with a trial vector whose objective function value is equal or lower, Lampinen's method replaces a vector in the current population when the trial vector dominates it. Lampinen's method is easy to apply to problems with multiple constraints (Sect. 4.3), those with multiple objectives (Sect 4.6) and multi-objective problems with multiple constraints (Sect 4.6). Among its principal advantages are that objectives and constraints do not need to be weighted. Details on Lampinen's method can be found in the sections indicated above.

In summary, DE's one-to-one selection offers numerous advantages beyond its simplicity. It does not require mapping objective function values to selection probabilities. It is elitist, easy to implement in parallel, compensates for increased acceptance rates at low Cr and has all the traditional advantages of tournament selection's versatility which include invariance to objective function transposition. DE selection is also flexible, allowing either target or base indices to be randomly specified by permutations, or the criterion of "less than or equal" to be replaced by "Pareto-dominant" when problems have multiple objective and/or multiple constraints.

2.8 Termination Criteria

Sometimes it is obvious when an optimization should be halted. For example, in constraint satisfaction problems (Sects. 4.3 and 4.5) the optimization is over when all constraints are satisfied, i.e., when a feasible vector is found. In multi-objective optimization (Sect. 4.6), however, objectives often conflict. Satisfying one objective leaves another unfulfilled, so it is not always clear when to stop searching for a better compromise. This section briefly describes some halting criteria and the scenarios in which they are appropriate.

2.8.1 Objective Met

In some optimization tasks, the objective function's minimum value is already known. For example, the goal when designing telescope optics is to reduce the geometric spot size of a star's image to a point. The wave nature of light, however, renders meaningless any improvement beyond certain well-known limits. Consequently, an optical system optimization can be halted when spot sizes fall below the limits set by the wave nature of light. The same is true of other error functions for which the tolerable error is given. This is also the method used when working with test functions whose minima are known. If the best-so-far vector's objective function value is within a specified tolerance of the global minimum, the optimization halts.

2.8.2 Limit the Number of Generations

Usually, the objective function minimum is not known in advance. Even for many test functions, only the best-known results are reported. In these cases, optimizations can be terminated after g_{max} generations. When testing optimizers with functions whose optima *are* known, setting g_{max} may halt optimizations that do not reach the objective function minimum within the specified tolerance. Finding a value of g_{max} that is large enough to give an optimizer enough time to find the optimum, but not so long that a second trial would be a better way to invest computer time, involves some guess work.

Alternatively, an optimization can be halted when Δg_{max} generations have passed without a trial vector being accepted. Again, some experimentation may be needed to find a good value for Δg_{max}. Long periods without improvement are perhaps more common in DE than other EAs, so it is important that Δg_{max} not be set too low.

2.8.3 Population Statistics

An optimization can also be terminated when a population statistic reaches a pre-specified value. For example, an optimization can be halted when the difference between the population's worst and best objective function values falls below some predetermined limit. This method needs to be applied with caution because it can cause an optimization to halt prematurely. For example, if the optimization is halted when the difference between the population's worst and best objective function values is less than, e.g.,

1.0×10^{-6}, the population's best objective function value might not yet be within 1.0×10^{-6} of the minimum value. Thus, the interruption is premature because DE may still be making progress even though the range of objective function values is small. When using this criterion, it is usually a good idea to make the difference between the population's worst and best objective function values several orders of magnitude lower than the tolerance set for locating the optimum. The same advice applies when monitoring the standard deviation of population vectors or the longest vector difference as termination criteria.

2.8.4 Limited Time

Sometimes only a limited amount of time is available for an optimization. In such cases, the optimization must terminate regardless of the state of the population or the number of generations. For example, in on-line optimization, only a small amount of time may available to adjust manufacturing process parameters (e.g., Sect. 7.12). Similarly, it may be that computer time is limited or simply that a deadline must be met. Monitoring and manual intervention can help determine whether the available time is best spent completing an ongoing optimization or running a new trial.

2.8.5 Human Monitoring

Because of the inherent uncertainties in knowing when an optimization is over, it usually helps to personally monitor time-consuming optimization tasks. The feedback from the best objective function value, number of trial vectors accepted per generation, the distribution of the population, etc., usually makes it clear when no more improvement is possible or when time might be better spent running a new trial. In addition, human monitoring allows the optimization to be altered in response to perceived opportunities.

2.8.6 Application Specific

Finally some applications will have their own termination criterion. In evolutionary art, for example, an optimization to find the most pleasing picture might end when interest in the exhibit wanes, or when a certain group of people have participated.

References

Ali MM, Törn A (2000) Optimization of carbon and silicon cluster geometry for Tersoff potential using differential evolution. In: Floudas CA, Pardalos PM (eds) Optimisation in computational chemistry and molecular biology. Kluwer Academic, Dordrecht, pp 1–15

Bäck T (1993) Optimal mutation rates in genetic search. In: Forrest F (ed) Proceedings of the fifth international conference on genetic algorithms. Morgan-Kaufmann, San Mateo, CA, pp 2–8

Bäck T (1996) Evolutionary algorithms in theory and in practice. Oxford University Press

Bäck T, Schwefel H-P (1993) An overview of evolutionary algorithms for parameter optimization. Evolutionary Computation 1(1):1–23

Bäck T, Schwefel H-P (1995) Evolution strategies I: variants and their computational implementation. In: Periaux J, Winter G (eds) Genetic algorithms in engineering and computer science. Wiley, Chichester, Chap. 6

Bäck T, Rudolph G, Schwefel H-P (1993) Evolutionary programming and evolution strategies: similarities and differences. In: Fogel DG, Atmar W (eds) Second annual conference on evolutionary programming, February. Evolutionary Programming Society, La Jolla, CA, pp 11–22

Baker JE (1987) Reducing bias and inefficiency in the selection algorithm. In: Grenfenstette JJ (ed) Proceedings of the first international conference on genetic algorithms and their applications. Lawrence Erlbaum, Hillsdale, NJ, pp 14–21

Beckman FS (1980) Mathematical foundations of programming. Addison-Wesley, Reading, MA

Beyer HG (1999) On the dynamics of EAs without selection. In: Banzaf W, Reeves C (eds) Foundations of genetic algorithms. Morgan Kaufmann, San Mateo, CA, pp 5–26

Blahut RE (1984) Fast algorithms for digital signal processing. Addison-Wesley, Reading, MA, p 329

Caruana RA, Eshelman LJ, Schaffer JD (1989) Representation and hidden bias II: eliminating defining length bias in genetic search via shuffle crossover. In: Sidharan NS (ed) Eleventh international joint conference on artificial intelligence, Morgan Kaufmann, San Mateo, CA, vol 1, pp 750–755

Chakraborti N, Misra K, Bhatt P, Barman N, Prasad R (2001) Tight-binding calculations of Si–H clusters using genetic algorithms and related techniques: studies using differential evolution. Journal of Phase Equilibria 22(5):525–530

Eiben AE, Smith JE (2003) Introduction to evolutionary computing. Springer, Berlin Heidelberg New York

Eshelman LJ, Caruana RA, Schaffer JD (1989) Biases in the crossover landscape. In: Schaffer JD (ed) Proceedings of the third international conference on genetic algorithms. Morgan Kaufmann, San Francisco, pp 10–19

Fogel DB (1991) System identification through simulated evolution: a machine learning approach to modeling. Ginn Press, Needham Heights, MA

Fogel LJ, Owens AJ, Walsh MJ (1966) Artificial intelligence through simulated evolution. Wiley, New York

Gamperle G, Mueller SD, Koumoutsakos P (2002) A parameter study for differential evolution. In: Grmela A, Mastorakis NE (eds) Advances in intelligent systems, fuzzy systems, evolutionary computation. WSEAS Press, Athens, pp 293–298

Goldberg DE (1989) Genetic algorithms in search, optimization and machine learning. Addison-Wesley, Reading, MA

Halton J, Weller G (1964) Algorithm 247: radical inverse quasi-random point sequence. Communications of the ACM, 7(12):701–702

Holland, JH (1973) Genetic algorithms and the optimal allocation of trials. SIAM Journal of Computing 2:88–105

Holland J (1992) Adaptation in natural and artificial systems. MIT Press, Cambridge, MA. First edition 1975, The University of Michigan Press, Ann Arbor

Kennedy J, Eberhart RC (1995) Particle swarm optimization. In: Proceedings of the international conference on evolutionary computation, Perth, Australia. Invited paper

Koza J (1992) Genetic programming: on the programming of computers by means of natural selection. MIT Press, Cambridge, MA

Lampinen J (2002) Multi-constrained nonlinear optimization by the differential evolution algorithm. In: Rajkumar Roy, Mario Köppen, Seppo Ovaska, Takeshi Furuhashi, Frank Hoffmann (eds) Soft computing and industry: recent advances. Springer, Berlin Heidelberg New York, pp 305–318 (Proceedings of the 6th online world conference on soft computing in industrial applications (WSC6), September 10–24, 2001. Available at: http://vision.fhg.de/wsc6)

Lampinen J, Zelinka I (2000) On stagnation of (the) differential evolution algorithm. In: Ošmera P (ed) Proceedings of MENDEL 2000, sixth international Mendel conference on soft computing, June 7–9, Brno, Czech Republic. Brno University of Technology, Faculty of Mechanical Engineering, Institute of Automation and Computer Science, Brno, pp 76–83. Available at: http://www.lut.fi/~jlampine/MEND2000.ps

Macready WG, Wolpert DH (1998) Bandit problems and the exploration/exploitation tradeoff. IEEE Transactions on Evolutionary Computing 2(1):2–22

Michalewicz Z (1996) Genetic algorithms + data structures = evolution programs, 3rd edn. Springer, Berlin Heidelberg New York

Mühlenbein H (1992) How genetic algorithms really work I: mutation and hill climbing. In: Schwefel H-P, Männer R (eds) Proceedings of the second international conference on parallel problem solving from nature, Springer, Berlin Heidelberg New York, pp 15–26

Mühlenbein H, Schlierkamp-Voosen D (1993) Predictive models for the breeder genetic algorithm. Evolutionary Computation 1(1):25–50

Peitgen H-O, Saupe D (eds) (1998) The science of fractal images. Springer, Berlin Heidelberg New York

Potter MA, DeJong KA (1994) A cooperative co-evolutionary approach to function optimization. In: Davidor Y, Schwefel H-P, Männer R (eds) Proceedings of parallel problems solving from nature 3. Springer, Berlin Heidelberg New York, pp 249–257

Price KV (1996) Differential evolution: a fast and simple numerical optimizer. In: Smith MH, Lee MA, Keller J, Yen J (eds) Proceedings of the 1996 biennial conference of the North American fuzzy information processing society – NAFIPS, June 19–22, Berkeley, CA, USA. IEEE Press, New York, pp 524–527

Price KV (1997) Differential evolution vs. the functions of the second ICEO. In: Proceedings of the 1997 IEEE international conference on evolutionary computation, Indianapolis, Indiana, USA. IEEE Press, New York, pp 153–157

Rudolph G (1996) Convergence of evolutionary algorithms in general search spaces. In: Proceedings of the third IEEE conference on evolutionary computation, IEEE Press, New York, pp 50–54

Saravanan N, Fogel DB (1997) Multi-operator evolutionary programming: a preliminary study on function optimization. In: Angeline PJ, Reynolds RG, McDonnell JR, Eberhart R (eds) Evolutionary programming 6: sixth international conference, Indianapolis, Indiana, USA, April. Springer, Berlin Heidelberg New York, pp 215–221

Salomon R (1996a) Re-evaluating genetic algorithm performance under coordinate rotation of benchmark functions: a survey of some theoretical and practical aspects of genetic algorithms. Biosystems 39(3):263–278

Salomon R (1996b) The influence of different coding schemes on the computational complexity of genetic algorithms in function optimization. In: Voigt H-M, Ebeling E, Rechenberg I, Schwefel H-P (eds) Proceedings of the fourth international conference on parallel problem solving from nature, Springer, Berlin Heidelberg New York, pp 227–235

Salomon R (1997) Raising theoretical questions about the utility of genetic algorithms. In: Proceedings of the sixth international conference on evolutionary programming. Lecture notes in computer science, vol 1213. Springer, Berlin Heidelberg New York, pp 275–284

Spears WM, DeJong KA (1991) An analysis of multi-point crossover. In: Rawlins G (ed) Foundations of genetic algorithms. Morgan Kaufmann, San Francisco, pp. 301–315

Storn R (1996) On the usage of differential evolution for function optimization. In: Smith MH, Lee MA, Keller J, Yen J (eds) Proceedings of the 1996 biennial conference of the North American fuzzy information processing society – NAFIPS, June 19–22, Berkeley, CA, USA. IEEE Press, New York, pp 519–523

Storn R, Price KV (1996) Minimizing the real functions of the ICEC'96 contest by differential evolution. In: Proceedings of the 1996 IEEE international conference on evolutionary computation, Nagoya, Japan. IEEE Press, New York, pp 842–844

Storn R, Price KV (1997) Differential evolution – a simple and efficient heuristic for global optimization over continuous spaces. Journal of Global Optimization 11:341–359

Syswerda G (1989) Uniform crossover in genetic algorithms. In: Schaffer JD (ed) Proceedings of the third international conference on genetic algorithms. Morgan Kaufmann, San Francisco, pp 2–9

Wright AH (1991) Genetic algorithms for real parameter optimization. In: Rawlins GJE (ed) Foundations of genetic algorithms. Morgan-Kaufmann, San Mateo, CA, pp 205–218

Yang J-M, Chen Y-P, Horng J-T, Kao C-M, Chen Y-P, Horng J-T, Kao C-Y (1997) Applying family competition to evolution strategies for constrained optimization. In: Angeline PJ, Reynolds RG, McDonnell JR, Eberhart R (eds) Evolutionary programming 6: sixth international conference, Indianapolis, Indiana, USA, April. Springer, Berlin Heidelberg New York, pp 201–211

Yao X, Liu Y (1997) Fast evolution strategies. In: Angeline PJ, Reynolds RG, McDonnell JR and Eberhart R (eds) Proceedings of the sixth international conference on evolutionary programming, Springer, Berlin Heidelberg New York, pp 151–161

Zaharie D (2002) Critical values for the control parameters of differential evolution algorithms. In: Matoušek R, Ošmera P (eds) Proceedings of MENDEL 2002, 8th international conference on soft computing, June 5–7, 2002, Brno, Czech Republic. Brno University of Technology, Faculty of Mechanical Engineering, Institute of Automation and Computer Science, Brno, pp 62–67

Zimmons P (n.d.) Polynomial fitting with differential evolution. Available at: http://www.cs.unc.edu/~zimmons/cs258/poly.html

3 Benchmarking Differential Evolution

3.1 About Testing

Testing can be a valuable tool for understanding how and why an algorithm performs as it does. For example, testing can measure how an algorithm's performance depends on objective function characteristics like dimension, number of local minima, degree of parameter dependence, dynamic range of parameters, constraints, quantization, noise, etc. Testing can also show which control parameter combinations are the most effective. This knowledge can be particularly useful, since finding an effective set of control parameter combinations is itself a multi-objective optimization problem in which fast and reliable convergence are conflicting objectives. In addition, test functions are a convenient way to compare one algorithm's performance to that of another. Furthermore, testing can lead to new insights that can be exploited to enhance an optimizer's performance.

Despite its value, testing can be misleading if results are not correctly interpreted. For example, the dimension of some test functions can be arbitrarily increased to probe an algorithm's scaling behavior. In all but the simplest cases, however, changing an objective function's dimension also changes its other characteristics as well (e.g., number of local minima, dynamic range of optimal function parameter values, etc.). Thus, an algorithm's response to a change in test function dimension must be understood in the context of the accompanying alterations to the objective function landscape.

Test beds that consist entirely of separable functions are another example in which test functions provide misleading clues about an algorithm's versatility. For many years, most of the functions used for testing GAs were separable (see Sect. 1.2.3). Consequently, GAs with low mutation rates performed well on these early test beds, leading to high expectations for their success as numerical optimizers. It is now clear that these early successes do not extend to parameter-dependent functions because GAs are limited by their lack of a scheme for correlating mutations (see Sects. 1.2.3 and 2.6.2) (Salomon 1996).

Ensuring that test bed functions are sufficiently diverse is just one challenge to creating a valid benchmark against which algorithms can be compared. Benchmarking is further complicated by the fact that a fair comparison must take into account not only an algorithm's speed and probability of convergence, but also the effort required to tune its control parameters. For example, in the First International Contest on Evolutionary Optimization (1st ICEO), probability and speed of convergence were combined into a single performance measure, but control parameter tuning was not considered (see Sect 3.2 for a description of this method) (Bersini et al. 1996). In the 2nd ICEO, "crafting effort" was taken into account so that given two algorithms with comparable speed and probability of convergence, the algorithm that required less tuning scored better (e.g., Price (1997), Yen and Lee (1997)).

Even if tuning issues can be overcome, it is sometimes hard to define a problem without favoring one approach over another. For example, floating-point optimizers require less computational effort to compute the optima of real-valued functions than do GAs using bit-strings (see Sect. 2.2). In addition, there can be other important factors that determine an optimizer's true value, like the computer resources that it requires. Furthermore, the lack of a sufficiently detailed account of the testing procedure can also lead to test results being misinterpreted.

Perhaps most importantly, the test bed has been shown to have fundamental limits as a benchmarking tool. In their landmark paper on the "No Free Lunch" (NFL) theorem, Wolpert and Macready (Wolpert and Macready 1997) proved that all optimizers meeting certain criteria perform the same on average if the sample of test functions over which their performance is averaged is large enough. In other words, if algorithm A outperforms algorithm B on one test bed, then there will be another test bed on which algorithm B outperforms algorithm A by a similar margin. For example, in the 1st ICEO, an optimizer specifically designed for separable functions won because the contest functions were separable. If the contest functions had been non-separable, this same method would have performed dismally and the second place method based on Latin squares might have won. If the test functions had been both high-dimensional and non-separable, the third place DE algorithm might have won because high-dimensional Latin squares are computationally expensive. Even when testing versions of DE, no single algorithm always wins.

The NFL theorem assumes that optimizers have no special knowledge about the objective function, i.e., that they are "black-box" algorithms. Special knowledge might be that the function is separable, symmetric, unimodal, etc. Restated, the NFL theorem says that if one optimizer consistently outperforms another, then it must be using special knowledge about the objective functions. This observation has encouraged the development of memetic algorithms, i.e., EA-based optimizers that include one or more phases of problem-specific heuristics and local optimization (Moscato 1989).

Although DE begins with no knowledge about an objective function, the population soon "learns" about the function by distributing itself along the function's contours. As the simple experiments with the hyper-ellipsoid in Sect. 2.6 suggest, a function's contours contain some of the most useful information for an optimizer because they usually reveal the best step size and orientation. DE exploits this knowledge and adapts step sizes accordingly. The question regarding DE's performance then becomes, "How prevalent are functions for which contour matching is an effective strategy?" Based on the many objective functions to which DE has been applied, most practical problems seem to have a structure that contour matching can exploit, even though DE is ultimately no better than a random walk when averaged over the universe of functions. Of course, knowing how DE works makes it easy to *invent* infinitely many functions for which contour matching is counterproductive, but, in practice, contour matching seems to be far more of a benefit than a drawback. The next section describes the criteria used in this chapter to measure optimizer performance.

3.2 Performance Measures

There are several ways to measure an algorithm's performance. The *convergence* or *progress plot* graphs the current best vector's objective function value as a function of time, i.e., as a function of the number of either generations or function evaluations. Rather than plotting a single trial's progress, it is better to conduct a series of trials and then plot the mean best objective function value. Alternatively, a progress plot can track the distance of the best-so-far vector from the optimum as a function of time. For this type of plot, the location of the global optimum must already be known. Again, the multi-trial average better represents the optimizer's performance than does the result of a single trial.

By themselves, progress plots do not give a complete picture of an optimizer's abilities. For example, if only one of ten trials becomes stuck in a local optimum, then the plot of the best vector's average performance can make it look as though none of the trials were successful. In practice, an algorithm's best performance is often more important than is its average or worst performance. For example, many design optimizations conduct multiple trials just to find the best solution possible.

To better estimate an algorithm's ability to locate the true global optimum, a trial can be classified as a "success" when the best vector's objective function value falls below a predetermined limit known as the *value-to-reach*, or "VTR". Trials that do not reach the VTR within a predetermined maximum number of evaluations, E_{max}, are treated as "failures". The VTR must be set low enough so that any vector with an objective function value that is less than or equal to the VTR falls within the basin of attraction to which the global minimum belongs. Consequently, this method is feasible only when a test function's global optimum is already known. In addition, this method presumes that once the population begins to inhabit the basin of attraction occupied by the global optimum, any further progress will be easy enough that a local optimizer can be used if need be. With the VTR and E_{max} as criteria for success, one can estimate the speed and the probability with which an optimizer locates the basin of attraction to which the global optimum belongs.

Ideally, an optimizer should be both fast and reliable, but high speed and high probability of convergence are usually conflicting objectives. Forcing an algorithm to converge more quickly usually increases the odds that it will prematurely converge. Just as conflicting objectives make it hard to decide which compromise solution to a multi-objective optimization problem is best, determining which one of two algorithms performs best, even on a single function, is difficult because high speed and high reliability are conflicting performance measures. One algorithm can be fast but unreliable, while the other might be very slow, but far more likely to succeed. To resolve this dilemma, the creators of the 1st ICEO proposed a combined measure, the "expected number of (function) evaluations per success", or ENES. The ENES is the total number of function evaluations taken over t trials, divided by the number of successful trials, s (Eq. 3.1). In Eq. 3.1, E_k is the number of objective function evaluations taken by the kth trial.

$$\text{ENES} = \frac{1}{s} \cdot \sum_{k=1}^{k=t} E_k, \quad s > 0. \tag{3.1}$$

If no trials are successful, the ENES is not defined.

The total number of function evaluations over t trials has two components: the total number of evaluations (up to and including E_{max}) taken by successful trials and E_{max} evaluations for all those trials that end in failure. Since E_{max} is manually set, the ENES does not always reflect the true performance of an algorithm. This study resolves the conflict between speed and reliability by measuring algorithm speed at a constant success probability. In particular, control parameter combinations are sought that can produce the lowest *average (number of function) evaluations per success*

(AES) in ten consecutive successful trials. The AES is not explicitly dependent on E_{max}, but setting E_{max} too low will prematurely terminate potentially successful trials. On the other hand, comparatively large values for E_{max} do not affect the AES like they do the ENES.

In the 2nd ICEO, contestants were allowed to tune their algorithms to solve each function, but a penalty was imposed based on how diverse the final set of control parameters was. For example, there was no penalty if a single set of control parameters was used for each function, whereas using different control parameter settings for each function incurred the highest penalty. This study takes a different approach that is based on phase portraits, primarily because F, Cr and P_F can be limited to the range [0, 1] and sampled uniformly. If successful random control parameter combinations are plotted as points, a phase portrait emerges that shows not only which combinations are effective, but also how difficult it is to find them. By contrast, the entropy-based diversity measure used in the 2nd ICEO relies on the skill of a researcher to find an effective set of control parameters and provides no clues about how hard this decision was for a particular function.

The next section compares four versions of DE, first using phase portraits both to find reliable control parameter combinations and to provide a measure of an algorithm's speed, then using progress plots of the mean best vector's performance. The progress plots show DE's time-dependent behavior and demonstrate that it has no trouble driving objective function values below the VTR. Together with the control parameter plots, a fairly clear picture of each algorithm's strengths and weaknesses emerges.

3.3 DE Versus DE

3.3.1 The Algorithms

This section compares four versions of DE that differ only in how new solutions are generated:

- DE/rand/1/bin (classic DE)
- DE/best/1/bin, with uniform jitter, $d = 0.001$
- DE/target-to-best/1/bin, $K = F$
- DE/rand/1/either-or, $K = 0.5 \cdot (F+1)$

In this shorthand notation, the first term after "DE" specifies how the base vector is chosen. For example, "best" means that the base vector is the current-best-so-far vector. Similarly, "rand" means that base vectors are randomly chosen, while "target-to-best" means that base vectors are chosen to lie on the line defined by the target vector and the best-so-far vector. The number that follows indicates how many vector differences contribute to the differential. In each case above, only one vector difference is used. The three DE versions that use uniform crossover are appended with the additional term "bin" for "binomial" (distribution). The term "either-or" indicates that trial vectors are either three-vector recombinants or randomly chosen population vectors to which a randomly chosen vector difference has been added. Pseudo-code for the three versions that use uniform crossover appears in Fig. 3.1, while pseudo-code for the DE/rand/1/either-or algorithm is given in Fig. 3.2. In all four versions, base, target and difference vector indices are all distinct.

```
...
for (j=0; j<D; j++)
{
    if (rand_j(0,1)<=Cr or j==jrand)
    {
        if (DE/rand/1/bin; classic DE)
        {
            u_j,i=x_j,r0+F*(x_j,r1-x_j,r2);
        }
        if (DE/best/1/bin, uniform jitter, d=0.001)
        {
            F_j=F+0.001*(rand_j(0,1)-0.5);
            u_j,i=x_j,best+F_j*(x_j,r1-x_j,r2);
        }
        if (DE/target-to-best/1/bin, K=F)
        {
            u_j,i=x_j,i+F*(x_j,best-x_j,i)+F*(x_j,r1-x_j,r2);
        }
    }
    else u_j,i=x_j,i;
}
...
```

Fig. 3.1. The generating loop for DE/rand/1/bin, DE/best/1/bin (with uniform jitter) and DE/target-to-best/1/bin. In each case, base, difference and target indices are distinct (not shown).

```
...
if (rand_i(0,1)<P_F)
{
    for (j=0; j<D; j++)
    {
        u_i=x_r0+F*(x_r1-x_r2);
    }
}
else
{
    for (j=0; j<D; j++)
    {
        u_i=x_r0+0.5*(F+1)*(x_r1+x_r2-2*x_r0);
    }
}
...
```

Fig. 3.2. Pseudo-code for the generating loop for the DE/rand/1/either-or algorithm used in these experiments

Each algorithm is initialized with a uniformly distributed random population that is restricted to a region of the search space by a set of initial parameter bounds that is provided along with each function. Once the optimization begins, however, bounds are *not* enforced. In addition, all four algorithms employ classic DE selection.

To be fair, each algorithm is permitted two control parameters, either F and Cr, or F and P_F, in addition to Np. To comply with this restriction, $K = F$ in DE/target-to-best/1/bin and $K = 0.5 \cdot (F + 1)$ in DE/rand/1/either-or. Both values have some limited empirical support. In DE/target-to-best/1/bin, increasing F as the base point approaches the best-so-far vector helps counteract the increased convergence pressure that relying on the best-so-far vector creates. Evidence for the validity of the relation $K = 0.5 \cdot (F + 1)$ is visible in the phase portraits for DE/rand/1/either-or in Sect. 3.3.3 where it can be seen that the left side of the distribution of points representing effective control parameter combinations is usually vertical. Deviations from this relation between F and K cause the left side of the distribution to slope in phase portraits for DE/rand/1/either-or.

3.3.2 The Test Bed

Table 3.1 shows that the test bed chosen for this chapter consists of two uni-modal and eight multi-modal, unconstrained functions. Using unconstrained test functions helps simplify the comparison because there are no bound resetting methods or constraint handling techniques to complicate the analysis. Functions are formally defined in the Appendix, as are several additional functions that are available for experimentation with the software that accompanies this book. In addition, the Appendix also lists each function's optimal value, $f(\mathbf{x}^*)$, and a value for ε such that VTR $= f(\mathbf{x}^*) + \varepsilon$. The Appendix also lists initial parameter bounds for each function.

Table 3.1. Test bed functions

Function	D	Modality	Separable?	Comments
Hyper-ellipsoid	10	uni-modal	yes	hard when rotated
Rosenbrock	10	uni-modal	no	a 2nd ICEO function
Ackley	10	multi-modal	yes	common in literature
Chebyshev	9	multi-modal	no	a 2nd ICEO function
Griewangk	10	multi-modal	no	gets easier at high D
Rastrigin	10	multi-modal	yes	highly multi-modal
Mod. Langerman	10	multi-modal	no	a 2nd ICEO function
Shekel	10	multi-modal	yes	a 2nd ICEO function
Whitley	10	multi-modal	no	Whitley's F8F2
Lennard-Jones	15	multi-modal	no	many-body problem

To ensure that all test functions exhibit parameter dependence, all trial vectors are evaluated in a coordinate system that is rotated 45° with respect to the coordinate system in which the functions are defined. In two dimensions, the rotation routine aligns the $+x_0$ axis to the diagonal between the $+x_0$ and $+x_1$ axes. In three dimensions, the $+x_0$ axis is rotated to align with the diagonal that lies between the $+x_0$, $+x_1$ and $+x_2$ axes, and so on for higher dimensions. This technique for transforming separable functions into parameter-dependent ones was pioneered by Ralf Salomon and is described in Salomon (1996). The next subsection presents a series of phase portraits to show which algorithms and control parameter settings are effective on these test bed functions.

3.3.3 Phase Portraits

To create a phase portrait, the two-dimensional control space (excluding Np) for each algorithm was subdivided into a ten-by-ten grid and ten points were randomly chosen from within each grid square. A point was plotted if the trial using the corresponding pair of control parameters was successful. This is the same technique that was explored in Sect. 2.6.4. The resulting distribution of points shows which control parameter combinations are effective on a particular function for a given DE version with the given Np. Figure 3.3 presents an example plot of the control parameter combinations that were effective for solving the ten-dimensional hyper-ellipsoid with DE/rand/1/bin.

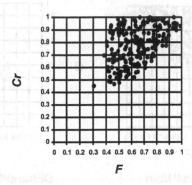

Fig. 3.3. A phase portrait for DE/rand/1/bin with points showing the control parameter combinations of F and Cr that were effective on the ten-dimensional rotated hyper-ellipsoid with $Np = 50$

The best control parameter settings are taken to be the coordinates of the *center* of the grid square with the lowest average number of function evaluations per success, or AES, for which all ten trials were successful (P = 1). A successful trial is one in which the objective function value of the current best vector becomes less than or equal to the VTR in less than or equal to E_{max} function evaluations. The title of Fig. 3.3 reports the DE version, Np (= 50) and Z (= 0.206), the fraction of all 1000 trials that were successful. This fraction measures how hard it is to find an effective set of control parameters for a given function–algorithm combination.

Ideally, finding the population size best suited for a given algorithm–function combination would allow an algorithm's very best performance to be determined. Because of the high computational effort of an exhaustive search for the optimal Np, this study explored algorithmic performance at a preset series of population sizes: Np = 50, 100, 200, 400, 800 and 1600. The population size ultimately chosen for a phase portrait is the one that gives the lowest AES when all ten randomly chosen control parameter combinations from a single grid square are successful. If no grid square produces ten consecutive trials for any $Np \leq 1600$, the AES and Np of the grid square giving the highest convergence probability, P, is used. If two or more grid squares have the same maximum probability, P < 1, then the Np of the square with the lowest AES is chosen for the plot. Table 3.2 summarizes the AES for each algorithm–function combination.

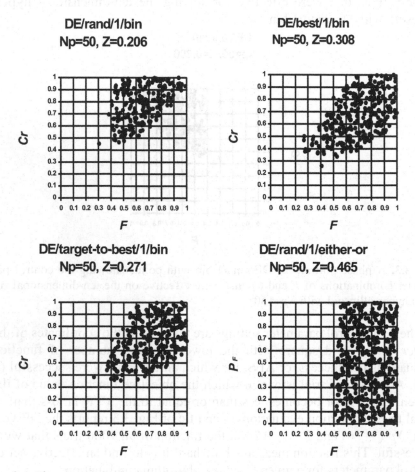

Fig. 3.4. Phase portraits for the ten-dimensional rotated hyper-ellipsoid

Ten-Dimensional Rotated Hyper-Ellipsoid

Each trial for this problem was limited to 200,000 objective function evaluations. Each algorithm produced its lowest AES at $Np = 50$. At this population size, DE/target-to-best/1/bin gave the fastest performance, while DE/rand/1/either-or had the most robust control space (Z = 0.465).

In each case, the best F was in the range $0.55 \leq F \leq 0.65$. For algorithms using crossover, $Cr > 0.4$ gave the best results, whereas the choice of P_F for the DE/rand/1/either-or algorithm was almost arbitrary when $F > 0.5$.

Ten-Dimensional Rotated Rosenbrock

Each trial was limited to 200,000 objective function evaluations. Once again, each algorithm performed best at $Np = 50$. On this function, both DE/best/1/bin and DE/target-to-best were faster than DE/rand/1/bin and DE/rand/1/either-or, which took about twice as long. Classic DE had the most robust control parameter space ($Z = 0.345$). In each case, the best F was in the range $0.65 \leq F \leq 0.75$. For algorithms using crossover, $Cr > 0.5$ was critical, while the choice for P_F was not important when $F > 0.6$.

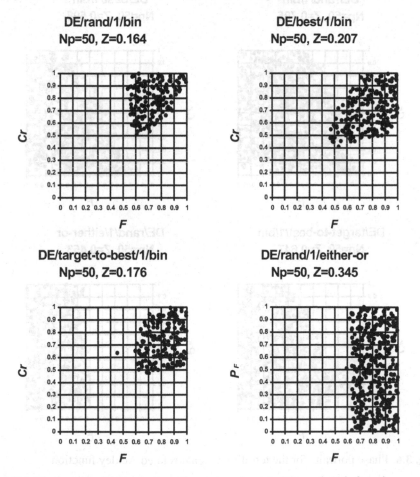

Fig. 3.5. Phase portraits for the ten-dimensional rotated Rosenbrock function

Ten-Dimensional Rotated Ackley

Even when rotated, the Ackley function proved to be very easy to solve. Once again, all four algorithms performed best when $Np = 50$. DE/best/1/bin was a little faster than DE/target-to-best/1/bin and about twice as fast as DE/rand/1/bin and four times faster than DE/rand/1/either-or. DE/rand/1/bin had the most robust control parameter space ($Z = 0.795$), although none of the algorithms was hard to tune for this function. The best F fell in the range $0.25 \leq F \leq 0.55$. Neither Cr nor P_F had to be well chosen when F was in the range $0.5 \leq F \leq 1.0$. E_{max} was 200,000 objective function evaluations.

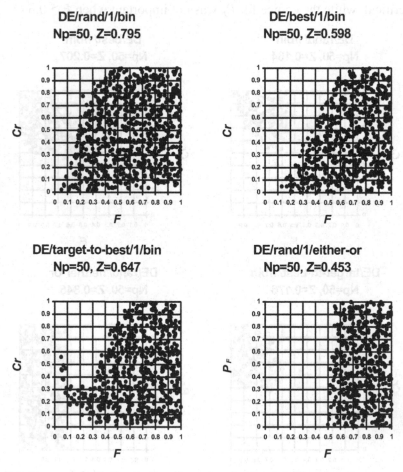

Fig. 3.6. Phase portraits for the ten-dimensional rotated Ackley function

Nine-Dimensional Rotated Storn's Chebyshev

Because this problem is already parameter dependent, rotation does not substantially affect DE's ability to solve it. Again, $Np = 50$ was more than large enough to solve this problem. DE/target-to-best/1/bin proved to be significantly faster than DE/best/1/bin on this function, but both methods incurred about the same value of Z. The highest Z was posted by DE/rand/1/either-or ($Z = 0.396$). The best F at this Np was in the range $0.65 \leq F \leq 0.75$. For algorithms using crossover, Cr had to be above about 0.5 while P_F was arbitrary when $F > 0.6$. E_{max} was 200,000 objective function evaluations.

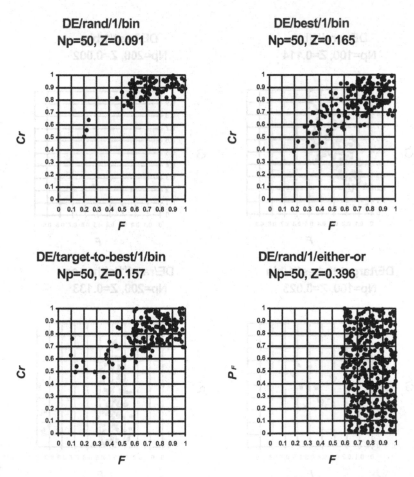

Fig. 3.7. Phase portraits for the nine-dimensional rotated Storn's Chebyshev function

Ten-Dimensional Rotated Griewangk

The ten-dimensional rotated version of Griewangk's function is the first problem in this test bed for which Np had to be larger than 50 to produce regular convergence. Both DE/rand/1/bin (Np = 100) and DE/rand/1/either-or (Np = 200) solved this problem, although DE/rand/1/either-or was more than ten times as fast and displayed a higher Z (= 0.133). Even when Np = 1600, neither DE/target-to-best/1/bin nor DE/best/1/bin could solve this function consistently within 2,000,000 function evaluations. The best F was relatively small, lying between 0.25 and 0.45. For this function, P_F is *not* arbitrary and a setting that favors three-point recombination over mutation, e.g., $P_F < 0.5$, is beneficial.

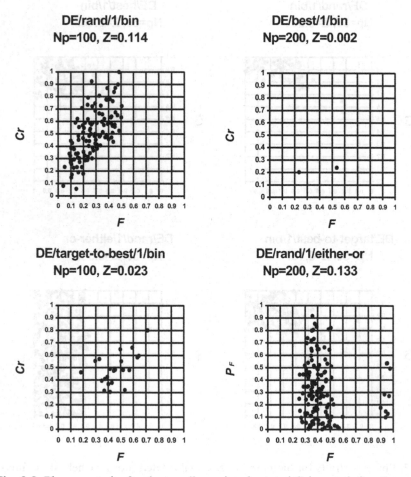

Fig. 3.8. Phase portraits for the ten-dimensional rotated Griewangk function

Ten-Dimensional Rotated Rastrigin

Both DE/best/1/bin and DE/target-to-best/1/bin struggled on this highly multi-modal function, posting $Z = 0.001$ and $Z = 0.003$, respectively. The AES for these two methods was also very high. DE/rand/1/bin was also unable to achieve regular convergence ($Z = 0.006$). By contrast, access to the three-point recombination axis lets DE/rand/1/either-or find solutions with regularity as long as $P_F < 0.5$ and $0.2 < F < 0.4$. At $Np = 400$, $F = 0.35$ was the best scale factor for DE/rand/1/either-or. E_{max} was 2,000,000 objective function evaluations.

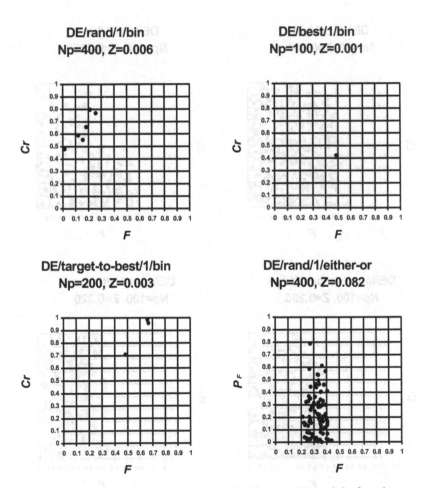

Fig. 3.9. Phase portraits for the ten-dimensional rotated Rastrigin function

Ten-Dimensional Rotated Modified Langerman

None of the four methods had any difficulty with this 2nd ICEO function. DE/rand/1/best was faster by virtue of its ability to use a smaller population ($Np = 50$) than could the other three methods ($Np = 100$), yet it also had the highest Z (= 0.415). Compared to the other three algorithms, DE/rand/1/either-or needed a higher F (= 0.75) and required about five times more objective function evaluations than did DE/rand/1/bin. Neither Cr nor P_F had to be chosen with care as long as F was in the right range. E_{max} was 2,000,000 objective function evaluations.

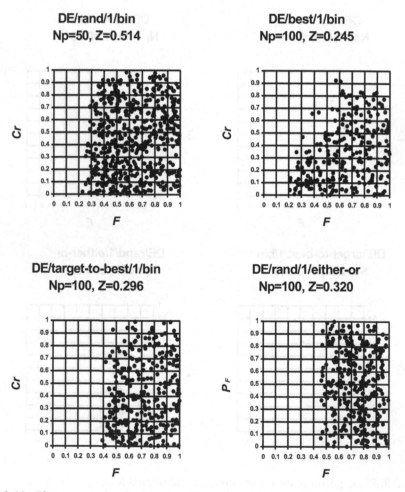

Fig. 3.10. Phase portraits for the ten-dimensional rotated Modified Langerman function

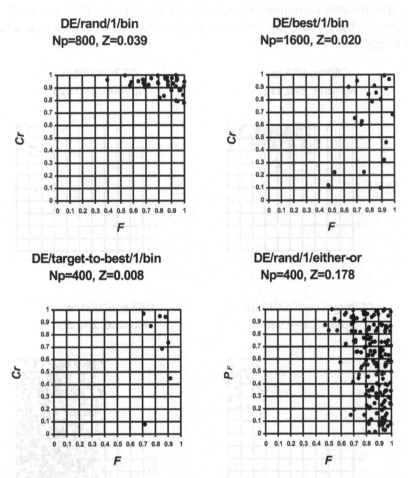

Fig. 3.11. Phase portraits for the ten-dimensional rotated Shekel's foxholes function

Ten-Dimensional Rotated Shekel's Foxholes

Like the rotated Rastrigin and Griewangk functions, the rotated Shekel's foxholes function also proved to be very hard for both methods that rely on the best-so-far vector. Neither DE/best/1/bin nor DE/target-to-best/1/bin was able to produce regular convergence in 2,000,000 function evaluations or with $Np \leq 1600$. Classic DE was more successful, achieving nine consecutive successes when $Np = 800$, $F = 0.95$, $Cr = 0.95$. By contrast, $Np = 400$ provided the either-or algorithm with enough diversity to solve this problem with regularity. Unlike the rotated Griewangk and Rastrigin functions that favored recombination ($P_F < 0.5$), the plot for DE/rand/1/either-or shows that preferentially searching the mutation axis ($P_F > 0.5$) is a better strategy.

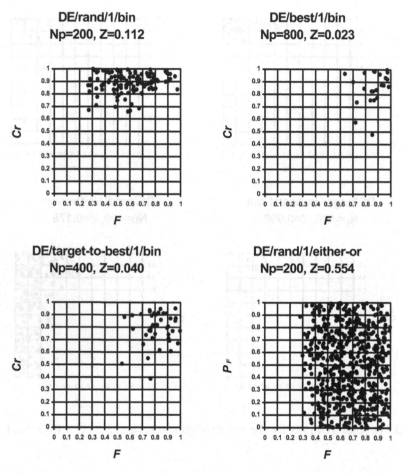

Fig. 3.12. Phase portraits for the ten-dimensional rotated Whitley function

Ten-Dimensional Rotated Whitley

This function is a composite of the one-dimensional Griewangk function and the ten-dimensional generalized Rosenbrock function. Its large-scale structure resembles Rosenbrock's function, while at a small scale, it displays a myriad of local optima due to the contribution from Griewangk's function. All methods achieved some level of success, but only DE/rand/1/bin and DE/rand/1/either-or could produce ten consecutive trials. DE/rand/1/either-or was not only the fastest algorithm, but also the most robust (Z = 0.544). E_{max} was 2,000,000 objective function evaluations.

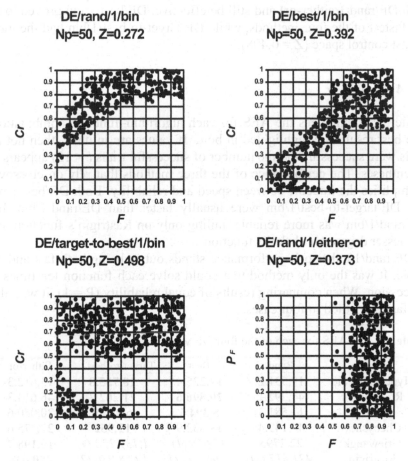

Fig. 3.13. Phase portraits of the fifteen-dimensional rotated Lennard-Jones function

Fifteen-Dimensional Lennard-Jones

This problem asks for the optimal arrangement of five atoms of the hypothetical "Lennard-Jonesium" in three-dimensional space. The optimal arrangement for three atoms is an equilateral triangle, while the minimum "energy" configuration for four atoms is a tetrahedron. Although this is a highly parameter-dependent "many-body" problem, none of the algorithms had difficulty finding solutions with $Np = 50$. E_{max} was limited to

2,000,000 objective function evaluations, although this proved to be unnecessarily high. Like the scale factor values reported in other cluster optimization studies with DE, F could be very small (except in conjunction with DE/rand/1/either-or) and still be effective. DE/best/1/bin proved to be the fastest of the four methods, while DE/target-to-best/1/bin had the most robust control space ($Z = 0.498$).

3.3.4 Summary

Table 3.2 summarizes the AES for each algorithm–function combination. The best results are highlighted in bold. If values are in italic, then not all trials were successful and the number of successful trials (< 10) appears in parentheses. The performance of the three methods that rely on crossover exemplifies the tradeoff between speed and reliability. Both DE/best/1/bin and DE/target-to-best/1/bin were usually faster than DE/rand/1/bin, but DE/rand/1/bin was more reliable, failing only on Rastrigin's function and to a lesser degree on Shekel's function.

DE/rand/1/either-or's performance stands out as being both fast and reliable. It was the only method that could solve each function ten times in succession. When comparing results of equal reliability ($P = 1$), it was also the fastest method in four cases.

Table 3.2. The AES for each of the four DE versions

Function	Rand	Best	Target-to-best	Either-or
Hyper-ellipsoid	19,531.2	12,225.4	**10,755.8**	13,612.3
Rosenbrock	44,292.1	**20,896.6**	21,262.2	46,614.3
Ackley	11,581.3	**5,394.6**	6,074.9	20,050.6
Chebyshev	44,142.4	18,332.9	**12,952**	27,775.6
Griewangk	22,2795	*31,678 (1)*	*1,139,120 (4)*	**19,198.7**
Rastrigin	*421,317 (2)*	*1,295,790 (1)*	*1,658,980 (2)*	**490,648**
Mod. Langerman	**56,156.6**	128,728	122,968	293,159
Shekel	*1,205,800 (9)*	*176,545 (2)*	*87,404 (2)*	**710,499**
Whitley	177,305	*369,544 (4)*	*453,481 (5)*	**70,380.8**
Lennard-Jones	34,592	**10,165.5**	21,549.7	59,116.9

Table 3.3 summarizes the Z values for each algorithm–function combination. The highest Z for each function, highlighted in bold, indicates the most robust control space. In all but three cases, DE/rand/1/either-or had the most robust control space. If functions had not been rotated to make parameters dependent, the methods that rely on crossover would have posted significantly higher Z values on the (otherwise) separable problems like Rastrigin's function.

Table 3.4 lists the control parameter settings derived from the phase portrait experiments. Settings in italic did not produce ten consecutive trials. In those cases where settings did produce ten consecutive trials, population sizes were fairly consistent, never differing by more than a factor of two (e.g., Modified Langerman and Griewangk functions). The largest population required by reliable methods (P = 1) was never more than 400.

Table 3.3. The fraction of 1000 random control parameter combinations that were successful (Z)

f(x)	Rand	Best	Target-to-best	Either-or
Hyper-ellipsoid	0.206	0.308	0.271	**0.465**
Rosenbrock	0.164	0.207	0.176	**0.345**
Ackley	**0.795**	0.598	0.647	0.453
Chebyshev	0.091	0.165	0.157	**0.396**
Griewangk	0.114	0.002	0.023	**0.133**
Rastrigin	0.006	0.001	0.003	**0.082**
Langerman	**0.514**	0.254	0.296	0.320
Shekel	0.039	0.020	0.008	**0.178**
Whitley	0.112	0.023	0.040	**0.554**
Lennard-Jones	0.272	0.392	**0.498**	0.373
Average	0.2313	0.1970	0.2119	**0.3299**

Table 3.4. Quasi-optimal control parameter values for each algorithm–function combination

f(x)	Rand			Best			Target-to-best			Either-or		
	Np	F	Cr	Np	F	Cr	Np	F	Cr	Np	F	PF
Helli.	50	0.55	0.95	50	0.65	0.85	50	0.65	0.85	50	0.55	0.35
Ros.	50	0.65	0.95	50	0.75	0.85	50	0.75	0.95	50	0.75	0.05
Ack.	50	0.25	0.55	50	0.45	0.55	50	0.55	0.75	50	0.45	0.45
Cheb.	50	0.65	0.95	50	0.75	0.95	50	0.75	0.95	50	0.65	0.45
Gri.	100	0.25	0.45	*200*	*0.25*	*0.25*	*100*	*0.45*	*0.45*	200	0.25	0.05
Ras.	*400*	*0.15*	*0.55*	*100*	*0.45*	*0.45*	*200*	*0.65*	*0.95*	400	0.35	0.05
Lang.	50	0.45	0.55	100	0.35	0.15	100	0.55	0.15	100	0.75	0.25
Shek.	*800*	*0.95*	*0.95*	*1600*	*0.75*	*0.65*	*400*	*0.85*	*0.95*	400	0.95	0.85
Whit.	200	0.45	0.95	*800*	*0.95*	*0.95*	*400*	*0.85*	*0.95*	200	0.45	0.95
L-J	50	0.45	0.95	50	0.45	0.65	50	0.95	0.95	50	0.95	0.45

When P = 1, the best scale factor ranged from $F = 0.25$ to $F = 0.95$. Allowing for the fact that points were randomly chosen within a grid square – not just sampled at its center as the values in Table 3.4 might suggest – expands this range to $0.2 < F < 1.0$. No one particular range of F seemed to be favored, which suggests that choosing the right value for the scale factor may require some effort.

Except for the Modified Langerman and Lennard-Jones functions, Cr was always greater than 0.4, and $Cr = 0.95$ was the most common value. This is to be expected, since Cr must be near 1.0 for the search to remain efficient when parameters become dependent (Sect. 2.6.2). Table 3.4 shows that P_F varies over the full spectrum of values, but it does not show that in all but two cases (the rotated Rastrigin and Griewangk functions), the choice of P_F was effectively arbitrary if F was chosen from within the right range.

In summary, the two methods that relied on the best-so-far vector were clearly faster on the easiest functions, but both became unreliable once the test functions became highly multi-modal. Not shown in the phase portraits is the fact that on the most difficult functions, large increases in Np had relatively little impact on both run times and convergence probabilities for DE/best/1/bin and DE/target-to-best/1/bin. Similarly, increasing Np did not always increase Z for these two methods. For example, in the case of the rotated Griewangk function, DE/best/1/bin never found more than two solutions ($Z = 0.002$) regardless of whether Np was 100 or 1600.

Classic DE (DE/rand/1/bin) was slower, but more robust than the two methods that relied on the best-so-far vector. This tradeoff reflects the usual condition in which speed and probability of convergence are conflicting objectives. The only function that posed a significant challenge to DE/rand/1/bin was the rotated Rastrigin's function.

DE/rand/1/either-or was both reliable and fast. In addition, its control space was very robust and P_F seldom had to be chosen with care. DE/rand/1/either-or performed well when other versions did not because of its ability to access the three-vector recombination axis. Furthermore, its rotationally invariant generating scheme keeps the search efficient when parameters become dependent. The next section examines DE's performance on five commonly cited thirty-dimensional test functions to give some idea of how DE performs compared to other optimizers.

3.4 DE Versus Other Optimizers

This section compares the four algorithms tested in the previous section to a variety of EAs and classical optimization techniques. In Sect. 3.4.1, DE is compared to as many as 16 other methods on a set of five, thirty-dimensional problems. Guidelines for version and control parameter selection are then given. Section 3.4.2 summarizes the results of four studies that compared DE to other optimizers with test beds dominated by uncon-strained problems. Section 3.4.3 then summarizes the results of four stud-ies that compared DE to other optimizers in constrained, multi-objective, mixed-variable and noisy problem domains. Finally, Sect. 3.4.4 presents a series of results comparing DE with other optimizers on real-world appli-cations.

3.4.1 Comparative Performance: Thirty-Dimensional Functions

This subsection compares DE's performance to that of 16 other optimizers using a test bed composed of five commonly cited thirty-dimensional test functions. Included in the test bed are the Rosenbrock, Ackley, Griewangk, Rastrigin and Schwefel functions. None of these functions are rotated and all but Rosenbrock (uni-modal) and Griewangk (multi-modal) are separa-ble. The Schwefel function is unique in this group because it is bound con-strained. Function descriptions are given in the Appendix.

Table 3.5 lists the EAs cited in these comparisons. Among the compet-ing optimizers are four genetic algorithms, (GAs), three evolution strate-gies (ESs), three versions of evolutionary programming (EP), three particle swarm optimization (PSO) algorithms, a simple evolutionary algorithm (SEA) and the evolutionary optimization (EO) algorithm.

Instead of reporting the number of objective function evaluations that an optimizer needs to reach the VTR, most published results simply state the mean best value after E_{max} function evaluations. Even though not all stud-ies use the same termination criteria or the same E_{max}, their results can be plotted as points on a progress plot of the mean objective function value. To this end, Figs. 3.14–3.18 plot the published results for the algorithms listed in Table 3.5. Not all studies provided results for all five, thirty-dimensional test functions, although most did. In addition, results for the HTGA, StGA and OGA are not plotted for technical reasons. Instead, re-sults for those three GAs are reported in Table 3.7.

The five progress plots also chart DE's performance as the ten-trial av-erage of the best-so-far vector's objective function value sampled at regu-lar intervals. Plots are provided for each of the four versions of DE exam-ined in the previous section. These DE performance curves are plotted as either dotted, dashed, or dash-dot lines, while the VTR is plotted as a solid horizontal line. Table 3.6 lists the control parameter settings used to gener-ate the DE progress plots in Figs. 3.14–3.18.

Table 3.5. Methods used for comparison

Symbol	Method	Reference
ALEP	Evolutionary programming with adaptive Levy mutations	(Lee and Yao 2004)
arPSO	Attractive and repulsive particle swarm optimization	(Vesterstrøm and Thomsen 2004)
CCGA	Cooperative co-evolutionary genetic algorithm	(van den Bergh and Englebrecht 2004)
CEP	Classical evolutionary programming	(Yao et al. 1999)
CEP/AM	Conventional evolutionary programming with adaptive mutations (b = 1.0×10^{-4})	(Chellapilla 1998)
CES	Classical evolution strategies	(Yao and Liu 1997)
CPSO-S6	Cooperative particle swarm optimization (S6)	(van den Bergh and Englebrecht 2004)
EO	Evolutionary optimization	(Angeline 1998)
FEP	Fast evolution programming	(Yao et al. 1999)
FES	Fast evolutionary strategies	(Yao and Liu 1997)
HTGA	Hybrid Taguchi-genetic algorithm	(Tsai et al. 2004)
OGA	Orthogonal genetic algorithm	(Leung and Wang 2001)
PSO	Particle swarm optimization	(Angeline 1998)
QEA/R	Quantum evolutionary algorithm with rotation	(Han and Kim 2004)
SEA	Simple evolutionary algorithm	(Vesterstrøm and Thomsen 2004)
StGA	Stochastic genetic algorithm	(Tu and Lu 2004)

Table 3.6. DE settings for data plotted in Figs. 3.14–3.18

$f(\mathbf{x})$	Rand			Best			Target-to-best			Either-or		
	Np	F	Cr	Np	F	Cr	Np	F	Cr	Np	F	P_F
Ros.	50	0.75	0.95	400	0.75	0.95	100	0.75	0.95	150	0.75	0.5
Ack.	50	0.5	0.2	50	0.5	0.2	50	0.5	0.2	250	0.5	0.5
Gri.	50	0.5	0.2	500	0.5	0.2	100	0.5	0.2	300	0.5	0.0
Ras.	50	0.5	0.2	800	0.5	0.2	100	0.5	0.2	800	0.35	0.0
Sch.	50	0.5	0.2	300	0.8	0.2	200	0.9	0.2	400	0.7	0.5

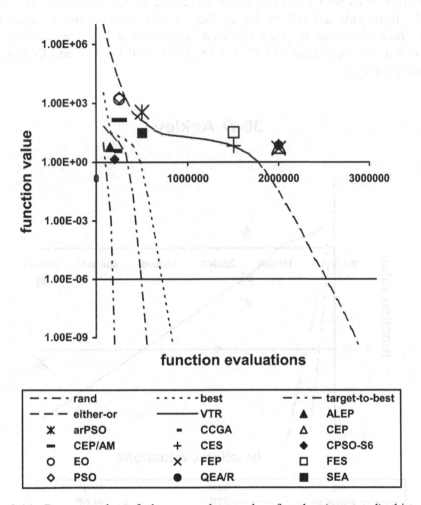

Fig. 3.14. Progress plot of the mean best value for the (unrotated) thirty-dimensional Rosenbrock function

Thirty-Dimensional Rosenbrock

As Fig. 3.14 shows, none of the four DE algorithms had any difficulty solving this parameter-dependent, uni-modal problem. To overcome this function's parameter dependence, Cr had to be set near 1.0. To produce regular convergence, F had to be set to at least 0.75 when $Np = 50$. Even though this function is uni-modal, none of the other algorithms drove the mean best vector's value below 1.0, let alone below the VTR ($= 1.0 \times 10^{-6}$). While CSPO–S6, CCGA and ALEP were all fast, none reached $f(\mathbf{x}) < 1.4$. The time scale defined by the number of function evaluations taken by DE/rand/1/either-or to reach the VTR suggests that E_{max} may have been too low for algorithms like CES, FES, CEP, FEP QEA/R and CCGA to reach the VTR.

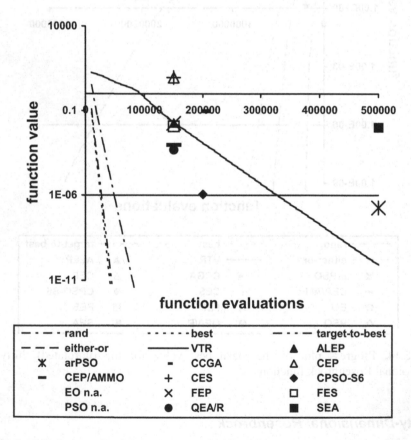

Fig. 3.15. Progress plot for the (unrotated) thirty-dimensional Ackley function

Thirty-Dimensional Ackley

Like its ten-dimensional rotated counterpart studied in the previous section, Ackley's function was easily solved by all four versions of DE. Again, DE/rand/1/either-or was significantly slower than the three versions of DE that use crossover, but it was still able to reach the VTR (= 1.0×10^{-6}). Only one other result (arPSO) exceeded the VTR, although CPSO–S6 came close, posting a mean best value of 1.12×10^{-6}.

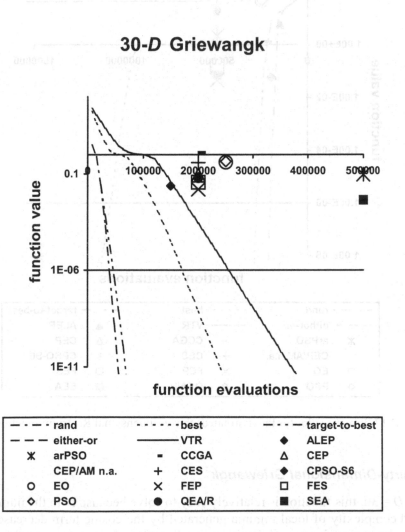

Fig. 3.16. Progress plot for the (unrotated) thirty-dimensional Griewangk function

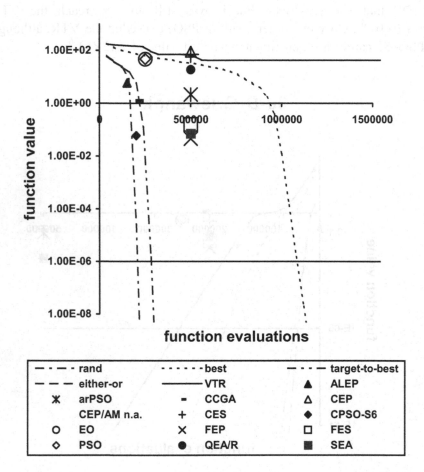

Fig. 3.17. Progress plot for the (unrotated) thirty-dimensional Rastrigin function

Thirty-Dimensional Griewangk

At $D = 30$, this function is relatively easy to solve because both the number and complexity of local minima generated by the cosine term decrease as the dimension increases (Whitley et al. 1996). Despite this, only DE reached the VTR (= 1.0×10^{-6}). To be fair, it must be pointed out that the goal in the other studies cited here was not to reach the VTR, but to find the mean best value after E_{max} generations. Thus it may be that given more time, some of these other methods may also have reached the VTR. DE/best/1/bin's performance was slowed by the fact that it needed a large population (Np = 500) to produce ten consecutive trials when F = 0.5.

Thirty-Dimensional Rastrigin

This function is characterized by having an enormous number of local minima. Because Rastrigin's function is separable, all three DE algorithms that rely on crossover were successful, although DE/best/1/bin was again slowed by the need for a high population to avert premature convergence. None of the other algorithms displayed here reached the VTR (= 1.0×10^{-6}). Fast evolutionary programming (FEP) gave the best non-DE result ($f(\mathbf{x})$ = 0.046). The DE/rand/1/either-or algorithm struggled on this function, although it did eventually reach the VTR at around 4.6×10^7 AES (not shown).

Thirty-Dimensional Schwefel

Not as many studies have reported results for this function, perhaps because it is bound-constrained. All three versions of DE that rely on crossover had no trouble solving this separable function with Cr = 0.2 and F = 0.5. Although the population sizes that produced ten consecutive trials differed, the progress plots for all three DE versions that use crossover are virtually indistinguishable. All four DE algorithms used the bounce-back method to reset out-of-bound trial parameters (see Sect. 4.3.1). As with Rastrigin's function, DE/rand/1/either-or could not exploit this function's decomposability, although it did eventually converge to the VTR (= $-418.983 + 0.01$) with regularity after about 2.0×10^7 function evaluations. Several other methods also solved this function, e.g., FES, FEP and QEA/R.

Several GAs developed for numerical optimization have proven very effective on this five-function test bed. Except for Schwefel's function, results for these GAs could not be graphed along with the other results because the reported mean best value, $f(\mathbf{x})$ = 0, cannot be plotted on a logarithmic scale. In the case of Schwefel's function, all three GAs were so close to the true optimum that they would have been indistinguishable if they had been plotted. Instead, both the average number of objective function evaluations taken and the mean best result for these three GAs are listed in Table 3.7. Results for Rosenbrock's function for both the OGA and HTGA are for D = 100, while all other results are for D = 30. The results for Schwefel's function in Table 3.7 have been divided by the function's dimension $(D = 30)$ to provide a normalized optimal function value.

30-*D* Schwefel

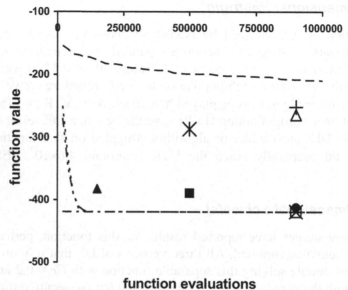

Fig. 3.18. Progress plot for the (unrotated) thirty-dimensional Schwefel function

Table 3.7. GAs compared

$f(\mathbf{x})$	HTGA		StGA		OGA	
	Evals.	Mean	Evals.	Mean	Evals.	Mean
Ros.	60,737	0.7	45,000	0.04435	167,863	0.7520
Ack.	16,632	0	10,000	3.52×10^{-8}	112,421	4.4×10^{-16}
Gri.	20,999	0	52,500	2.4×10^{-17}	134,000	0
Ras.	16,267	0	28,500	4.4×10^{-13}	224,710	0
Sch.	16,3468	−418.982	1500	−418.983	302,166	−418.981

Each of the algorithms in Table 3.7 used a different termination criterion, making it difficult to provide a comparable DE result. For example, the OGA terminated trials once the objective function value of the best-so-far vector was unchanged for fifty generations after the first 1000 generations. Authors reporting results for the HTGA found the 1000-generation limit too long and terminated trials once the best result equaled the mean best result found by the OGA. Trials for the StGA were halted after a preset number of generations. Table 3.8 lists the number of objective function evaluations that DE/rand/1/bin took to optimize these functions. Np was adjusted in increments of five until fifty consecutive trials were successful. Trials were terminated once the objective function value of the best-so-far vector reached the VTR. As the steep slopes in Figs. 3.14–3.18 show, DE very quickly closes in on the optimum once it reaches the VTR. Whatever extra time it takes DE to drive the value of the mean best vector from the VTR to the values listed in Table 3.7 (e.g., "0") increases the number of function evaluations by only a small percentage.

Although they differed in speed, each GA performed very well on all but Rosenbrock's function. Like the versions of DE that use crossover, these GAs are successful on this test bed in part because they use a low mutation rate. In each of these three GAs, the probability that a bit was mutated (inverted) was $p_m = 0.02$. Although parameters in Griewangk's functions are mildly dependent, results obtained with DE in Table 3.8 show that a low mutation rate strategy is effective on this function nevertheless.

Table 3.8. The number of evaluations it took DE/rand/1/bin to reach the VTR

$f(\mathbf{x})$	Evaluations	VTR	NP	CR	F
Rosenbrock	115,137	1.0×10^{-6}	60	0.9	0.8
Ackley	18,741	1.0×10^{-6}	20	0.2	0.5
Griewangk	14,446.3	1.0×10^{-6}	20	0.2	0.5
Rastrigin	118,936	1.0×10^{-6}	35	0.2	0.5
Schwefel	20,690.7	-418.982	45	0.2	0.5

The three GAs' inability to solve the simple uni-modal Rosenbrock function suggests that their incremental strategy, although effective on separable functions, would – for the reasons given by Salomon – be less effective on functions with dependent parameters unless mutations are correlated (see Sects. 1.2.3 and 2.6.2).

Selecting Effective Control Parameters: Rules of Thumb

If time and computational effort are not an issue, then either classic DE (DE/rand/1/bin) or DE/rand/1/either-or will be a good first choice. DE/rand/1/bin has a proven record of success, whereas DE/rand/1/either-or – although potentially more effective on parameter-dependent functions – is new and relatively untested on real-world problems. If the objective function is known to be separable or if it exhibits a low degree of parameter dependence, then classic DE will be more effective than DE/rand/1/either-or. While $Cr = 0.2$ should be the default crossover setting for separable functions or those that exhibit limited parameter dependence, $Cr = 0.9$ or 0.95 will ensure that optimization remains efficient in the presence of dependent parameters. When $Cr = 0.2$, F may be as small as 0.3, but $F = 0.5$ will be a better first choice. As Cr increases, however, F usually needs to increase as well. For example, if $Cr = 0.9$, then $F \geq 0.8$ will be more likely to give regular convergence than will $F = 0.5$.

Like F, Np may also have to be increased as Cr increases. For example, when the base vector was randomly chosen, $Np = 50$ was sufficient to solve the four, thirty-dimensional functions that used $Cr = 0.2$. For DE/rand/1/either-or, however, Cr is implicitly equal to 1 and Np had to be between 5 and 16 times as large as the populations that were effective when $Cr = 0.2$. Setting $Np = 5 \cdot D \cdot Cr$ is usually a good low-end default setting, but for highly multi-modal, parameter-dependent functions, Np may need to be $10 \cdot D$ or higher.

If it is not known whether or not the objective function is decomposable, or if a low-Cr strategy with DE/rand/1/bin fails to give satisfactory results, then DE/rand/1/either-or is probably the most viable option. Effective values for F will likely be found in the range (0.3, 1.0). The mutation probability, P_F, will probably not be difficult to choose, although selecting a good value for P_F becomes more important as F grows smaller. In general, $P_F = 0.5$ seems to be a good, if non-optimal, first choice. For DE/rand/1/either-or, Np will be about the same size as that required by DE/rand/1/bin when $Cr = 1$. While $Np = 5 \cdot D$ may be adequate for simple functions, populations of $10 \cdot D$ or larger may be required to achieve regular convergence on highly multi-modal, parameter-dependent functions.

If the objective function is not highly multi-modal and if function evaluations are very time-consuming, then DE/best/1/bin with a small amount of jitter may be the best choice. In particular, this method has proven very effective for designing digital filters. If DE/best/1/bin is chosen, then F and/or Np may have to be relatively large to maintain diversity and forestall premature convergence. The guidance given for selecting Cr for classic DE applies here as well: $Cr = 0.2$ for separable functions or those that exhibit limited parameter dependence, and $Cr = 0.9$ or 0.95 otherwise.

As might be expected, DE/target-to-best/1/bin exhibits performance that is intermediate between that of classic DE and DE/best/1/bin. Performance would improve if F and K were both independent and well chosen, but this extra "tuning" further complicates the process of finding an effective set of control parameters. Similarly, DE/rand/1/either-or would have performed at least as well as did classic DE on the thirty-dimensional Rastrigin and Schwefel functions if the target vector were crossed with either a mutant with probability P_F, or a three-vector recombinant with probability $1-P_F$.

Very small values of F seem particularly effective when searching for the minimum energy configuration of atoms. This unusual case emphasizes that these "rules of thumb" are only intended to serve as guidelines and that experimenting with control parameter combinations is strongly encouraged if these general rules do not appear to be effective.

3.4.2 Comparative Studies: Unconstrained Optimization

This section explores four studies that compared DE to a variety of well-known optimizers. The first paper (Storn and Price 1997) compared DE to two annealing methods, two EAs and to the method of stochastic differential equations (Aluffi-Pentini et al. 1985), although the test beds were not very challenging. The next paper (Ali and Törn 1999) compared DE to several versions of the controlled random search and to two floating-point GAs. While the first test bed consisted of unchallenging low-dimensional problems, the second test bed was substantially more difficult. The third paper (Vesterstrøm and Thomsen 2004) employed an extensive test bed consisting of functions having up to 100 dimensions for a comparison of DE to two versions of the particle swarm optimization algorithm and to a simple EA. The authors of Paterlini and Krink (2004) also compared DE to particle swarm optimization and to a floating-point GA. Their test bed was a set of partitional clustering problems that ranged from easy to difficult.

Storn and Price

In this early study (Storn and Price 1997), the authors compared DE/rand/1/bin to two annealing methods, two EAs and to the method of stochastic differential equations (SDE) (Aluffi-Pentini et al. 1987). The first comparison pitted DE against both the annealed Nelder–Mead algorithm (ANM) (Press et al. 1992) and adaptive simulated annealing (ASA) (Ingber 1993). Each method, including DE/rand/1/bin, was tuned to give its best performance. The test bed was a modified De Jong test suite (Storn and Price 1997) that included Corana's parabola ($D = 4$) (Corana et al. 1987), Griewangk's function ($D = 10$) (Griewangk 1981), Zimmerman's function (a constrained problem) (Zimmermann 1990) and the Chebyshev polynomial fitting problem ($D = 5$ and $D = 9$) (Storn and Price 1997). The ANM method regularly solved only four of the test bed's ten functions (sphere ($D = 3$), Rosenbrock ($D = 2$) and step ($D = 5$) from the modified De Jong test bed and Zimmerman ($D = 2$)). Adaptive simulated annealing did better than ANM, but it could not reach the VTR for either the Griewangk function or the Chebyshev problems. DE was the only method to optimize all functions with regularity, i.e., for all twenty trials.

Population sizes for DE/rand/1/bin varied from $Np = 5$ (sphere ($D = 3$)) to $Np = 100$ (Chebyshev ($D = 9$)). In most cases, $0 \leq Cr \leq 0.2$, but the best value for Cr for the Chebyshev, Zimmerman and Rosenbrock functions was between 0.9 and 1, inclusive. Both the Chebyshev and Rosenbrock functions have nonlinear terms that give rise to parameter dependence and Zimmerman's function has a single nonlinear constraint. Most functions were solved with $F = 0.9$, except for the Chebyshev problem ($F = 0.6$) and Zimmerman's problem ($F = 0.5$).

The second comparison in (Storn and Price 1997) compared DE to the breeder genetic algorithm (BGA) (Mühlenbein 1993) and an EA with "soft" genetic operators (EASY) (Voigt 1995). The five-function test bed included the hyper-ellipsoid ($D = 30$ and $D = 100$) (Storn and Price 1997), Rastrigin ($D = 20$ and $D = 100$) (Mühlenbein et al. 1991), Griewangk ($D = 20$ and $D = 100$) (Griewangk 1981), Ackley ($D = 30$ and $D = 100$) (Ackley 1987) and Katsuura ($D = 10$ and $D = 30$) (Katsuura 1991) functions. Performance data for the BGA on the hyper-ellipsoid and Katsuura's functions was not available, but in all other cases, all 3 methods solved each version of every function. DE was the fastest method on all but Rastrigin's function, which was more quickly optimized by the BGA. Despite the high dimension of these functions, DE solved them all with population sizes no larger than $Np = 25$. In addition, $F = 0.5$, $Cr = 0.1$ were used in each case except for Rastrigin's function ($Cr = 0$). DE's success with small population sizes and low values for Cr reflects the fact that all test bed functions were separable. Consequently, both competing EAs also performed well because they used low mutation rates, e.g., $p_m = 1/D$ in the case of EASY.

In a final study, (Storn and Price 1997) compared DE to the method of stochastic differential equations (SDE) (Aluffi-Pentini et al. 1985) on a very simple, fifteen-member test bed composed of functions having from one to ten dimensions. Both algorithms successfully optimized all test bed functions, but DE was faster than SDE in every case, often by a factor of ten or more. The DE control parameter setting $Np = 20$, $F = 0.5$, $Cr = 0$

was effective in all but two cases. On one non-separable function, Cr was set to 1 and the population size had to be increased to $Np = 30$. On another function, F was set to 1 and Np was raised to $Np = 40$. Setting F or Cr to just less than 1 in these two cases would probably have allowed Np to be smaller, since both $F = 1$ and $Cr = 1$ dramatically reduce the number of potential trial vectors.

Ali and Törn

The authors of this study (Ali and Törn 1998) compared four versions of DE, four versions of the Controlled Random Search (CRS) (Price 1977) algorithm and two GAs based on the real-coded GA proposed in Hu et al. (1997). Benchmark functions were organized into two test beds. The first test bed consisted of nine, relatively simple, low-dimensional test functions, while the second, more difficult test bed included functions developed for the 2nd International Contest on Evolutionary Optimization. Optimizers were rated based on both the number of function evaluations, the total CPU time that they took to reach the VTR and on the fraction of trials that were successful. Ali and Törn then ranked the algorithms from best to worst in each of these three categories. On the easy test bed, the best algorithm depended on which of the three criteria was applied. DE solved all problems, but versions of the CRS algorithm required fewer function evaluations and less CPU time than did DE. One of the GAs was faster than DE in terms of the number of function evaluations, but not when CPU times were compared.

The roles were reversed, however, when algorithms were compared based on their rate of success. DE algorithms captured the top five spots followed by four CRS algorithms in places 6, 8, 9 and 10 with a GA taking the seventh ranked position. To be fair, the comparisons should have been conducted at the same probability of success, but this is often difficult to achieve in practice.

On the second, more difficult test bed, versions of DE were not only more successful than any CRS or GA algorithm, but also faster. DE algorithms took the first three spots in each category, requiring fewer evaluations, less CPU time and achieving a greater rate of success than the competing algorithms. The very best performing versions on this test bed were two of the modified DE algorithms proposed in Ali and Törn (1998).

Vesterstrøm and Thomsen

The primary focus of this comparison (Vesterstrøm and Thomsen 2004) was to compare DE (DE/rand/1/exp), a simple evolutionary algorithm

(SEA) (Thomsen 2003) and two particle swarm optimization algorithms (PSO) (Vesterstrøm and Riget 2002) and (arPSO) (Vesterstrøm and Riget 2002). Results for both the SEA and arPSO are plotted in Figs. 3.14–3.18 in Sect. 3.4.1. The test bed consisted of 23 functions, most of which were taken from Yao and Liu (1997). In addition, the test bed included two noisy functions and one whose landscape contained flat plateaus. In all experiments, DE's control parameter settings were $Np = 100$, $Cr = 0.9$ and $F = 0.5$. Despite Np and F being smaller than would normally be recommended with $D = 30$ and $Cr = 0.9$, DE gave a lower mean value than competing algorithms in 17 of the 23 problems having dimension thirty or less.

DE's mean best result after 500,000 function evaluations for two of the six problems for which it was not the best performer differed from the best performing algorithm (SEA) only beyond the 5th decimal place, e.g., −1.03163 for the SEA and −1.03162 for DE. On the thirty-dimensional versions of both Schwefels' uni-modal ridge function (Schwefel 1995) and the one-max problem (Stanhope and Daida 1997), PSO gave a lower mean best value than did DE, although DE's final results for these two functions, 2.02×10^{-8} and 3.85×10^{-8}, respectively, were excellent. Only on two functions did DE show a significant difference from the best result. One of these was a simple four-dimensional function that caused DE and both PSO algorithms to stagnate. For a function of such low dimension, it seems certain that DE would give a better final solution if some control parameter tuning had been attempted. DE's other inferior result was a uni-modal, thirty-dimensional function to which uniformly distributed evaluation noise was added.

Vesterstrøm and Thomsen also explored optimizer performance on a one hundred-dimensional subset of Yao and Liu's test bed. For these experiments, E_{max} was set to 5,000,000 function evaluations. Even without changing its control parameter settings, DE performed even better than on the first test bed, posting the lowest or equal mean best value for twelve of the thirteen functions. The only one hundred-dimensional function on which DE did not do as well or better than the other algorithms was the noisy uni-modal function. In general, the authors called DE's performance with respect to the optimizers analyzed "outstanding" and judged DE to be simple, robust, reliable and easy to tune.

Paterlini and Krink

A recent study (Paterlini and Krink 2004) compared the ability of four optimizers to solve a set of six simulated and four real-world partitional clustering problems. In addition to DE, the methods evaluated included the k-means method, random search, a floating-point encoded GA and the

PSO algorithm. The goal of partitional clustering is to determine the optimal partitioning for a data set, i.e., to maximize the similarities within a group while minimizing the dissimilarities between different groups. The GA tested in this comparison used tournament selection, Gaussian mutations and arithmetic crossover. The PSO algorithm was based on the original method described in Kennedy and Eberhart (1995), while DE/rand/1/exp was chosen to represent DE. The random search was not incremental, but consisted of E_{max} evaluations of vectors chosen with random uniformity from within the search space. DE, PSO and the GA were all tuned, but the authors noted that both the GA and PSO methods took "much more time" to tune than did DE.

Data clusters can be defined in many ways and several clustering criteria were explored. For the simplest data sets, all three algorithms reached the same mean value and the clustering criterion was not an important factor. The more difficult clustering problems showed the superiority of EAs over the simple random search and the method of k-means. On these problems, DE was both the most accurate in terms of mean objective function value and the most robust in terms of being able to reproduce a result.

DE's status as an effective algorithm for identifying clusters was reinforced by its performance on the four real-world data sets. While all methods performed well on the simplest problem, DE was better in two cases and not significantly worse (based on a 95% confidence level) than PSO in the other. (Both methods found the optimum.) In general, the more difficult the clustering problem was (e.g., the more overlapping clusters there were), the better DE performed compared to the competing algorithms. The authors concluded that DE was "clearly and consistently superior compared to GAs and PSO for hard clustering problems".

3.4.3 Performance Comparisons from Other Problem Domains

Multi-Constrained, Nonlinear Optimization

In Lampinen (2002), DE/rand/1/bin was used in conjunction with Lampinen's Pareto-dominance inspired selection criterion (Sect. 4.3) to optimize a set of nonlinear benchmark functions that were subject to multiple, nonlinear constraints. All functions were taken from the test beds in Michalewicz and Schoenauer (1996) and Koziel and Michalewicz (1999). The ten functions studied by Lampinen include one linear objective function with both linear and nonlinear constraints and one nonlinear objective function with linear constraints. The remaining eight functions were nonlinear and subject to one or more nonlinear constraints. In addition, three functions were subject to equality constraints, e.g., $\gamma(x) = 0$, which Lampinen converted into a pair of inequality constraints: $0.001 \leq \gamma(x)$ and $\gamma(x) \leq 0.001$.

When compared to a uniform random search, DE found the first feasible solution between 3% and 99% faster, indicating that Lampinen's criterion was effectively guiding the population toward feasible regions. DE's improvement over the random search was the highest for the most difficult functions. Lampinen then compared DE's performance to the very best result reported in Joines and Houck (1994), Koziel and Michalewicz (1999) and Michalewicz (1995). Lampinen chose to set $Cr = 0.9$ and $F = 0.9$ for all ten problems, but spent some effort tuning the population size because problems varied in dimension from $D = 2$ to $D = 50$. Except for one problem that required $Np = 120$, the remaining nine functions were solved using $15 \leq Np \leq 40$. In each case, 1000 trials were run to test DE's reliability. For all test problems, DE with Lampinen's criterion found the best-known solution except in two cases for which DE's solution was better than the best previously known solution. In addition, all 1000 trials for each function found the reported optimum. Precision for the solutions to the three functions with equality constraints was limited by the tolerance chosen for converting an equality constraint into two inequality constraints.

Mixed-Variable Optimization

Another study (Lampinen and Zelinka 1999) used a popular test bed of mechanical design problems to compare DE to ten different optimizers. The problems are to design a gear train (four integer variables), a pressure vessel (two discrete and two continuous variables) and a coil spring (one integer, one discrete and one continuous variable). In addition, each variable in the gear train problem is subject to both upper and lower bound constraints. The pressure vessel design problem includes six inequality constraint functions, while the coil spring design is subject to eight inequality constraints. In this early study, constraints were implemented with traditional penalty functions, not Lampinen's Pareto-dominance-based selection criterion (Sect. 4.3). A more recent study (Lampinen and Storn 2004) solved the functions in this test bed using Lampinen's criterion and compared the results to those reported for twenty other methods.

The popularity of this test bed made it possible to compare DE to a wide range of optimization methods that included classical approaches, several genetic algorithms, evolutionary programming (EP), evolution strategies (ES), simulated annealing (SA) and several unique methods. Of the seven methods reporting results for the gear train design problem, only DE and two other methods – the meta-GA (Wu and Chow 1995) and the modified

GA in Lin et al. (1995) – found the optimal solution. In addition, multiple runs with DE found the four alternative solutions to this problem.

Variations in the test problem implementations in other studies forced Lampinen and Zelinka to test DE on three different versions of the pressure vessel design problem. Both Sandgren (1990) and Fu et al. (1991) reported results for Case A, which treats all parameters as continuous. DE's solution to Case A was both different and significantly better than those found by the other two methods, e.g., $f(x)$ = 7,790.588 for Fu et al. and $f(x)$ = 7,019.031 for DE. Case B of the pressure vessel design problem was formulated according the original problem statement. Four algorithms reported results for this version and DE once again produced the best result. The sequential linearization algorithm in (Loh and Papalambros 1991) gave a result that was almost as good, but one of their published parameter values violates a constraint. Case C was investigated so that DE could be compared to several algorithms that used a different numerical value in one of the constraint functions. Once this adjustment was taken into account, DE once again gave the best result ($f(x)$ = 7,006.358). The two-level parallel ES (Thierauf and Cai 1997) gave a result that was almost as good ($f(x)$ = 7,006.9), but neither of the other two methods (Li and Chow 1994; Cao and Wu 1997) were competitive.

The coil spring problem also had both continuous and mixed-variable versions. In the continuous case, DE improved on Sandgren's value of 2.6353 by posting an objective function value of 2.61388. DE also returned the best result on the mixed-variable version of the coil spring problem, beating the best result reported by Sandgren and two other methods (Chen and Tsao 1993; Wu and Chow 1995).

Multi-Objective Optimization

Kukkonen and Lampinen (Kukkonen and Lampinen 2004) compared DE with other optimizers on multi-objective benchmark functions. In addition to DE, the comparison included the non-denominated sorting genetic algorithm (NSGA–II) (Deb et al. 2002) and the strength Pareto evolutionary algorithm (SPEA) (Zitzler and Thiele 1999). In particular, NSGA–II was chosen for its good performance in previously published tests (Deb et al. 2002). Two versions of DE were tested. Generalized differential evolution (GDE) uses a Pareto-based selection criterion for handing constraints and multiple objectives (see Sect. 4.6). GDE2 adds a mechanism to improve the extent and diversity of GDE's approximation to the Pareto-front.

The test bed was composed of five, bi-objective benchmark functions described in Zitzler et al. (2000). For all test problems, the DE control parameter settings were Np = 100, Cr = 0.05 and F = 0.1. Control parameter values for both F and Cr were determined by preliminary testing with values from the set {0.05, 0.1, 0.2, 0.3, 0.4}. The authors report that values for F outside this range tended to result in rapid convergence to a single point on the Pareto-front. In addition, each of the 100 trials was limited to 250 generations, but NSGA–II and GDE2 used twice as many function evaluations per generation. Closeness to the Pareto-front was measured by both an error ratio and by a generational distance. In addition, Kukkonen and Lampinen also measured diversity in the solutions with spacing, spread and maximum spread metrics. These metrics were applied to both versions of GDE and NSGA–II, but results for the SPEA were plotted for visual reference only.

Results showed that GDE gave a good approximation to the Pareto-front, but that GDE2 improved both the extent and diversity of the solutions. Overall, NSGA–II performed best in most statistical categories on four of the five benchmark functions, although GDE2 outperformed NSGA–II on the test function ZDT6 that tests an optimizer's response to non-uniformity in the Pareto-front. Despite handling multiple objectives directly using Pareto-dominant based selection, both versions of DE performed comparably to both NSGA–II and SPEA, although this conclusion is based on limited experimentation.

Optimizing Noisy Functions

In a comparison with PSO and a simple EA, the authors of Krink et al. (2004) explored DE/rand/1/exp's performance on objective functions with noisy evaluations. The EA used binary tournament selection, two-vector arithmetic recombination and a Gaussian mutation operator. Control parameter settings for each algorithm were chosen based on preliminary experiments, but they were not tuned for each problem. In DE's case, the setting chosen was $Np = 50$, $Cr = 0.8$, $F = 0.5$. The test bed included the Schaffer ($D = 2$), sphere ($D = 5$), Griewangk ($D = 50$), Rastrigin ($D = 50$) and Rosenbrock ($D = 50$) functions. Trials for the sphere and Schaffer functions were limited to 100,000 function evaluations, whereas the remaining three, fifty-dimensional functions were allowed 500,000 function evaluations.

For their experiment, the authors first optimized the non-noisy version of each function, and found that DE outperformed the other two algorithms when measured by the mean objective function value at E_{max} function evaluations. Next, zero-mean Gaussian noise having a variance of 1 was added to each vector's objective function value. To minimize the effects of the noise, trials were conducted in which vectors were reevaluated 1, 5, 20,

50 and 100 times, with each evaluation contributing to the total allowed number of function evaluations.

Performance was compared based on the mean value of the final population. DE gave the best result for both the noisy Rastrigin and Rosenbrock functions, but only because both other methods stagnated before reaching the optimum. Once, however, the variance of the population's objective function values were on a par with the variance of the evaluation noise, DE's performance was comparable to that of PSO, but not as good as that of the EA.

Summary

DE was particularly effective when handling constraints directly. In several cases, DE found a result that was better than the previously best known solution, even though functions were defined with mixed variables. DE was also competitive optimizing functions with multiple objectives using a simple Pareto-based selection scheme. Functions with added evaluation noise, however, proved to be more difficult for DE than for an EA with Gaussian mutation. DE dominated the comparison, however, when noise was absent.

3.4.4 Application-Based Performance Comparisons

Some of the greatest challenges for an optimizer are found in the realm of real-world applications. Unlike test functions, real-world applications seldom display regularities. Furthermore, noise, constraints, and both mixed and dependent variables are common in many real-world problems. Practical applications also offer an opportunity to discover how hard it is for researchers who are not necessarily optimization experts to adapt, implement and tune an optimizer. As such, real-world applications are a demanding proving ground for optimizers. Below is a brief survey of studies that have compared DE to other optimizers on real-world problems. In several cases, DE is compared to traditional methods that are industry standards.

Multi-Sensor Fusion

In Joshi and Sanderson (1999) and Sect. 7.4 of this book, the authors use DE to integrate information from both visual and tactile sensors to categorize an object's shape and determine its orientation (pose). Tactile sensor data is taken from a robotic hand that grasps the objects, while a camera records the object's shape and extracts vertex/edge features. The goal is to integrate the tactile and visual cues to improve the robot's ability to manipulate the object. The authors used a minimum representation size criterion to help select a model for the sensor data. Finding the minimum representation involves a tradeoff between model size (number of parameters) and residual error. In addition, the problem contains both discrete and continuous variables.

In their study, Joshi and Sanderson compared DE/best/2/bin's performance to that of a binary encoded GA that used one-point crossover, a mutation probability of $p_m = 0.05$ and a population size of 100. For DE, the control parameter settings were $Np = 100$, $Cr = 0.8$ and $F = 0.8$. Settings for both algorithms were chosen by trial and error to minimize the size of the representation at the end of the search. When compared to the GA, DE found a much smaller representation size in less time with fewer evaluations and also showed smaller interpretation errors. Details can be found in Sect. 7.4 of this book.

Earthquake Relocation

Also appearing in Chap. 7 (Sect. 7.5) is a study by Růžek and Kvasnička based on Růžek and Kvasnička (2001) in which the authors explored DE's ability to find an earthquake's epicenter from seismographic data recorded at multiple stations. This four-dimensional inverse problem is hard because it is nonlinear and, in some cases, multi-modal as well. For their study, the authors chose DE/rand/1/bin and applied it to problems with both synthetic and real data. Considerable effort was spent exploring how different control parameter settings affected DE's performance. The author's found that DE was very robust in this regard and settled on $Np = 30$, $Cr = 0.8$, $F = 0.5$ for all experiments. Tests with synthetic data showed that DE's results were reliable, fast and accurate. In addition, adding uniform parameter noise to synthetic data did not jeopardize DE's ability to locate the global optimum. Real-world seismic data from events located in the Gulf of Corinth showed that DE always gave a significantly better result than did the industry standard HYPO71 location program. Section 7.5 contains details.

Active Compensation in RF-Driven Plasmas

Langmuir probes are a diagnostic tool for measuring the properties of low-pressure plasma. The act of measurement, however, disturbs the plasma from its normal state, but if a radio-frequency signal with the right phase, amplitude and waveform is applied to the probe, these distortions can be actively cancelled. As a result, plasma used for circuit etching can produce cleaner shapes. In this study (Sect. 7.12), Zelinka and Nolle used DE to adjust fourteen waveform variables to actively compensate for nonlinear interactions between plasma and a Langmuir probe. After some experimentation, they found the best DE control parameter settings to be $Np = 50$, $Cr = 0.8$, $F = 0.5$. When compared to simulated annealing, DE's results were more precise and more consistent. Consult Sect. 7.12 for more details.

DC Operating Point Analysis for Nonlinear Circuits

Finding a circuit's quiescent, or DC, operating point is the starting point for performing other types of circuit analysis. In Crutchley and Zwolinski (2003), the classical Newton–Raphson (NR) method is compared to two versions of DE (DE/rand/1/bin and DE/target-to-best/1/bin) and to three ES algorithms. The versions of the ES algorithm are a $(\mu + \lambda)$-ES with standard adaptive mutations (ESA), a $(\mu + \lambda)$-ES with correlated mutations (ES) and a tournament selection based ES (TSEA). Neither the ES nor the TSEA used adaptive mutations. After some experimentation, the authors chose $Np = 10 \cdot n$, $Cr = 0.5$, $F = 0.4$ for DE/rand/1/bin and $Np = 10 \cdot n$, $Cr = 0.3$ $F = 0.8$, $K = 0.9$ for DE/target-to-best/1/bin, where n is the number of circuit nodes.

The test bed consisted of nine benchmark circuits, two of which have multiple solutions. Results included the mean error per node compared to the Newton–Raphson method (assumed to be the most accurate), the number of solutions found, the number of generations taken and the CPU time consumed. DE/rand/1/bin and TSEA showed comparable relative errors in all but one case, for which TSEA was better. For circuits having a single solution, TSEA required fewer generations and less CPU time in all but one case, for which DE/rand/1/bin was faster. Only the two DE algorithms and ES, however, were able to find multiple solutions in a single run. Although the NR method was by far the fastest method in all cases, it must be manually reinitialized to find all possible solutions. In brief, the authors argued that the best algorithms in terms of accuracy, speed and ability to find multiple solutions were the two versions of DE.

Identifying Induction Motor Parameters

DE's ability to find optimal parameters for two induction motor models was examined in Ursem and Vadstrup (2004). One of the models incorporated nonlinear magnetic saturation effects while the other motor model did not. In a previous study, eight stochastic search algorithms were tested on these same problems. The methods previously tested were the steepest descent local search, simulated annealing, a simple EA, a diversity-guided EA, ES with simple mutations (ES1), ES with correlated mutations (ES2), standard PSO and diversity-guided PSO. In this study, DE was compared to the best of these optimizers – the two ESs and the diversity-guided EA.

DE's control parameters were tuned by trial and error, with $Np = 100$, $Cr = 0.5$, $F = 0.5$ being chosen for both problems. The authors noted that compared to other algorithms, DE's control parameters required significantly fewer trials to "tune". For the five-dimensional induction motor problem without magnetic saturation, both DE and DGEA found the "exact" solution for all twenty trials, but DE was three to four times faster than both DGEA and ES2.

All four algorithms in this study found the exact solution to the five-dimensional problem more than once, but only DE was able to find the exact result for the eight-dimensional model with magnetic saturation. DE found the exact result not only for all twenty trials in the original experiment, but also when eighty additional trials were conducted. On this problem, the ES with correlated mutations (ES2) performed worse than did the ES with simple mutations (ES1). The authors concluded that DE was robust, easy to tune, fast, accurate and simple to implement.

Estimation of Heat Transfer Parameters in a Trickle-Bed Reactor

Trickle-bed (chemical) reactors are widely used in the petroleum industry and to a lesser extent in the chemical and pharmaceutical industries. Two-dimensional models have been developed that consist of coupled partial differential equations of the parabolic type. In Babu and Sastry (1999), the authors use the method of orthogonal collocation to transform the set of partial differential equations into a function minimization problem. In a typical trickle-bed reactor, the heat transfer parameters, effective radial thermal conductivity and wall-to-bed heat transfer coefficients are unknown and must be estimated. Once these values have been determined, the all-important temperature profile of the reactor bed can be numerically determined. In their experiment, the authors measured the radial temperature profile in a working trickle-bed reactor and then sought the model's parameter values that could reproduce the measured profile. In all, 232 data points were obtained that covered a wide range of flow rates and packings.

The authors chose DE/target-to-best/1/bin to compete with the classical radial temperature profile (RTP) method, which uses Powell's method. DE control parameters were $Np = 20$, $Cr = 0.9$, $F = 0.7$, $K = 0.7$. Compared to the RTP method, DE was two and a half to three times faster and its estimates were much more accurate. In addition, DE was very robust, converging to the global optimum regardless of the initial population, whereas the RTP method needed an initial guess close to the solution.

Aerodynamic Optimization

Aerodynamic shape optimization involves finding the most efficient shape for bodies moving through air. In Rogalski et al. (1999), the authors used DE/target-to-best/1/bin, the Nelder–Mead downhill simplex algorithm and simulated annealing to design three fan blades. All three problems were attacked with the inverse method, where the goal is to produce a shape that exhibits the prescribed surface pressure distribution. The authors used Bezier curves to model both a blade's thickness (width) profile and the camber curve that gives the blade its characteristic arch. Several constraints were applied using penalty functions. A program that simulates airflow computed the pressure distribution around each proposed shape. The difference between the computed pressure distribution and the target distribution was then used to compute an error function based on the L2 norm that served as an objective function. In all, each problem involved finding fifteen real-valued coefficients. The control parameters for DE were $Np = 150$, $Cr = 1$, $F = K = 0.85$.

For the first problem, all three optimizers were able to accurately represent the target shape. In terms of residual error, DE was twice as good as its nearest competitor, although it took almost fifty times as long to converge. A high-pressure region near the nose of the second fan blade made finding the target shape more difficult than in the first case. For this second design problem, DE was again the slowest method, but it was the only one to accurately model the target shape. Similarly, DE was the only algorithm of the three to accurately model the third target shape.

Image Registration

Thomas and Vernon. Image registration is a fundamental image-processing task that matches two or more images. In Thomas and Vernon (1997), the authors begin their experiment by generating a 255-by-255 pixel target image. The second image is a copy of the first that has been rotated and to which noise has been added to both the x and y coordinates of each pixel. Control points are identified in each image and the error in mapping the control points serves as an objective function that must be minimized.

Control points can also be used to solve a set of simultaneous linear equations so that a least squared error (LSE) solution can be found. On data sets generated by the identity transformation with noise, DE matched the performance of the LSE method exactly. On data sets generated by nonlinear transformations, LSE was slightly better than DE in four of nine cases, whereas DE was better in the remaining five. In two of these cases, DE's result was significantly better that LSE's and in each of these two cases, one parameter value was much larger than that found by LSE (e.g., 224.72 for DE vs. 2.14 for LSE). DE/rand/1/bin appears to be the method employed in this study, with $Np = 160$, $F = 0.4$. A value for Cr was not provided.

Salomon et al. In Sect. 7.6, Salomon et al. evaluate DE as a tool for three-dimensional medical image registration. This study did not consider comparative performance, but focused instead on how well DE/target-to-best/1/bin (called "DE/*rand*-to-best/1/bin" by the authors) performs when implemented in parallel. The authors found that DE was not only fast and accurate, but also scaled almost linearly with the number of processors. See Sect. 7.6 for details.

Optimization of Carbon and Silicon Cluster Geometry

Ali and Törn. Cluster optimization is a many-body problem that exhibits a very high degree of parameter interaction. In Ali and Törn (2000), the authors sought the minimum binding energies of both carbon and silicon clusters consisting of up to fifteen atoms and 39 variables. The binding energy between atoms was simulated with the Tersoff potential.

DE was modified to include both a derivative-based local optimization technique and an auxiliary population, S_a, of Np vectors. In this algorithm, known as topographical DE (TDE), a trial vector that does not improve on the target vector subsequently competes with the corresponding vector in S_a. If the trial vector wins this second competition, it replaces the inferior vector in S_a.

For this series of experiments, the authors chose $Np = 10 \cdot D$, but other DE settings were not provided. For silicon clusters up to six atoms, TDE's result could be compared to the best result found by the eight optimizers studied in Ali et al. (1997). In two of these cases, TDE found even better minima than the best known values.

Chakraborti. In Sect. 7.1 of this book, N. Chakraborti uses DE to discover the minimum energy configurations of silicon–hydrogen clusters whose interactions are based on the "tight-binding" model. When compared to the simple GA and simulated annealing, DE gave an equal or better result in all but one case. The author also concluded that DE's elitist selection criterion was an asset for this type of problem and that DE could resolve closely spaced minima without resorting to "niching" strategies. Consult Sect. 7.1 for additional details.

Optimizing Neural Networks

Fischer et al. In Fischer et al. (1999), the authors use DE to optimize a neural network (NN) having three inputs, a single hidden layer with a fixed number of hidden units and a single output unit. Keeping the topology of the network fixed restricts the problem to one of determining network weights. To train the network, the authors chose the Austrian inter-regional telecommunications data set because its multiple local minima are known to pose a difficult challenge to gradient-based learning algorithms. The objective function measured the squared error between the network's output and the actual training data. The goal was to make good predictions about the intensity of telecommunications traffic between two locations.

The authors chose DE/best/1/bin, $Cr = 1$ and experimented with a range of value for both F and Np. They decided on $F = 0.9$ for further experiments even though neighboring values were also effective. Values for Np ranging from 50 to 100 were explored with $F = 0.9$. The authors found that increasing Np beyond 200 did not improve the out-of-sample average. $Np = 400$ was used for subsequent experiments in which the number of hidden nodes was varied. A topology with eight hidden nodes proved best. For comparison, the authors also trained weights with a multi-start, conjugate-gradient (CG) back-propagation method. Both methods were given the same time to train the networks, with each DE trial taking as long as six CG trials. While the CG method showed a better in-sample performance, DE exhibited statistically better performance in the more important out-of-sample category.

Plagianakos et al. In Plagianakos et al. (2001), the authors investigated DE's ability to train neural networks that use discrete activation functions. Most NNs use continuous activation functions, like the well-known sigmoid function, but discrete, e.g., binary activation functions are well suited for inherently binary tasks. In addition, discrete networks are computationally simple to understand and provide a starting point for investigations into networks with continuous activation functions. Furthermore, networks with discrete activation functions are cheaper to implement in hardware. When the activation function becomes discrete, however, back-propagation methods that rely on gradients are not effective, so a direct search algorithm like DE is ideally suited for this problem.

In their investigation, Plagianakos et al. used DE/target-to-best/1/bin, with $F = K$, to solve a set of three benchmark problems. No control parameter settings were reported. Problems included the "exclusive or" (XOR) classification problem (known to be sensitive to the initial choice of weights), the three-bit parity problem (hard because members of different classes differ by a single bit) and controlling a lathe cutting process. DE's results were compared to those found by four other algorithms: GLO (Gorwin et al. 1994), T (Tom 1990), GZ (Goodman and Zeng 1994) and MVGA (Magoulas et al. 1997). On the XOR problem, DE was successful 100% of the time, while the nearest competitor (GLO) was successful 84% of the time. Both DE and GLO did well on the three-bit parity problem, scoring 100% and 96%, respectively, but the GZ algorithm did not have any successes. On the lathe control problem, DE (93%) and MVGA (34%) scored successes. The authors concluded that DE was a promising method even when compared with other methods that require the gradient approximations of the error function and train networks by progressively altering the shape of the sigmoid function.

3.5 Summary

The results summarized in this chapter echo several themes. One is that while DE may not always be the fastest method, it is usually the one that produces the best result, although the number of cases in which it is also faster is significant. DE also proves itself to be robust, both in how control parameters are chosen and in the regularity with which it finds the true optimum. In addition, when compared to one-point optimizers like Powell's method, DE is relatively immune to differences in initial populations. Because it is a direct search method, DE is versatile enough to solve problems whose objective functions lack the analytical description needed to compute gradients. As a bonus, DE is also very simple to implement and modify. As these researchers have found, DE is a good first choice when approaching a new and difficult global optimization problem is defined with continuous and/or discrete parameters.

References

Ackley DH (1987) A connectionist machine for genetic hillclimbing. Kluwer, Boston, MA, USA

Ali MM, Törn A (1998) Evolution based global optimization techniques and the controlled random search algorithm: Proposed modifications and numerical studies. Submitted to the Journal of Global Optimization, 1998, Kluwer Academic Publishers, The Netherlands

Ali MM, Törn A (2000) Optimization of carbon and silicon clusters geometry for Tersoff potential using differential evolution. In: Floudas CA, Pardalos PM (eds) Optimization in computational and molecular biology. Kluwer Academic Publishers pp 1–15

Aluffi-Pentini F, Parisi V, Zirilli F (1985) Global optimization and stochastic differential equations. Journal of Optimization and Theory and Applications 47(1):1–16

Angeline PJ (1998) Evolutionary optimization versus particle swarm optimization. In: Porto VW, Saravanan N, Waagen D, Eiben AE (eds) Evolutionary programming VII. Springer, Berlin pp 601–610

Babu BV, Sastry KKN (1999) Estimation of heat transfer parameters in a trickle-bed reactor using differential evolution and orthogonal collocation. Computers and Chemical Engineering 23:327–339

Bersini H, Dorigo M, Langerman S, Seront G, Gambardella L (1996) Results of the first international contest on evolutionary optimization (1st ICEO). In: Proceedings of the 1996 international conference on evolutionary computation, Nagoya, Japan, May 20-22. IEEE Press

Cao YJ, Wu QH (1997) Mechanical design optimization by mixed-variable evolutionary programming. In: Proceedings of the 1997 conference on evolutionary computation. IEEE Press pp 443–446

Chellapilla K (1998) Combining mutation operators in evolutionary programming. IEEE Transactions on Evolutionary Computation 2:91–96

Chen JL, Tsao YC (1993) Optimal design of machine elements using genetic algorithms. Journal of the Chinese Society of Mechanical Engineers 14(2):193–199

Corana A, Marchesi M, Martini C, Ridella S (1987) Minimizing multimodal functions for continuous variables with the "simulated annealing algorithm". ACM Transactions on Mathematical Software, March 1987, pp 272–280

Crutchley DA, Zwolinski M (2003) Globally convergent algorithms for DC operating point analysis for nonlinear circuits. IEEE Transactions on Evolutionary Computation 7(1):2–10

Deb K, Pratap A, Agarwal S, Meyarivan T (2002) A fast and elitist multi-objective genetic algorithm: NSGA-II. IEEE Transactions on Evolutionary Computation 6:182–197

Fischer MM, Reismann M, Hlavackova-Schindler K (1999) Parameter estimation in neural spatial interaction modelling by a derivative free global optimization method. In: Proceedings of IV international conference on geocomputation, Mary Washington College, Fredericksburg, VA, USA, July 25–28, 1999 Available via Internet: http://www.geovista.psu.edu/sites/geocomp99/Gc99/007/gc_007.htm

Fu J-F, Fenton RG, Cleghorn WL (1991) A mixed integer-discrete-continuous programming method and its application to engineering design optimization. Engineering Optimization 17(4):263–280

Goodman R, Zeng Z (1994) A learning algorithm for multi-layer perceptrons with hard-limiting threshold units. In: Proceedings of the IEEE Neural Networks for Signal Processing, pp 219–228

Gorwin EM, Logar AM, Oldham WJB (1994) An iterative method for training multilayer networks with threshold functions. IEEE Transactions on Neural Networks 5:507–508

Griewangk AO (1981) Generalized descent for global optimization. JOTA 34:11–39

Han K-H, Kim J-H (2004) Quantum-inspired evolutionary algorithms with a new termination criterion, Hε gate, and two-phase scheme. IEEE transactions on Evolutionary Computation 8(2):156–169

Hu YF, Mcguire KC, Cokljat D, Blake RJ (1997) Parallel controlled random search algorithms for shape optimization. In: Emerson DR, Ecer A, Periaux J, Satofuka N (eds) Parallel computational fluid dynamics. North-Holland, pp 345–352

Ingber L (1993) Simulated annealing: Practice versus theory. Journal of Mathematical and Computer Modeling 18(11):29–57

Joines JA, Houck CR (1994) On the use of non-stationary penalty functions to solve nonlinear constrained optimization problems. In: Proceedings of the first IEEE conference on evolutionary computation, June 27–29. IEEE Press vol 2, pp 579–584

Joshi R, Sanderson AC (1999) Minimal representation multisensor fusion using differential evolution. IEEE Transactions on systems, man and cybernetics – part A: Systems and Humans 29(1):63–76

Katsuura H (1991) Continuous nowhere differential functions – an application of contraction mappings. The American Mathematical Monthly 5(98)

Kennedy J, Eberhart RC (1995) Particle swarm optimization. In: Proceedings of the 1995 IEEE international conference on neural networks, 4. IEEE Press, Piscataway, NJ, USA pp 1942–1948

Krink T, Filipie B, Fogel GB (2004) Noisy optimization problems – a particular challenge for differential evolution? In: Proceedings of the 2004 Congress on evolutionary computation vol 1, pp 332–339

Kozeil S, Michalewicz Z (1999) Evolutionary algorithms, homomorphous mappings and constrained parameter optimization. Evolutionary Computation 7(1):19–44

Kukkonen S, Lampinen J (2004) An extension of generalized differential evolution for multi-objective optimization with constraints. In: Proceedings of PPSN 2004, the 8th International conference on parallel problem solving from nature, September 18–22   2004, Birmingham, UK, pp 752–761. Springer, ISBN: 3-540-23092-0

Lampinen J (2002). A constraint handling approach for the differential evolution algorithm. In: Proceedings of the 2002 IEEE world congress on computational intelligence – WCCI 2002, 2002 Congress on evolutionary computation – CEC 2002, Honolulu, Hawaii, May 12-17, 2002. IEEE Press, 6 pages. ISBN 0-7803-7281-6

Lampinen J, Storn R (2004) Differential evolution. In: Onwubolu GC, Babu BV (eds) New optimization techniques in engineering. Studies in fuzziness and soft computing, vol 141, Chapter 6. Springer, pp 123–166. ISBN 3-540-20167-X

Lampinen J, Zelinka I (1999) Mechanical engineering design optimization by differential evolution. In: Corne D, Dorigo M, Glover F (eds) New ideas in optimization. McGraw-Hill, Maidenhead, UK pp 127–146

Lee C-Y, Yao X (2004) Evolutionary programming using mutations based on the Levy probability distribution. IEEE Transactions on Evolutionary Computation 8(1):1–13

Li HL, Chow CT (1994) A global approach for nonlinear mixed discrete programming in design optimization. Engineering Optimization 22:109–122

Leung Y-W, Wang Y (2001) An orthogonal genetic algorithm with quantization for global numerical optimization. IEEE Transactions on Evolutionary Computation 5(1):41–53

Lin SS, Zhang C, Wang H–P (1995) On mixed-discrete nonlinear optimization problems: A comparative study. Engineering Optimization 23(4):287–300

Loh HT, Papalambros PY (1991) A sequential linearization approach for solving mixed-discrete nonlinear design optimization problems. Transactions of the ASME, Journal of Mechanical Design 113(3):325–334

Loh HT, Paplambros PY (1991a) Computational implementation and tests of a sequential linearization algorithm for mixed-discrete nonlinear design optimization. Transactions of the ASME, Journal of Mechanical Design 113(3):335–345

Margoulas GD, Vrahatis MN, Grapsa TN, Androulackis GS (1997) A training method for discrete multilayer neural networks. In: Ellacot SW, Mason JC, Anderson IJ (eds) Mathematics of neural networks: Models, algorithms and applications, chapter 41. Kluwer Academic Publishers

Michalewicz Z (1995) Genetic algorithms, numerical optimization and constraints. In: Proceedings of the sixth international conference on genetic algorithms, Pittsburgh, July 15–19 pp 151–158

Michalewicz Z, Schoenauer M (1996) Evolutionary algorithms for constrained parameter optimization problems. Evolutionary Computation 4(1):1–32

Moscato PA (1989) On evolution, search, optimization, genetic algorithms and martial arts: Towards memetic algorithms. Technical report, Caltech concurrent computation program report 826, Caltech, Pasadena, California

Mühlenbein H, Scomisch D, Born J (1991) The parallel genetic algorithm as function optimizer. Parallel Computing 17:619–632

Mühlenbein H, Schlierkamp-Vosen D (1993) Predictive models for the breeder genetic algorithm, I. Continuous parameter optimization. Evolutionary Computation 1(1):25–49

Paterlini S, Krink T (2004) Differential evolution and particle swarm optimization in partitional culstering. In: Proceedings of the 2004 Congress on Evolutionary Computation (CEC 2004), IEEE Press, Piscataway, NJ, USA

Press WH, Teukolsky SA, Vetterling WT, Flannery BP (1992) Numerical recipes in C. Cambridge University Press

Price KV (1997) Differential evolution vs. the contest functions of the 2nd ICEO. In: Proceedings of the 1997 IEEE international conference on evolutionary computation, April 13-16, Indianapolis, IN, USA. IEEE Press, pp 153–157

Price WL (1977) Global optimization by controlled random search. Computer Journal 20:367–370

Plagianakos VP, Magoulas GD, Nousis NK, Vrahatis MN (2001) Training multi-layer networks with discrete activation functions. In: Proceedings of the INNS-IEEE international joint conference on neural networks, July 14–19, 2001, Washington DC, USA

Rogalsky T, Derksen RW, Kocabiyik S (1999) Differential evolution in aerodynamic optimization. In: Proceedings of the 46th annual conference of the Canadian aeronautics and space institute, May 2-5, 1999, pp 29–36. Available via Internet: http://home.cc.umanitoba.ca/~umrogal1/publications.html

Růžek B, Kvasnička M (2001) Differential evolution algorithm in the earthquake hypocenter location. Pure and Applied Geophysics 158:667–693

Salomon R (1996) Reevaluating genetic algorithm performance under coordinate rotation of benchmark functions: A survey of some theoretical and practical aspects of genetic algorithms. Biosystems 39(3):263–278

Sandgren E (1990) Nonlinear integer and discrete programming in mechanical design optimization. Transactions of the ASME, Journal of Mechanical Design 112(2):223–229

Schwefel H-P (1995) Evolution and optimum seeking. Wiley

Stanhope SA, Daida JM (1997) An individually variable mutation rate strategy for genetic algorithms. In: Angeline PJ, Reynolds RJ, McDonnell JR, Eberhart R (eds) Evolutionary programming VI; Lecture notes in computer science 1213. Springer, pp 235–245

Storn R, Price KV (1997) Differential evolution – A simple and efficient heuristic for global optimization over continuous spaces. Journal of Global Optimization 11:341–359

Thierauf G, Cai J (1997) Evolution strategies – parallelization and application in engineering optimization. In: Topping BHV (ed) Parallel and distributed processing for computational mechanics. Saxe-Coburg Publications, Edinburgh

Thomas P, Vernon D (1997) Image registration by differential evolution. In: Proceedings of the first Irish machine vision and image processing conference IMVIP-97, Magee College, University of Ulster, pp 221–225. PostScript file available from http://www.cs.may.ie/~pthomas/

Thomsen R (2003) Flexible ligand docking using evolutionary algorithms: Investigating the effects of variation operators and local search hybrids. Biosystems 72(1–2):57–73

Tom DJ (1990) Training binary node feed forward neural networks by back-propagation of error. Electronics Letters 26:1745–1746

Tsai J-T, Liu T-K, Chou J-H (2004) Hybrid Taguchi-genetic algorithm for global numerical optimization. IEEE Transactions on Evolutionary Computation 8(4):365–377

Tu Z, Lu Y (2004) A robust stochastic genetic algorithm for global numerical optimization. IEEE Transactions on Evolutionary Computation 8(5):456–470

Ursem RK, Vadstrup P (2004) Parameter identification of induction motors using differential evolution. Applied Soft Computing 4(1): 49–64

Van den Bergh F, Englebrecht AP (2004) A cooperative approach to particle swarm optimization. IEEE Transactions on Evolutionary Computation 8(3):225–239

Vesterstrøm JS, Riget J (2002) Particle swarms: Extensions for improved local, multi-modal and dynamic search in numerical optimization. Master's thesis, EVALife, Dept. of Computer Science, University of Aarhus, Denmark

Vesterstrøm J, Thomsen R (2004) A comparative study of differential evolution, particle swarm optimization, and evolutionary algorithms on numerical benchmark problems. In: Proceedings of the 2004 congress on evolutionary computing, vol 2, pp 1980–1987

Voigt H-M (1995) Soft genetic operators in evolutionary computation and bio-computation. In: Lecture Notes in Computer Science 899. Springer, Berlin, pp 123–141

Whitley D, Mathias K, Rana S, Dzubera J (1996) Evaluating evolutionary algorithms. Artificial Intelligence 85:1–32

Wolpert DH, Macready WG (1997) No free lunch theorems for optimization. IEEE transactions on evolutionary computation, IEEE Press, 1(1):67–82

Wu S-J, Chow P-T (1995) Genetic algorithms for nonlinear mixed discrete-interger optimization problems via meta-genetic parameter optimization. Engineering Optimization 24(2): 137–159

Yao X, Liu Y (1997) Fast Evolution Strategies. In: Angeline PJ, Reynolds RJ, McDonnell JR, Eberhart R (eds) Evolutionary programming VI. Springer, Berlin, pp 151–161

Yao X, Liu Y, Lin GM (1999) Evolutionary programming made faster. IEEE Transactions on Evolutionary Computation 3:82–102

Yen J, Lee B (1997) A simplex genetic algorithm hybrid. In: Proceedings of the 1997 IEEE conference on evolutionary computation, Indianapolis, Indiana, April 13-16. IEEE Press, pp 175–180

Zimmermann W (1990) Operations research. Oldenbourg

Zitzler E, Thiele I (1999) Multi-objective evolutionary algorithms: A comparative case study and the strength Pareto approach. IEEE Transactions on Evolutionary Computation 4:257–271

Zitzler E, Deb K, Thiele L (2000) Comparison of multi-objective evolutionary algorithms: Empirical results. Evolutionary Computation 8:173–195

4 Problem Domains

4.1 Overview

Up until now, this book has focused primarily on unconstrained, and to a lesser degree, bound constrained continuous optimization. This chapter explores how to apply DE in several different, less idealized problem domains. Among the topics discussed are how to optimize functions with discrete or mixed-type parameters as well as those that are subject to bound, inequality and/or equality constraints. In addition, the challenges associated with optimizing both noisy functions and those with multiple objectives are discussed. This chapter also explores the possibility of applying DE to combinatorial problems like the Traveling Salesman Problem, or TSP. The next section, however, looks at how DE handles quantized functions and parameters.

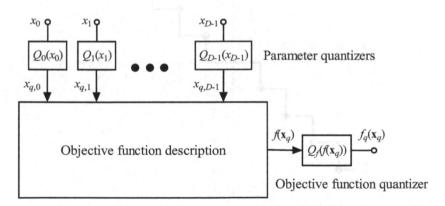

Fig. 4.1. An objective function whose parameters and output are both quantized

4.2 Function and Parameter Quantization

Real-world optimization often involves functions and/or parameters that vary discretely. A continuous range of values becomes discretely distributed when operated on by a *quantizing function, Q*. The diagram in Fig. 4.1 shows the most general case in which the function is rendered discrete by the quantizing function Q_f and parameters are quantized by the functions $Q_j, j = 0, 1, …, D-1$.

4.2.1 Uniform Quantization

Uniform quantization transforms a continuous range of values into a set of evenly spaced values, like the integers. Uniform quantization (Rabiner and Gold 1975) is based on the quantizing function

$$Q(y) = \frac{floor(k \cdot y)}{k}. \tag{4.1}$$

The "floor" function returns the integer part of its argument, e.g., floor(4.13) = 4. As Fig. 4.2 illustrates, when $k = 1$, $Q(y)$ returns the integer part of y.

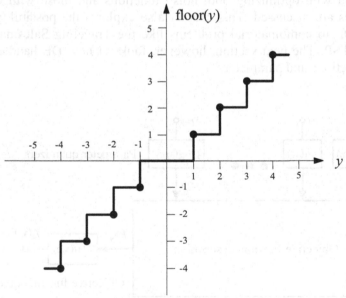

Fig. 4.2. Uniform quantization: floor(y) returns the function value indicated by the dots that mark discontinuities.

Selecting components for a gear train is an example of an optimization task where objective function parameters are uniformly quantized (Lampinen and Storn 2004). The goal is to select a combination of gears that minimizes the (absolute) difference between the actual and target rotation rates. Since both the number of gears and the number of teeth on a gear are integers, parameters are uniformly discrete. Additionally, the objective function value itself is also discrete because for a fixed input rate of rotation, a finite number of gear combinations produce only a limited number of isolated output rates. Unlike parameter values, however, objective function values are not uniformly distributed in this example.

4.2.2 Non-Uniform Quantization

Discrete values need not be regularly spaced and many applications optimize non-uniformly quantized variables (Kondoz 1994). *Non-uniform quantization* maps the continuum to a set of isolated and irregularly spaced real values. Figure 4.3 illustrates non-uniform quantization with an example that also displays a saturation characteristic, i.e., $Q(y)$'s output has a limiting magnitude.

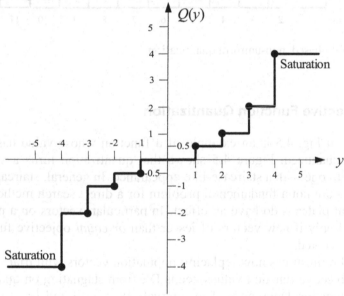

Fig. 4.3. Non-uniform quantization

Parameters with non-uniformly quantized values are common in many mechanical and electronic design problems, often because only a limited set of components is commercially available. For example, electronic resistors in the E12 series have the following discrete normalized values: 1.0, 1.2, 1.5, 1.8, 2.2, 2.7, 3.3, 3.9, 4.7, 5.6, 6.8, 8.2. Not all resistance values are available and those that are, are not uniformly spaced.

Transformation tables can be an effective way to non-uniformly quantize continuous values when a quantizing function is impractical. For example, if objective function parameters must be prime numbers, then a quantizing function will first have to "floor" a floating-point number into an integer and then search for the closest prime value.

Locating the nearest prime value involves a search process that can be computationally intensive if primes are not stored in advance. A table-based approach is both faster and simpler (Fig. 4.4). Instead of varying the parameters x_j and quantizing them to the closest prime value, it is much easier to vary surrogate parameters $z_j \in [1.5, 12.5]$ and use floor(z_j) as the table index to primes already stored in memory. In Fig. 4.4, for example, index 5 leads to table entry 11. When the objective function is evaluated, table entries take the place of the corresponding parameters, x_j.

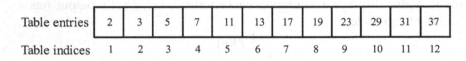

Table entries	2	3	5	7	11	13	17	19	23	29	31	37
Table indices	1	2	3	4	5	6	7	8	9	10	11	12

Fig. 4.4. Table-based, non-uniform quantization

4.2.3 Objective Function Quantization

The graph in Fig. 4.5 is an example of a function whose value has been uniformly quantized. Figure 4.5 shows that quantization turns a smooth function into one with a staircase-like appearance. In general, staircase discontinuities are not a fundamental problem for a direct search method like DE, but flat plateaus do have an effect. In particular, vectors on a plateau will spread only if new vectors of lesser than *or equal* objective function value are accepted.

Figure 4.6 illustrates how replacing population vectors with trial vectors of equal objective function values keeps DE from stagnating on quantized objective function landscapes. The intervals $[x_{0,L}, x_{0,U}]$ and $[x_{1,L}, x_{1,U}]$ in Fig. 4.6 define a plateau where the objective function value is constant. DE's vector generating scheme inevitably places multiple trial vectors outside the hull that encloses the original population. If trial vectors are not allowed to replace competitors of equal objective function value, the population stagnates and remains confined to the original hull. If, however, trial vectors replace population vectors of equal objective function value, the population expands very quickly unless F is below Zaharie's limit (Zaharie 2002). If F is too small, the population will likely converge even if trial vectors replace target vectors of equal objective function value. For DE with $N_j(0,1)$ Gaussian jitter, Zaharie found that in the absence of selective pressure, the population's variance increases as long as

$$2 \cdot F^2 \cdot Cr - \frac{2 \cdot Cr}{Np} + \frac{Cr^2}{Np} < 0. \tag{4.2}$$

Other DE models behave similarly and experiments suggest that as long as F is above the Zaharie limit, the population should diverge (see Sect. 2.5).

Figure 4.7 shows how the presence of plateaus affects DE's ability to optimize both the ten-dimensional sphere

$$f(\mathbf{x}) = \sum_{j=0}^{9} x_j^2 \tag{4.3a}$$

and its quantized counterpart

$$f_q(\mathbf{x}) = \frac{\text{floor}\left(2 \cdot \sum_{j=0}^{9} x_j^2\right)}{2}. \tag{4.3b}$$

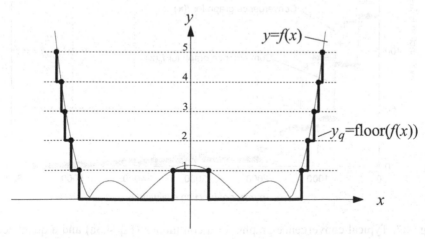

Fig. 4.5. A function, $f(x)$, and its uniformly quantized version, $y_q = \text{floor}(f(x))$

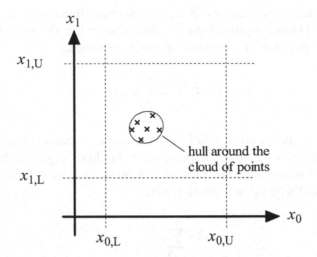

Fig. 4.6. A cloud of points will quickly spread over the plateau defined by the intervals $[x_{0,L}, x_{0,U}]$ and $[x_{1,L}, x_{1,U}]$ if DE's selection rule replaces target vectors with trial vectors of equal value.

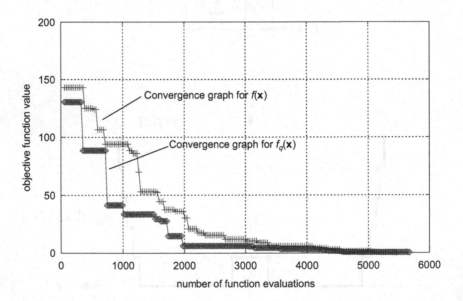

Fig. 4.7. Typical convergence graphs for a continuous (Eq. 4.3a) and a quantized (Eq. 4.3b) ten-dimensional sphere. The initialization interval was $[-10,10]$. The algorithm was DE/rand/1/bin with $Np = 30$, $F = 0.85$ and $Cr = 1$.

The convergence graphs in Fig. 4.7 plot the current best vector's objective function value *versus* the number of function evaluations. Both plots were generated with classic DE (DE/rand/1/bin), $Np = 30$, $F = 0.85$ and $Cr = 1$. The quantized sphere's convergence graph exhibits larger vertical jumps and less incremental improvement than its continuous counterpart. Similarly, the number of function evaluations spent without making improvement is higher in the quantized version than in the continuous case. Intuitively, this behavior makes sense because the population's best function value does not improve while DE explores a plateau. Even though quantizing the objective function affected DE's convergence profile, it did not impact DE's effectiveness.

In reality, examples where parameters are continuous but the objective function output is discrete are rare. More often, parameters are quantized but the objective function is not – a possibility that also can be treated as if the reverse was true – with continuous parameters serving as input for a quantized objective function. The next section shows how this transformation adds diversity to the spectrum of vector differences, thereby reducing the chance that DE will stagnate.

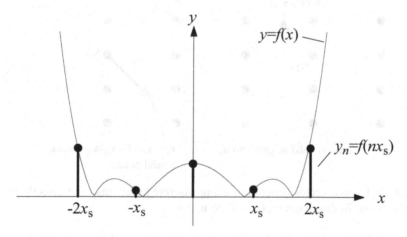

Fig. 4.8. Uniform parameter quantization. Not all parameter values are allowed. n is an integer.

4.2.4 Parameter Quantization

Quantizing the parameters of a real-valued objective function transforms a continuous optimization task into a discrete one. Although the objective function itself is not quantized, function values are nevertheless discrete simply because they can only be sampled at allowed parameter values (Fig. 4.8). Limiting parameters to discrete values restricts the distribution of vector differences, which in turn limits DE's ability to explore the problem space.

When parameters are discrete, differential mutation rarely creates a trial point that coincides with an allowed discrete value. Only in special cases, like that of uniform quantization and $F = 1$, do trial values coincide with allowed values. As Fig. 4.9 shows, even for uniformly distributed discrete variables, $F \neq 1$ creates child vectors that are not allowed. Consequently, a criterion must be established when dealing with discrete variables that can locate the allowed parameter value that is nearest to the proposed value. Finding the nearest allowable vector in D-dimensional space is probably the best approach, but that task is significantly harder than finding the nearest discrete value for each parameter independently, which is the method used for the experiments in this book.

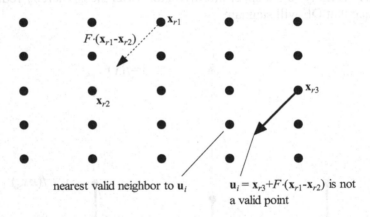

Fig. 4.9. Difference vector generation using discrete points may not directly lead to a valid point. In this diagram, $Cr = 1$, so $\mathbf{u}_i = \mathbf{v}_i$.

A second difficulty posed by discrete parameters is that even if trial vectors are mapped to the closest allowed value, some points may not be accessible unless F is carefully chosen. For example, suppose that a variable is restricted to the even integers and a single odd value. Assuming that the odd point does not already belong to the population, the difference between any two instances of this quantized, discrete parameter will be a multiple of $2F$. As a result, F must be a special value (e.g., $0.5 < F < 0.75$) to place a trial point closer to the odd point than to an even one.

Instead of requiring F to assume a special value, it is more effective to use real values in place of discrete parameters when randomly initializing and generating vectors. In the example above, randomly initializing a discrete parameter with a floating-point value creates many differences smaller than 2 that can place a trial point near the lone odd point almost irrespective of F. Forming differentials with distributed real values generates a richer spectrum of differences that both reduces the risk that discrete values are inaccessible and relaxes the constraints on an effective F. For this reason, discrete parameters should be represented as floating-point values even when the problem is inherently discrete (Lampinen and Zelinka 1999).

When working with discrete parameters, the objective function is evaluated once DE's floating-point parameter values are quantized to, but not overwritten by, their nearest allowed discrete values. Figure 4.10 illustrates how copies of vectors generated in the continuous domain are quantized before being input into the objective function.

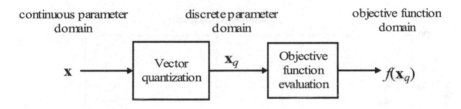

Fig. 4.10. An objective function's discrete arguments are copies of floating-point parameters quantized to their nearest allowed value.

When real-valued trial vectors are evaluated at their nearest allowable discrete value, DE is in effect optimizing a staircase function like the example of objective function quantization in Fig. 4.5. As Fig. 4.11 illustrates, the distribution of minima differs substantially between y and y_n. For y_n, only two minima exist instead of the previous four, i.e., two minima are masked. This is in contrast to the quantized objective function in Fig. 4.5 where the location of the minima is broadened, but all minima still exist. Simply put, objective function quantization does not mask minima, but parameter quantization can.

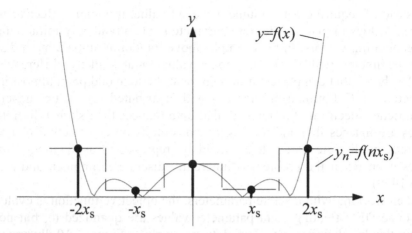

Fig. 4.11. If parameter quantization is applied after differential mutation but before evaluation, then the objective function is, in effect, stepped (dashed line).

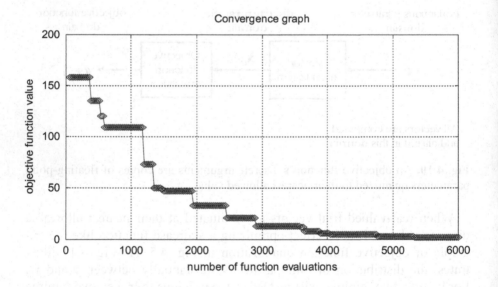

Fig. 4.12. Convergence graph for the objective function in Eq. 4.4. The DE variant was DE/rand/1/bin, $Np = 30$, $F = 0.85$, $Cr = 1$. The initialization interval was $[-10,10]$.

Applying the "floor" operation to the sphere of Eq. 4.4 is a simple example of a function with quantized parameters. The corresponding convergence graph in Fig. 4.12 suggests that as long as parameters are generated and maintained using real values, quantization is not an impediment. Equation 4.4, however, is a very simple case:

$$f_{pq}(x) = \sum_{i=0}^{9} (floor(x_i))^2 \qquad (4.4)$$

Equation 4.4 exhibits a single minimum at $x = 0$ that is not masked. In real-world applications, minima can be masked if discrete parameter values are too far apart. Equations 4.5–4.7 outline a constraint satisfaction problem like those that can be solved with the demo version of the FIWIZ digital filter design program (Storn 2000) that accompanies this book on CD.

The problem is solved when a_n and b_n are such that $H(z)$ is

$$H(z) = \frac{U(z)}{D(z)} = \frac{\sum_{n=0}^{N} a_n \cdot z^{-n}}{1 + \sum_{m=1}^{M} b_m \cdot z^{-m}} = A_0 \frac{\prod_{n=0}^{N-1} (z - z_{0n})}{\prod_{m=0}^{M-1} (z - z_{pm})} \qquad (4.5)$$

with

$$z = \exp(\imath \cdot 2\pi\Omega) = \cos(2\pi\Omega) + \imath \cdot \sin(2\pi\Omega), \quad \imath = \sqrt{-1} \qquad (4.6)$$

and

$$\Omega = \frac{f}{f_s} \qquad (4.7)$$

is constrained to lie within the tolerances shown in Fig. 4.13 as Ω ranges from 0.0 to 0.5. The variables f and f_s in Eq. 4.7 denote the natural and sampling frequencies, respectively (Storn 1999b; Mitra and Kaiser 1993). The parameters a_n and b_n are quantized to simulate digital filter processor word-length limitations. In this constraint satisfaction problem, "success" means that $H(z)$ satisfies the tolerance scheme.

Table 4.1 shows how the level of quantization for the parameters a_n and b_n affects DE's ability to solve this problem. Listed along with the quantization step size are P, the fraction of trials that were successful, and the corresponding average number of function evaluations that DE required. Although the test set is small, the trend is clear. Results show that increasing the quantization step size slowed DE and made it less reliable. The coarser the quantization, the more difficult it becomes for DE to find a solution until eventually (7-bit step), all minima are masked and a solution is no longer possible. Results are ten-trial averages. DE settings were the same for all designs: DE/best/1/bin with uniform jitter, $Np = 30$, $F = 0.85$, $Cr = 1$. The number of parameters was $D = N + M + 1$ with $N = 4$, $M = 4$. The increment by 1 accounts for the parameter A_0 in Eq. 4.5.

Additional applications where objective functions with discrete parameters have been successfully optimized can be found in Storn (1997) and Lampinen and Zelinka (1999). Table 4.2 summarizes the similarities and differences between objective function and parameter quantization.

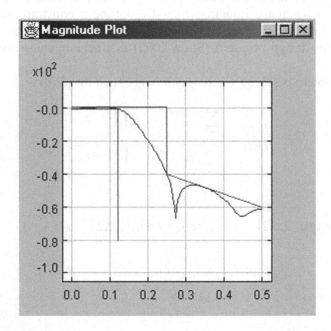

Fig. 4.13. A tolerance scheme constraining the function in Eq. 4.5. The x–axis in this plot denotes Ω while the y–axis plots $20 \cdot \log_{10} \cdot |H(\exp(\Omega 2\pi\Omega)|$.

Table 4.1. The effect that different quantization levels have on DE's ability to solve the design problem set forth in Eqs. 4.5–4.7.

Quantization level	Average evaluations	P
None	4230	1
16-bit (step size 2^{-15})	5970	1
10-bit (step size 2^{-9})	6016	0.9
8-bit (step size 2^{-7})	8317	0.7
7-bit (step size 2^{-6})	–	0

Table 4.2. Objective function and parameter quantization compared

	Objective function	Parameters
Number of function values	Restricted	Restricted
Function plateaus?	Yes, for $f_q(\mathbf{x})$	Yes, for $f(\mathbf{x})$ no for $f(\mathbf{x}_q)$
Minima	Broadened	May be hidden
Stagnation risk unless…	$f(\mathbf{u}_{i,g}) \le f(\mathbf{x}_{i,g})$ for $\mathbf{u}_{i,g}$ to survive	$\mathbf{x}$ is used instead of $\mathbf{x}_q$ for the differential

4.2.5 Mixed Variables

Mixed-variable problems, also known as mixed-discrete programming problems, contain both continuous and discrete parameters. As outlined in the previous sections, DE handles such tasks by representing all parameters internally as floating-point values and quantizing the discrete parameter values to the nearest allowed point. Lampinen and Zelinka (1999) described the first practical applications of DE to mixed-variable optimization.

4.3 Optimization with Constraints

Perhaps the majority of real-world optimization tasks involve finding a solution that not only is optimal, but also satisfies one or more constraints. There are several ways in which an optimization problem can be constrained. A general formulation for constrained optimization is

$$\text{Find } \mathbf{x} = (x_0, x_1, ..., x_{D-1})^{\mathrm{T}}, \quad \mathbf{x} \in \Re^D \tag{4.8}$$

to minimize: $f(\mathbf{x})$,

subject to:

inequality constraints: $\gamma_m(\mathbf{x}) \le 0, \quad m = 1, 2, ..., M$,

equality constraints: $\varphi_n(\mathbf{x}) = 0, \quad n = 1, 2, ..., N$,

and boundary constraints: $x_{j,\mathrm{L}} \le x_j \le x_{j,\mathrm{U}}, \quad j = 0, 1, ..., D-1$.

Strictly speaking, boundary constraints are inequality constraints, but they are listed separately because they occur frequently and are easier to handle than general inequality constraints.

Constraints typically make optimization harder for DE because they can create forbidden regions on the objective function landscape that restrict the free movement of vectors. It often happens, depending on how they are handled, that constraints divide the search space into several disjoint islands. On the other hand, constraints can eliminate local minima that might otherwise trap vectors, thereby reducing the chance that DE will prematurely converge. For an up-to-date survey on constraint handling techniques used with EAs, see Coello (2002).

4.3.1 Boundary Constraints

Boundary constraints are very common in real-world applications, often because parameters are related to physical components or measures that have natural bounds. For example, neither passive electronic resistance nor the length of a mechanical object can be negative. Even when the problem itself is unconstrained, bounds may be imposed by the limits set by the particular data type. For example, the limited number of bits dedicated to a fixed-point number in a digital signal processor sets a bound on the range of values that can be represented. Fortunately, handling boundary constraints in DE is particularly straightforward and several schemes have been applied with success.

In DE, each population vector is crossed with a randomly generated mutant vector. Since the current population of vectors already satisfies all bound constraints, only contributions from mutant vectors potentially violate parameter limits. Consequently, bounds need to be checked only when a mutant parameter is selected for the trial vector. For simplicity, however, the following examples test all trial parameters, not just those donated by the mutant.

There are two distinctly different general techniques for handling out-of-bounds parameters. *Resetting* schemes modify out-of-bounds parameters so that the trial vector satisfies all constraints. By contrast, *penalty* methods drive solutions from restricted areas through the action of an objective function-based criterion. The simplest of these penalty methods is the "brick wall".

Brick Wall Penalty

If *any* trial parameter exceeds a bound, the brick wall strategy sets the offending vector's objective function value high enough to guarantee that it will not be selected. If the optimum lies near bounds, the brick wall penalty can slow progress because generating a solution that has no out-of-bounds parameters may be improbable. The pseudo-code in Fig. 4.14 outlines this strategy as it applies to DE.

```
...
violate_flag = FALSE;
for (j=0; j<D; j++)
{
    if ((u_j<x_{j,L})||(u_j>x_{j,U}))  //if parameter exceeds bounds
    {
        violate_flag = TRUE;
    }
}
if (violate_flag = TRUE) return_value = HIGH_VALUE;
else return_value = objective_function(u);
...
```

Fig. 4.14. Pseudo-code for the brick wall penalty

Adaptive Penalty

Unlike the brick wall penalty, the adaptive penalty increases the objective function value by an amount that depends on the number of bounds that a trial vector violates. For example, an objective function can be incremented by a penalty whenever a parameter exceeds a bound. An alternative scheme (Fig. 4.15) imposes an additional penalty that depends not only on the number of violations, but also on their magnitude (Storn 1996a, 1996b, 2000).

```
...
return_value = objective_function(u);
penalty      = 0;
for (j=0; j<D; j++)
{
    if (u_j<x_j,L) //if parameter exceeds lower bound
    {
        penalty = penalty + CONST_PENALTY + FACTOR*(x_j,L-u_j);
    }
    if (u_j>x_j,U) //if parameter exceeds upper bound
    {
        penalty = penalty + CONST_PENALTY + FACTOR*(u_j-x_j,U);
    }
}
return_value = return_value + penalty;
...
```

Fig. 4.15. Pseudo-code for an adaptive penalty

Random Reinitialization

Penalty methods do not reset out-of-bounds parameters. If bounds are easily exceeded, then vectors that satisfy all bound constraints will be rare and progress will be slow. Resetting methods convert out-of-bounds parameter values into allowed values. The most unbiased approach, random reinitialization, replaces a parameter that exceeds its bounds by a randomly chosen value from within the allowed range (Lampinen and Zelinka 1999). Because it radically changes a parameter's value, reinitialization can disrupt progress toward solutions that lie near bounds. Equation 4.9 shows how to reinitialize an out-of-bounds trial parameter:

$$u_{j,i,g} = x_{j,L} + \text{rand}_j(0,1) \cdot \left(x_{j,U} - x_{j,L}\right) \text{ if } \left(u_{j,i,g} < x_{j,L}\right) \vee \left(u_{j,i,g} > x_{j,U}\right) \quad (4.9)$$

Bounce-Back

Like random reinitialization, the bounce-back method replaces a vector that has exceeded one or more of its bounds by a valid vector that satisfies all boundary constraints. In contrast to random reinitialization, the bounce-back strategy takes the progress toward the optimum into account by selecting a parameter value that lies between the base parameter value and the bound being violated. The base vector x_{r0} is the vector in DE's mutation scheme to which the random vector differential is added. As the population moves toward its bounds, the bounce-back method generates vectors that will be located even closer to the bounds. Figure 4.16 presents pseudo-code for the bounce-back strategy, while Fig. 4.17 illustrates the process in a two-dimensional search space.

```
...
x_r0 = base_vector;
u_i  = child_vector;
...
for (j=0; j<D; j++)
{
    if (u_{j,i}<x_{j,L}) //if child parameter exceeds lower bound
    {
        u_{j,i} = x_{j,r0} + rand(0,1)*(x_{j,L}-x_{j,r0});
    }
    if (u_{j,i}>x_{j,U}) //if child parameter exceeds upper bound
    {
        u_{j,i} = x_{j,r0} + rand(0,1)*(x_{j,U}-x_{j,r0});
    }
}
...
```

Fig. 4.16. Pseudo-code for bounce-back parameter constraint handling

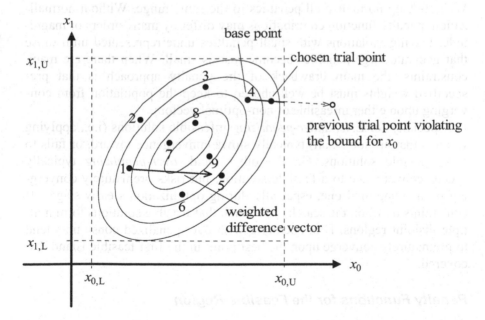

Fig. 4.17. Bounce-back bound resetting replaces an out-of-bounds trial parameter with one located between the base vector and the bound exceeded.

A simple, yet effective deterministic variant of bounce-back resetting forces an out-of-bounds trial parameter to the point midway between the bound violated and the base vector, e.g., $u_{j,i,g} = (x_{j,r0,g} + x_{j,U})/2$ when the upper bound is violated. Setting parameter values equal to the bounds they violate should be avoided because it lowers the diversity of the difference vector population.

4.3.2 Inequality Constraints

Inequality constraints require a solution to contain parameter values that satisfy one or more constraint functions. Most often, inequality constraints are implemented as *penalty functions*. Like the adaptive penalty for bound constraints, penalty functions increase the objective function value when constraints are violated. One common way to integrate constraint violations into an optimization task is to multiply each penalty by a weight, w_m, and add the result to the objective function, $f(\mathbf{x})$:

$$f'(\mathbf{x}) = f(\mathbf{x}) + \sum_{m=1}^{M} w_m \cdot p_m(\mathbf{x}).$$

(4.10)

Weights help normalize all penalties to the same range. Without normalization, penalty function contributions may differ by many orders of magnitude, leaving violations with small penalties underrepresented until those that generate large penalties become just as small. When there are many constraints, the main drawback of the penalty approach is that pre-specified weights must be well chosen to keep the population from converging upon either infeasible or non-optimal vectors.

Especially for EAs, *under-penalizing* infeasible solutions (i.e., applying weights that are too small) typically slows convergence toward, or fails to find, feasible solutions. On the other hand, *over-penalizing* typically speeds convergence to a feasible solution, but risks prematurely converging on a suboptimal one, especially during optimization's early stages. If constraints partition the search space such that feasible solutions form multiple disjoint regions, EAs operating with over-penalized constraints tend to prematurely converge upon the best point in the first feasible island discovered.

Penalty Functions for the Feasible Region

Classical penalty functions are *barrier functions* (Carrol 1961)

$$p_m(\mathbf{x}) = \frac{-1}{\gamma_m(\mathbf{x})} \tag{4.11}$$

or *log barrier functions* (Frisch 1955)

$$p_m(\mathbf{x}) = -\ln(-\gamma_m(\mathbf{x})) \tag{4.12}$$

where $\gamma_m(\mathbf{x})$ is from Eq. 4.8. For example, Fig. 4.18 (left) illustrates a simple one-dimensional constraint, $\gamma_1(x) = x - 1 < 0$, and the corresponding barrier function (right). A point from outside the feasible area has little chance to tunnel through the singularity at $x = 1$.

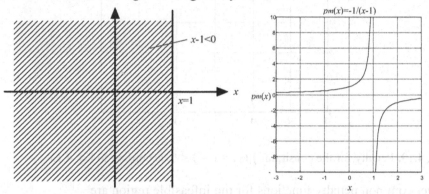

Fig. 4.18. Graphical illustration of the constraint $x - 1 < 0$ (left) and the corresponding barrier function, $p_m(x)$ (right).

The biggest drawback associated with barrier penalty functions is that they do not work when a vector violates a constraint. Instead, they require vectors to be resituated so that they fall within a feasible region.

Penalty Functions for Infeasible Regions

One penalty that often works well when trial solutions lie outside the constraint region is

$$p_m(\mathbf{x}) = \begin{cases} \gamma_m^2(\mathbf{x}) & \text{for } \gamma_m(\mathbf{x}) > 0, \\ 0 & \text{otherwise.} \end{cases} \tag{4.13}$$

When $\gamma_1(x) = x - 1 < 0$, the classical barrier penalty is infinite at the bound $x = 1$, but when computed according to Eq. 4.13, the penalty is zero. The effect of Eq. 4.13 is to steer the population within an infeasible region toward a feasible one, rather than obstruct it with an insurmountable barrier

(Fig. 4.19). Penalizing with the absolute value of $\gamma_m(\mathbf{x})$ instead of its square more gently steers vectors toward feasible areas.

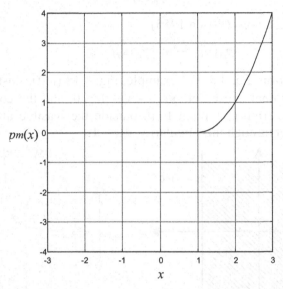

Fig. 4.19. Penalty for the constraint $\gamma_1(x) = x - 1 < 0$, according to Eq. 4.13

Other common penalty functions for the infeasible region are

$$p_m(\mathbf{x}) = \begin{cases} \cosh(\gamma_m(\mathbf{x})) - 1 & \text{for } \gamma_m(\mathbf{x}) > 0, \\ 0 & \text{otherwise.} \end{cases} \qquad (4.14)$$

and

$$p_m(\mathbf{x}) = \begin{cases} c_1 \cdot \gamma_m(\mathbf{x}) + c_2 & \text{for } \gamma_m(\mathbf{x}) > 0, \text{ with } c_1, c_2 > 0, \\ 0 & \text{otherwise.} \end{cases} \qquad (4.15)$$

Constraint Satisfaction: An Example

In constraint satisfaction problems, optimization terminates once constraints are satisfied. The following constraint satisfaction problem resembles those that arise in signal processing (filter design) and kinematics (trajectory design). The function

$$h(x_0, x_1, \tau) = \frac{1}{1 + x_0 \cdot \tau + (x_1 \cdot \tau)^2} \quad \text{for } \tau \in [0, 20] \qquad (4.16)$$

has two parameters x_0 and x_1 as well as the running variable τ. In addition, let $h(x_0, x_1, \tau)$ be subject to the constraint functions:

$$\gamma_1(x_0, x_1, \tau) = 1.04 - h(x_0, x_1, \tau) \text{ for } \tau \in [0, 10) \tag{4.17}$$

$$\gamma_2(x_0, x_1, \tau) = h(x_0, x_1, \tau) - 0.4 \text{ for } \tau \in [10, 20] \tag{4.18}$$

$$\gamma_3(x_0, x_1, \tau) = 0.8 - h(x_0, x_1, \tau) \text{ for } \tau \in [0, 5] \tag{4.19}$$

Figure 4.20 more clearly shows the nature of the optimization task specified by Eqs. 4.16–4.19.

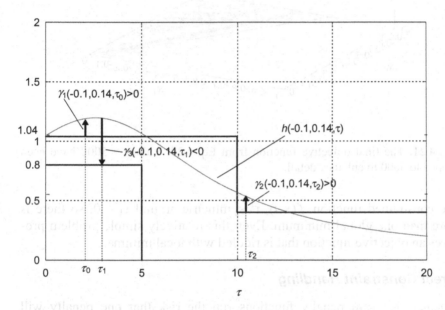

Fig. 4.20. The curved line must fall within the upper and lower bounds over the specified range. Points along the τ–axis illustrate each constraint function.

Combining Eq. 4.10 and Eq. 4.13 together with Eqs. 4.15–4.17 and taking S finely quantized samples on the τ-axis yields the objective function

$$f(x) = \sum_{m=1}^{M} w_m \cdot \sum_{s=0}^{S-1} \begin{cases} \left(\gamma_m(x, \tau_s)\right)^2 & \text{for } \gamma_m(x, \tau_s) > 0, \\ 0 & \text{otherwise;} \end{cases}, \quad M = 3. \tag{4.20}$$

Figure 4.21 plots $f(x)$ with all weights set equal to 1 ($w_m = 1$).

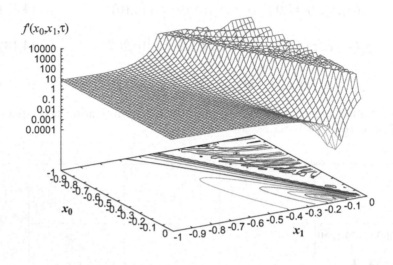

Fig. 4.21. The final objective function from Eq. 4.20. Values > 5000 have been clipped to 5000 to enhance detail.

The constrained function, $f(\mathbf{x}, \tau)$, is symmetric around $x_1 = 0$, so there is more than one global minimum. Even this relatively simple problem produces an objective function that is riddled with local minima.

Direct Constraint Handling

Schemes that sum penalty functions run the risk that one penalty will dominate unless weights are correctly adjusted. In addition, the population can converge upon an infeasible region if its objective function values are much lower than those in feasible regions. It can even happen that no set of weights will work. Because weight selection tends to be a trial and error optimization problem in its own right, simpler *direct constraint handling* methods have been designed that do not require the user to "tune" penalty weights.

Among the early direct handling techniques are those due to Kjellström and Taxen (1981) and Kreutzer (1985). These *constraint relaxation techniques* loosen constraints just enough so that all vectors in a population satisfy all constraints. Constraints are subsequently tightened over time.

This idea was extended by Storn (1999a) as CADE (Constraint Adaptation with Differential Evolution) to enhance DE's range of application.

Lampinen has devised a similar method (Lampinen 2001) that shows improved convergence speed when compared to CADE. In contrast to standard DE, each population vector is assigned not just one, but an array of objective values. The array contains not only each vector's objective function's value, but also its constraint function values, $\gamma_m(\mathbf{x}_i)$, $m = 1, \ldots,$ M; $i = 1, \ldots, Np$. Figure 4.22 provides an overview.

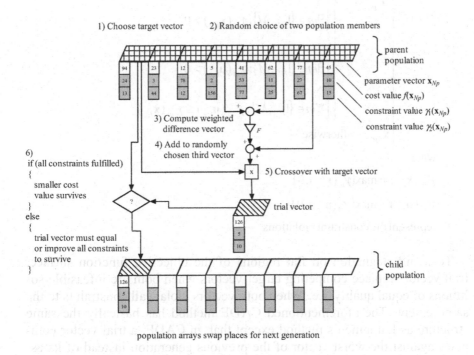

Fig. 4.22. Direct constraint handling with DE after Lampinen (2001). This example exhibits two constraints.

In simple terms, Lampinen's criterion selects the trial vector $\mathbf{u}_{i,g}$ if:

- $\mathbf{u}_{i,g}$ satisfies all constraints and has a lower or equal objective function value than $\mathbf{x}_{i,g}$, or
- $\mathbf{u}_{i,g}$ is feasible and $\mathbf{x}_{i,g}$ is not, or
- $\mathbf{u}_{i,g}$ and $\mathbf{x}_{i,g}$ are both infeasible, but $\mathbf{u}_{i,g}$ does not violate any constraint more than $\mathbf{x}_{i,g}$.

More formally, Lampinen's selection criterion states:

$$
\mathbf{x}_{i,g+1} = \begin{cases} \mathbf{u}_{i,g} & \text{if} \begin{cases} \begin{cases} \forall m \in \{1,...,M\}: \gamma_m(\mathbf{u}_{i,g}) \le 0 \wedge \gamma_m(\mathbf{x}_{i,g}) \le 0 \\ \wedge \\ f(\mathbf{u}_{i,g}) \le f(\mathbf{x}_{i,g}) \end{cases} \\ \vee \\ \begin{cases} \forall m \in \{1,...,M\}: \gamma_m(\mathbf{u}_{i,g}) \le 0 \\ \wedge \\ \exists m \in \{1,...,M\}: \gamma_m(\mathbf{x}_{i,g}) > 0 \end{cases} \\ \vee \\ \begin{cases} \exists m \in \{1,...,M\}: \gamma_m(\mathbf{u}_{i,g}) > 0 \\ \wedge \\ \forall m \in \{1,...,M\}: \gamma'_m(\mathbf{u}_{i,g}) \le \gamma'_m(\mathbf{x}_{i,g}) \end{cases} \end{cases} \\ \mathbf{x}_{i,g} & \text{otherwise} \end{cases}
\tag{4.21}
$$

where

$$\gamma'_m(\mathbf{x}_{i,g}) = \max(\gamma_m(\mathbf{x}_{i,g}),0)$$
$$\gamma'_m(\mathbf{u}_{i,g}) = \max(\gamma_m(\mathbf{u}_{i,g}),0)$$

represent the constraint violations.

To avoid stagnation on flat regions of the objective function surface, trial vectors replace competing target vectors when both are infeasible solutions of equal quality, i.e., when both vectors violate all constraints to the same extent. The aforementioned CADE method has basically the same structure as Lampinen's method except that, in CADE, a trial vector competes against the worst vector of the previous generation instead of its assigned target vector. Consequently, CADE tends to converge more slowly than Lampinen's method because CADE is more likely to accept small improvements in a trial vector.

If the objective function is unconstrained ($M = 0$, $N = 0$) or both vectors are infeasible, then Lampinen's criterion simply compares objective function values, just like DE's usual selection rule. When either one or both vectors are infeasible, however, objective function values are not compared. Consequently, it is not necessary to evaluate the objective function as long as one or more constraints are violated. This not only saves time, but also prevents *over-satisfying* constraints since there is no selective pressure to drive vectors into infeasible regions with low objective function values. Instead, selection drives vectors in the direction where constraint violations decrease (Lampinen 2002). For example, trying to make $\gamma_m(\mathbf{u}_{i,g})$ smaller than $\gamma_m(\mathbf{x}_{i,g})$ once $\gamma_m(\mathbf{x}_{i,g}) \le 0$ over-satisfies the constraints. As the second term in the if-condition for $\mathbf{u}_{i,g}$ in Eq. 4.21 shows, the trial vector's constraint function $\gamma_m(\mathbf{u}_{i,g})$ needs to be less than or equal to the target's constraint function $\gamma_m(\mathbf{x}_{i,g})$ only if $\gamma_m(\mathbf{x}_{i,g})$ is positive, i.e., only if the target violates the constraint. If $\gamma_m(\mathbf{x}_{i,g})$ is zero or negative, i.e., if the target constraint is fulfilled, then the trial vector only needs to be less than or equal to 0. The fact that Lampinen's method does not over-satisfy constraints is a significant benefit.

Like DE selection, Lampinen's criterion only needs to determine which of two solutions is better, so solutions can be compared even if objectives are not numerical. When both solutions are feasible, or one is feasible and the other is not, this determination is straightforward. The situation is less clear when competing vectors are both infeasible. The part of Lampinen's criterion that decides which of two infeasible solutions is better is based on the idea of Pareto-dominance in the constraint function space, $\gamma'(\mathbf{x}) = \max(0, \gamma(\mathbf{x}))$ (see Sect. 4.6). If a feasible solution for the problem exists, the Pareto-optimal front in the effective constraint function space is a single point ($\gamma'(\mathbf{x}) = 0$).

In addition to being simple, Lampinen's constraint handling approach can reduce the computational effort spent evaluating vectors. Not only does the objective function not have to be evaluated when one or both vectors are infeasible, but a vector can also be rejected before all its constraint violations have been computed, thus saving time. Figure 4.23 depicts an efficient implementation of Lampinen's constraint handling method that exploits these two time-saving features.

Figure 4.24 provides an example of how the order in which functions are evaluated affects the total number of evaluations.

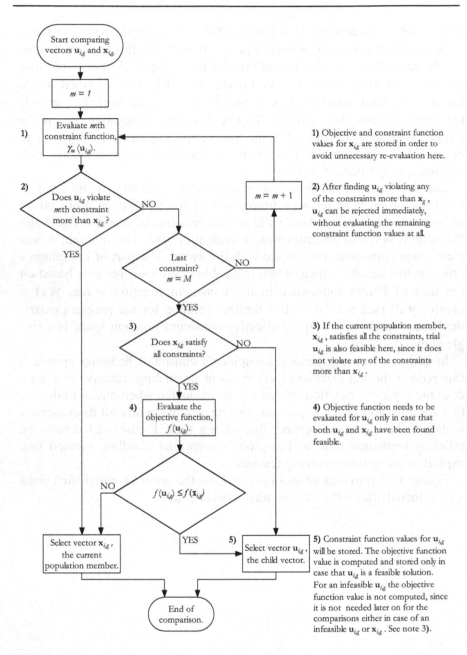

Fig. 4.23. The recommended way to implement Lampinen's selection scheme (Lampinen 2001, 2002)

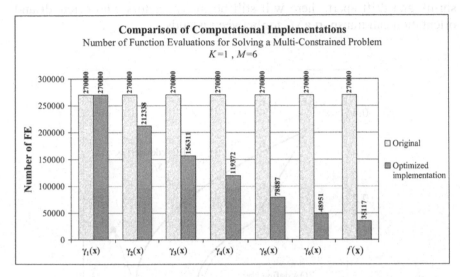

Fig. 4.24. The number of function evaluations to solve a multi-constrained problem, Problem no. 106 in Hock and Schittkowski (1981), with and without the optimized implementation described in Fig. 4.23. This problem has a single objective and is subject to six constraint functions. See Lampinen (2002) for details of the solution.

Direct Constraint Handling: An Example

Visualizing the shape of the region containing feasible vectors provides an insight into how DE operates when constraints are handled directly. In general, the shape of a *region of acceptability*, or ROA, in which all vectors satisfy a given set of constraints, is not known and may require considerable computational effort to determine. For the problem described by Eqs. 4.16–4.19, however, ROAs can be computed analytically. Figure 4.25 shows two ROAs corresponding to domains whose bounds are defined by infeasible trial and target vectors that satisfy a relaxed set of constraints. Based on Lampinen's criterion, the trial vector $\mathbf{u}_{i,g}$ will be selected because it falls inside the ROA of the target vector, $\mathbf{x}_{i,g}$.

The trial vector's chance of success depends on the shape of the target vector's ROA. As constraints are tightened, the ROA shrinks until it eventually splits into two disjoint islands (Fig. 4.26). Multiple feasible regions, like multiple local optima, make optimization more difficult. Nevertheless, DE's differential mutation scheme can handle the situation of split islands because its vector differentials adapt to the changing ROA. As islands shrink and drift apart, there will still be many vectors whose length and orientation can transport a vector between islands.

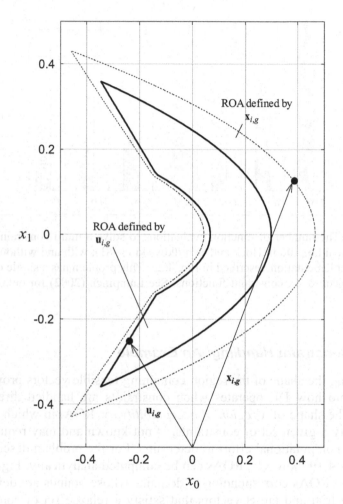

Fig. 4.25. Example for ROAs defined by the target and trial vectors. Since it falls inside the ROA defined by the target vector, $x_{i,g}$, the trial vector $u_{i,g}$ wins in this example.

Another example in which constraints partition the problem space into disjoint islands looks for the parts of an ellipse that are not shared by an overlapping circle. More precisely, the problem is to:

Find   (4.22)

$$\mathbf{x} = \{x_0, x_1\}, \quad \mathbf{x} \in \Re^D$$

subject to constraints

$$\gamma_1(\mathbf{x}) = 7 - \sqrt{x_0^2 + x_1^2} \leq 0$$

$$\gamma_2(\mathbf{x}) = -8 + \sqrt{x_0^2 + (2x_1)^2} \leq 0$$

and subject to boundary constraints

$$-10 \leq x_j \leq 10 \quad j = 1,2$$

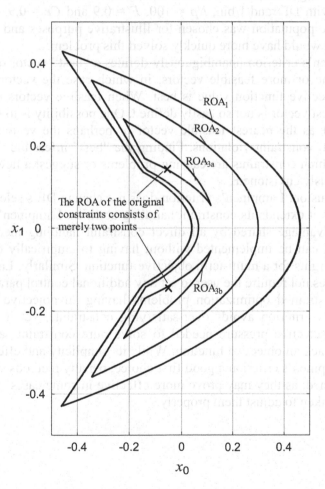

Fig. 4.26. The sequence ROA$_1$, ROA$_2$, ROA$_{3a}$ and ROA$_{3b}$ illustrates how the ROA changes as constraints are tightened.

Solutions lie inside the ellipse defined by $\gamma_2(\mathbf{x})$ and outside the circle defined by $\gamma_1(\mathbf{x})$ (see Fig. 4.27). In practice, multiple disjoint sets of feasible vectors can arise when constraints are nonlinear, as they are in Eq. 4.22. Figure 4.27 shows how the population evolves under Lampinen's criterion not only in the search space, but also in the constraint function space.

Since both the population and the feasible region(s) in Fig 4.27 are comparatively large, random initialization will usually generate at least one feasible solution. Despite this, Lampinen's method does not depend on the initial population containing any feasible solutions. Instead, the selective pressure it exerts drives vectors toward both feasible regions so that after only 40 generations, all vectors satisfy both constraints. Results were obtained with DE/rand/1/bin, $Np = 100$, $F = 0.9$ and $Cr = 0.9$. The relatively large population was chosen for illustrative purposes and a smaller population would have more quickly solved this problem.

Lampinen's criterion unambiguously defines a "best" vector only when there is one or more feasible vectors, in which case the vector with the lowest objective function value is best. When feasible vectors cannot be found, a best vector is not so easily defined. One possibility is to define the best vector as the nearest feasible vector or perhaps the vector with the lowest total constraint violations. Finding the "best" infeasible vector can indicate which constraints are causing problems or suggest a new solution not previously envisioned.

In conclusion, Lampinen's criterion does not change DE's selection rule as much as it extends its constraint handling abilities. Lampinen's method has the advantage shared by all direct constraint handling methods that constraints can be implemented without having to empirically determine penalty weights for a multi-term objective function. Similarly, Lampinen's method does not require the user to set any additional control parameters.

For constrained optimization problems having an objective function, Lampinen's criterion avoids over-satisfying constraints, yet it also provides the selective pressure needed to solve pure constraint satisfaction tasks that lack an objective function. While its simplicity and effectiveness make Lampinen's criterion a good first choice, penalty methods should not be abandoned, as they may prove more effective in some cases, especially if time is taken to adjust them properly.

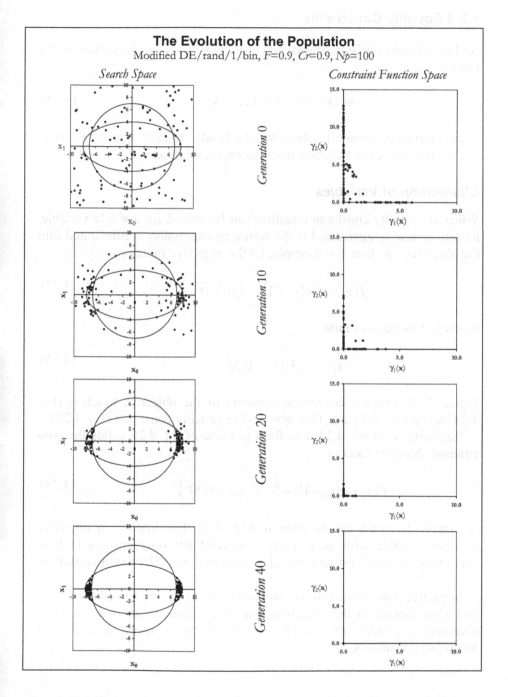

Fig. 4.27. DE's population evolves until the constraints in Eq. 4.21 are satisfied.

4.3.3 Equality Constraints

As Eq. 4.8 previously indicated, equality constraints can be written in the form

$$\varphi_n(\mathbf{x}) = 0, \quad n = 1,2,\ldots,N \tag{4.23}$$

If circumstances permit, the best way to handle equality constraints is to use them to eliminate variables from the objective function.

Elimination of Variables

When an equality constraint equation can be solved for a single variable, its satisfaction is guaranteed if the resulting expression is substituted into the objective function. For example, let the objective function

$$f(x_0, x_1) = (x_0 - 1)^2 + (x_1 - 2)^2 \tag{4.24}$$

be subject to the constraint

$$x_1 = 3 - 3 \cdot (x_0 - 0.5)^2. \tag{4.25}$$

Figure 4.28 plots the concentric contours of the objective function (Eq. 4.24) along with the curve that satisfies the equality constraint (Eq. 4.25).

Replacing x_1 in Eq. 4.24 with the right side of Eq. 4.25 yields the constrained objective function

$$f'(x_0) = (x_0 - 1)^2 + \left(1 - 3 \cdot (x_0 - 0.5)^2\right)^2 \tag{4.26}$$

the graph of which can be seen in Fig. 4.29. Eliminating an objective function variable with an equality constraint not only ensures that all vectors satisfy the constraint, but also reduces the problem's dimension by 1.

In pactice, not all equality constraint equations can be solved for a term that also appears in the objective function. When it cannot be used to eliminate a variable, an equality constraint can be recast as a pair of inequality constraints.

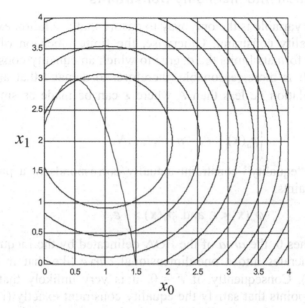

Fig. 4.28. The objective function's concentric contour lines (Eq. 4.24) and the line representing the parabolic constraint function (Eq. 4.25)

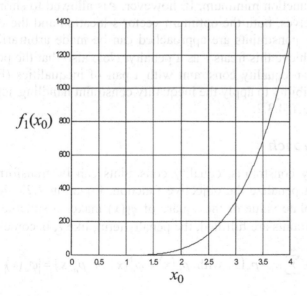

Fig. 4.29. A plot of the one-dimensional function in Eq. 4.26 whose minimum is 1.07131

Transformation into Inequality Constraints

Eliminating a variable is the only way to ensure that a solution exactly satisfies an equality constraint. Otherwise, the finite precision of floating-point number formats limits the degree to which an equality constraint can be satisfied. It is more reasonable, therefore, to demand that an equality constraint violation be less than ε, where ε can be made as small as desired:

$$|\varphi_n(\mathbf{x})| < \varepsilon, \quad n = 1,2,\dots,N. \tag{4.27}$$

This relaxed "equality" constraint actually corresponds to a pair of inequality constraints:

$$\varphi_n(\mathbf{x}) < \varepsilon \text{ and } \varphi_n(\mathbf{x}) > -\varepsilon. \tag{4.28}$$

As ε approaches 0, the *area* of the ROA delineated by the inequality constraints degenerates into a one-dimensional *curve* like that in Fig. 4.28 (Storn 1999a). Consequently, if $\varepsilon = 0$, it is very unlikely that DE will generate any points that satisfy the equality constraint exactly (i.e., within the floating-point format's limit of precision). As a result, population vectors are inevitably infeasible. Even if all parent vectors fall exactly on the equality constraint line, the children they create would almost never be feasible (unless the constraint was linear), rendering futile the search for the objective function minimum. If, however, ε is allowed to shrink as the population evolves, both the optimum vector's location and the degree to which equality constraints are approached can be made arbitrarily small. One way to achieve this treats ε as a penalty. Note also that the possibility of expressing an equality constraint with a pair of inequalities (Eq. 4.28) offers the possibility to apply the inequality constraint handling approaches described in Sect. 4.3.2.

Penalty Approach

Like inequality constraints, equality constraints can be transformed into cost terms that penalize the objective function. Equation 4.29 shows that either the absolute value or the square of $\varphi_n(\mathbf{x})$ makes a suitable penalty. When all constraints are fulfilled, the penalty term, like ε, becomes zero:

$$f'(\mathbf{x}) = f(\mathbf{x}) + \sum_{n=1}^{N} w_n \cdot p_n(\mathbf{x}) \text{ with } p_n(\mathbf{x}) = \varphi_n^2(\mathbf{x}) \text{ or } p_n(\mathbf{x}) = |\varphi_n(\mathbf{x})| \tag{4.29}$$

Using the constraint violation's absolute value, the penalized objective function for the problem set forth in Eqs. 4.24 and 4.25 becomes

$$f'(x_0, x_1) = (x_0 - 1)^2 + (x_1 - 2)^2 + w_1 \cdot \left| 3 - 3(x_0 - 0.5)^2 - x_1 \right| \qquad (4.30)$$

Figure 4.30 plots the contour lines for this function for $w_1 = 1$. DE has no particular difficulties solving this problem.

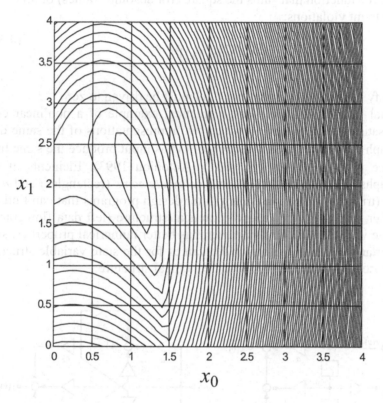

Fig. 4.30. Contour plot of Eq. (4.30)

Equality Constraint Satisfaction

Constraint satisfaction problems are characterized by their lack of an objective function. If all constraints are equality constraints, the problem reduces to solving a system of equations, $\varphi_n(\mathbf{x}) = 0$. When all constraint equations are linear, the simplex method (Hillier and Lieberman 1997) or interior point methods (Arbel 1993) like Karmarkar's algorithm (Karmarkar 1984) are both fast and reliable. (The simplex method referred to here is distinctly different from the simplex algorithm proposed by Nelder and Mead.) Once equality constraints become nonlinear, linear programming methods are not applicable. The fact that there is no general approach to solving systems of nonlinear equations makes DE a plausible alternative because it does not require nonlinear equations to be treated any differently than linear ones.

A pure constraint satisfaction problem (Eq. 4.27) can be recast into an "objective" function that sums the squares (or absolute values) of all equality constraint violations:

$$f'(\mathbf{x}) = \sum_{n=1}^{N} (\varphi_n(\mathbf{x}))^2. \tag{4.31}$$

To satisfy the equality constraints, $f'(\mathbf{x})$ must be driven to zero.

Digital filter design provides a practical example of a nonlinear constraint satisfaction problem in which two representations of the same data flowgraph that have the time sequence $s(v)$ as input produce the same time sequence $y(v)$ as output (Fig. 4.31) (Antoniou 1993). Elements of the flowgraphs are *adders* (circles), *unit delay elements* (rectangles) and *multipliers* (triangles). There are many filter design programs that can find the multiplier values for the second canonical structure (left data flowgraph). The state variable structure, however, has better numerical properties, so it is important to derive the multiplier values for the state variable structure from those computed for the second canonical structure.

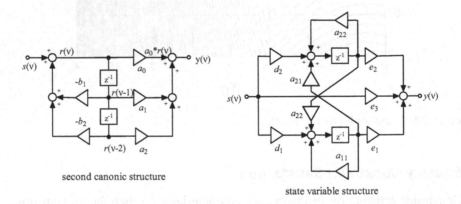

second canonic structure

state variable structure

Fig. 4.31. Two data circuits (flowgraph structures) that yield the same output $y(v)$ for the same input, $s(v)$

The following system of nonlinear equations describes the design problem:

$$\varphi_0 = -a_{11} - a_{22} - b_1 \qquad\qquad (4.32)$$
$$\varphi_1 = a_{11} \cdot a_{22} - a_{21} \cdot a_{12} - b_2$$
$$\varphi_2 = e_3 - a_0$$
$$\varphi_3 = -e_3 \cdot (a_{11} + a_{22}) + e_1 \cdot d_1 + e_2 \cdot d_2 - a_1$$
$$\varphi_4 = e_3 \cdot (a_{11} \cdot a_{22} - a_{21} \cdot a_{12}) + e_2 \cdot (d_1 \cdot a_{21} - d_2 \cdot a_{11})$$
$$+ e_1 \cdot (d_2 \cdot a_{12} - d_1 \cdot a_{22}) - a_2$$

For simplicity, no variables were eliminated from Eq. 4.32. Equation 4.31 was used instead.

To apply DE, each of the nine *state* variable multipliers a_{11}, a_{22}, a_{12}, a_{21}, d_1, d_2, e_1, e_2, e_3 is assigned to a vector component $x_j, j = 0, 1, \ldots, 8$. The objective function based on Eq. 4.31 will be zero if $\varphi_n = 0$ for $n = 1, 2, 3, 4$, 5. Since there are five known values and nine unknowns, the problem is under-specified and four of the nine unknowns can be preset, e.g., by setting both a_{11} and a_{22} to 0 and by setting e_3 to 0 and d_1 to 1. Presetting variables, however, is not entirely arbitrary. A poor presetting may render the resulting system of equations unsolvable (Hoefer 1987).

For example, let the numerical values input for the second canonical structure be

$a_0 = 0.015608325640383633$
$a_1 = 0.009443684818916196$
$a_2 = 0.01980858856448345$
$b_1 = -1.20703125$
$b_2 = 0.75250244140625.$

When there are no restrictions on coefficients in the state variable structure, DE found the following solution for state variable coefficients:

$d_1 \;=\; -0.050119$
$d_2 \;=\; 0.302834$
$e_1 \;=\; -0.128561$
$e_2 \;=\; 0.072084$
$e_3 \;=\; 0.015893$
$a_{11} = 0.387510$
$a_{12} = -0.622358$
$a_{21} = 0.698416$
$a_{22} = 0.819760.$

The convergence graph in Fig. 4.32 shows that DE/rand/1/bin with $Np = 30$, $F = 0.75$, $Cr = 1$ had no trouble finding a solution to this nonlinear system of equations.

Direct Equality Constraint Handling

Lampinen's method for direct constraint handling can also be extended to equality constraints by taking the absolute value of $\varphi_n(\mathbf{x})$ and selecting a trial vector that is not worse in any respect than the target vector:

$$
\mathbf{x}_{i,g+1} = \begin{cases} \mathbf{u}_{i,g} & \text{if } \begin{cases} \begin{cases} \forall n \in \{1,...,N\} : \left|\varphi_n(\mathbf{u}_{i,g})\right| \le \varepsilon \wedge \left|\varphi_n(\mathbf{x}_{i,g})\right| \le \varepsilon \\ \wedge \\ f(\mathbf{u}_{i,g}) \le f(\mathbf{x}_{i,g}) \end{cases} \\ \vee \\ \begin{cases} \forall n \in \{1,...,N\} : \left|\varphi_n(\mathbf{u}_{i,g})\right| \le \varepsilon \\ \wedge \\ \exists n \in \{1,...,N\} : \left|\varphi_n(\mathbf{x}_{i,g})\right| > \varepsilon \end{cases} \\ \vee \\ \begin{cases} \exists n \in \{1,...,N\} : \left|\varphi_n(\mathbf{u}_{i,g})\right| > \varepsilon \\ \wedge \\ \forall n \in \{1,...,N\} : \left|\varphi_n(\mathbf{u}_{i,g})\right| \le \left|\varphi_n(\mathbf{x}_{i,g})\right| \end{cases} \end{cases} \\ \mathbf{x}_{i,g} & \text{otherwise} \end{cases}
\tag{4.33}
$$

where $0 \le \varepsilon$

The constant ε in Eq. 4.33 may be set close to the floating-point precision limit.

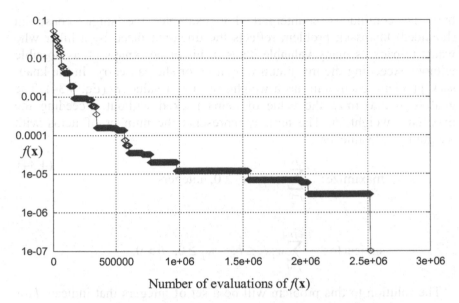

Fig. 4.32. Example for the convergence graph of the problem outlined in Fig. 4.31

4.4 Combinatorial Problems

In combinatorial problems, parameters can assume only a finite number of discrete states, so the number of possible vectors is also finite. The term "combinatorial optimization" is often equated with "discrete optimization on finite sets" even when a problem is the result of quantizing a highly constrained continuous problem (Babu and Munawar 2001; Du and Pardalos 1998; Press et al. 1992; Reeves 1993). Such problems are combinatorial (Du and Pardalos 1998; Corne et al. 1999) in a *wide sense* because they are viewed as "rearrangement problems" (Pahl and Damrath 2000). For example, picking the best optical glass for a telescope objective is a wide-sense combinatorial problem because a glass's optical properties (index of refection, dispersion, etc.) are continuous variables that are rendered discrete only because a limited number of glass types are commercially available. DE has solved wide-sense combinatorial problems (Babu and Munawar 2001; Storn 2000) in which discrete parameters are numerical and arithmetic operations are defined (see Sect. 4.2.5).

Many of the most familiar combinatorial problems, like the traveling salesman problem, the knapsack problem, the shortest-path problem, etc. (Syslo et al. 1983), are *strict-sense* combinatorial problems because they have no continuous counterpart. For example, the single constraint (bounded) knapsack problem reflects the dilemma faced by a hiker who wants to pack as many valuable items in his or her knapsack as possible without exceeding the maximum weight he or she can carry. In the knapsack problem, each item has a weight, w_j, and a value, c_j (Eq. 4.34). The goal is to maximize the value of items packed without exceeding the maximum weight, b. The term x_j represents the number of items with weight w_j and value, c_j:

$$\text{maximize:} \quad \sum_{j=0}^{j=D-1} c_j x_j, \quad x_j \geq 0, \text{ integers} \tag{4.34}$$

$$\text{subject to:} \quad \sum_{j=0}^{j=D-1} w_j x_j \leq b, \quad w_j \geq 0, \quad b > 0.$$

The solution to this problem will be a set of integers that indicate *how many* items of each type should be packed. As such, the knapsack problem is a strict-sense combinatorial problem because its parameters are discrete, solutions are constrained and it has no continuous counterpart (only a whole number of items can be placed in the knapsack).

In other strict-sense combinatorial problems, parameter values are discrete because they are symbolic. For example, in the board game of Scrabble, players are given seven randomly selected letters. Each letter has an associated numerical score. The game's objective is to find the combination of letters with the maximum score, subject to the constraint that letters form a real word that shares one or more letters with a word already on the board. In this strict-sense combinatorial problem, the objective function (word score) is numerical, but parameters are discrete, non-numerical letters whose combinations are highly constrained by dictionary entries and the existing board state.

For a problem like Scrabble, a parameter state is a letter whose meaning is understood in the context of language, not a number that measures quantity. Unless a symbolic combinatorial problem can be reformulated into one whose parameters measure quantity, DE is not likely to be effective because its mutation operator, like all numerical optimizers, relies on arithmetic operators (add, subtract, multiply, divide, modulo, etc.) rather than general data manipulation operations (swap, replace, move, etc.).

DE may prove effective on strict-sense, knapsack-like problems whose parameters measure quantity and whose constraints are not particularly restrictive, but its ability to optimize strict-sense problems also depends on the number and nature of any constraints. Strong constraints like those imposed in the traveling salesman problem make strict-sense combinatorial problems notoriously difficult for any optimization algorithm. In DE's case, the high proportion of infeasible vectors caused by constraints prevents the population from thoroughly exploring the objective function surface. To minimize the problems posed by infeasible vectors, algorithms can either generate only feasible solutions, or "repair" infeasible ones. The remainder of this chapter explores one approach that generates only feasible solutions and three others that rely to varying degrees on repair mechanisms. Each method proposes an analog of DE's differential mutation operator to solve the traveling salesman problem.

4.4.1 The Traveling Salesman Problem

The Traveling Salesman Problem (TSP) is a fairly universal, strict-sense combinatorial problem into which many other strict-sense combinatorial problems can be transformed (Syslo et al. 1983; Dolan and Aldous 1993). Consequently, many findings about DE's performance on the TSP can be extrapolated to other strict-sense combinatorial problems.

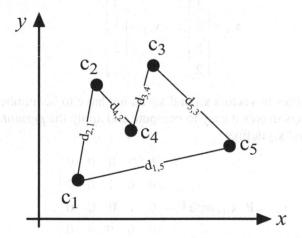

Fig. 4.33. An example of a five-city TSP. The tour indicated is one possible solution.

In the TSP, a salesman wants to minimize his travel expenses by finding the shortest route that visits each city in his territory just once. More generally, let there be M cities c_i, $i = 1, 2, ..., M$, each of which is at distance $d_{i,j} = d_{j,i}$ to some other city c_j, j not equal to i. The total distance for the tour is

$$\Theta = \sum_{i=1}^{M} d_{i,j} , \; j \neq i \qquad (4.35)$$

Figure 4.33 shows an example of a five-city tour.

4.4.2 The Permutation Matrix Approach

The basic idea behind DE is that two vectors define a difference that can then be added to another vector as a mutation. The same idea transfers directly to the realm of permutations, or the permutation group. Just as two vectors in real space define a difference vector that is also a vector, two permutations define a mapping that is also a permutation. This generalization of DE can be applied to the TSP because if cities are labeled with natural numbers, a valid tour is just a permutation of the sequence (1, 2, ..., M). For example, x_{r1} and x_{r2} in Eq. 4.36 encode tours, each of which is a permutation (M = 5):

$$\mathbf{x}_{r1} = \begin{pmatrix} 1 \\ 3 \\ 4 \\ 5 \\ 2 \end{pmatrix}, \quad \mathbf{x}_{r2} = \begin{pmatrix} 1 \\ 4 \\ 3 \\ 5 \\ 2 \end{pmatrix}. \qquad (4.36)$$

Labels for cities in vectors x_{r1} and x_{r2} do not have to be numbers, but using numerical tags makes it easy to compute and apply the *permutation matrix*, **P**, that x_{r1} and x_{r2} define:

$$\mathbf{x}_{r2} = \mathbf{P} \cdot \mathbf{x}_{r1}, \text{with } \mathbf{P} = \begin{pmatrix} 1 & 0 & 0 & 0 & 0 \\ 0 & 0 & 1 & 0 & 0 \\ 0 & 1 & 0 & 0 & 0 \\ 0 & 0 & 0 & 1 & 0 \\ 0 & 0 & 0 & 0 & 1 \end{pmatrix}. \qquad (4.37)$$

Like DE's difference vector, the randomly derived permutation matrix can permute a third randomly chosen vector into a mutant. Because all operations are permutations, mutants are always feasible solutions, i.e., all are valid tours that visit each city just once.

Figure 4.34 outlines an algorithm that scales the effect of the permutation matrix. Setting $\delta = 1$ leaves the permutation matrix unchanged, while

$\delta = 0$ reduces it to diagonal form. Intermediate values perform a fraction of the permutation defined by **P**.

```
...
for (i=1; i<M; i++)//search all columns of P
{
    if (element p(i,i) of P is 0) //1 not on diagonal
    {
        if (rand() > δ)  //if random number ex [0,1] exceeds δ
        {
            j=1; //find row where p(j,i) = 1
            while(p(j,i) != 1) j++;
            swap_rows(i,j);
        }
    }
}
...
```

Fig. 4.34. Algorithm to apply the factor δ to the difference permutation, **P**

In practice, this approach tends to stagnate because moves derived from the permutation matrix are seldom productive. In addition, this method is unable to distinguish rotated but otherwise equal tours. Because they display a unique binary signature, equal tours can be detected by other means, although this possibility is not exploited in the algorithm described in Fig. 4.34. This scheme is similar to the one described in Ruettgers (1997) who used it with some success on scheduling problems.

In part, a permutation takes a value from one parameter and copies it into a parameter with a different index. Like ordinary vector addition, traditional DE only combines values from parameters having the same index. The following two approaches resemble traditional DE because they both use vector addition, although their ultimate effect is to shuffle values between parameters, i.e., generate permutations.

4.4.3 Relative Position Indexing

Another way to guarantee that mutant trial tours are (almost) always valid is to transform parameters into the floating-point interval [0,1], perform mutation and then convert mutant parameters back into the integer domain using "relative position indexing" (Lichtblau 2002a). The first step in this simulation of DE mutation is to divide each parameter by the vector's largest element, in this case, $M = 5$.

$$\mathbf{x}_{r1,f} = \frac{\mathbf{x}_{r1}}{5} = \begin{pmatrix} 0.2 \\ 0.6 \\ 0.8 \\ 1 \\ 0.4 \end{pmatrix} \quad \text{and} \quad \mathbf{x}_{r2,f} = \frac{\mathbf{x}_{r2}}{5} = \begin{pmatrix} 0.2 \\ 0.8 \\ 0.6 \\ 1 \\ 0.4 \end{pmatrix}. \tag{4.38}$$

The subscript, f, denotes a vector's floating-point representation. After a third vector is chosen and normalized

$$\mathbf{x}_{r3} = \begin{pmatrix} 5 \\ 2 \\ 1 \\ 4 \\ 3 \end{pmatrix} \rightarrow \mathbf{x}_{r3,f} \begin{pmatrix} 1 \\ 0.4 \\ 0.2 \\ 0.8 \\ 0.6 \end{pmatrix}, \tag{4.39}$$

the mutation is applied:

$$\mathbf{v}_f = \mathbf{x}_{r3,f} + F \cdot \left(\mathbf{x}_{r1,f} - \mathbf{x}_{r2,f} \right)^{F=0.85} = \begin{pmatrix} 1 \\ 0.4 \\ 0.2 \\ 0.8 \\ 0.6 \end{pmatrix} + 0.85 \cdot \begin{pmatrix} 0 \\ -0.2 \\ 0.2 \\ 0 \\ 0 \end{pmatrix} = \begin{pmatrix} 1 \\ 0.23 \\ 0.37 \\ 0.8 \\ 0.6 \end{pmatrix}. \tag{4.40}$$

The floating-point mutant vector, $\mathbf{v}_f$, is then transformed back into the integer domain by assigning the smallest floating value (0.23) to the smallest integer (1), the next highest floating value (0.37) to the next highest integer (2), and so on to obtain

$$\mathbf{v}_f = \begin{pmatrix} 1 \\ 0.23 \\ 0.37 \\ 0.8 \\ 0.6 \end{pmatrix} \rightarrow \mathbf{v} = \begin{pmatrix} 5 \\ 1 \\ 2 \\ 4 \\ 3 \end{pmatrix}. \tag{4.41}$$

This backward transformation, or "relative position indexing", always yields a valid tour except in the unlikely event that two or more floating-point values are the same. When such an event occurs, the trial vector must be either discarded or repaired.

This method looks attractive at first sight and the results are reasonable, albeit not competitive with special-purpose TSP-solvers (Lichtblau 2002a, 2002b). A closer look, however, reveals that DE's mutation scheme together with the forward and backward transformations is, in essence, a shuffling generator. In addition, this approach does not reliably detect identical tours because the difference in city indices has no real significance. For example, vectors with rotated entries, e.g., (2, 3, 4, 5, 1) and (1, 2, 3, 4, 5), are the same tour, but their difference, e.g., (1, 1, 1, 1, –4), is not zero.

4.4.4 Onwubolu's Approach

Like Lichtblau, Onwubolu and Babu (2004) label cities with integral numerical indices, transform vectors into the real domain, manipulate them and then transform them back into the integer domain. Onwubolu and Babu defined the forward transformation of city indices into the continuous domain as

$$x_i' = -1 + x_i \cdot (1 + \varepsilon) \tag{4.42}$$

where ε is a small number. The backward transformation is defined as

$$x_i = \mathrm{round}\left[(1 + x_i') \cdot (2 - \varepsilon)\right] \tag{4.43}$$

where the round() function rounds the argument to the nearest integer.

Like the previous two methods, this approach impedes DE's self-steering mechanism because it fails to recognize rotated tours as equal. In addition, Onwubolu and Babu's method usually generates invalid tours that must be repaired. Even though competitive results are reported in Onwubolu and Babu (2004) and Onwubolu (2003), there is reason to believe that the success of this approach is primarily a consequence of prudently chosen local heuristics and repair mechanisms, not DE mutation.

4.4.5 Adjacency Matrix Approach

When tours are encoded as city vectors, the difference between rotated but otherwise identical tours is never zero. Rotation, however, has no effect on a tour's representation if it is encoded as an adjacency matrix. An adjacency matrix is a symmetric, $M \times M$ matrix matrix in which the entry in row i and column j is the number of connections from city i to city j (Dolan and Aldous 1993). The tour in Fig. 4.33 generates the adjacency matrix:

$$\mathbf{A}_1 = \begin{pmatrix} 0 & 1 & 0 & 0 & 1 \\ 1 & 0 & 0 & 1 & 0 \\ 0 & 0 & 0 & 1 & 1 \\ 0 & 1 & 1 & 0 & 0 \\ 1 & 0 & 1 & 0 & 0 \end{pmatrix}. \tag{4.44}$$

For example, city c_1, which corresponds to both row 1 and column 1 of $\mathbf{A}_1$, is connected to city c_2 and to city c_5 because there are ones in the second and fifth columns of row 1, as well as in the second and fifth rows of column 1.

Because the TSP allows each city to be visited only once, the elements of $\mathbf{A}_1$ must be either ones or zeros. More particularly, there must be exactly two ones in each row and in each column. In addition, the main diagonal of the adjacency matrix must be zero because there is no route from a city to itself. An adjacency matrix that satisfies the above requirements constitutes a valid TSP matrix.

Like the permutation matrix, the adjacency matrix is a semi-numerical, logical operator whose binary elements can be computed by comparing either numeric or non-numeric city symbols. Since matrix entries are zeros and ones, differences must be taken in the finite field GF(2), or Galois Field 2. All values in GF(2) are either 0 or 1 and all arithmetic operations are performed modulo 2, meaning that 2 is added to, or subtracted from, each computed result until only 0 or 1 remains. Equation 4.45 shows that addition and subtraction are the same in GF(2):

$$\begin{aligned} (0+0)\bmod 2 &= 0 \\ (0+1)\bmod 2 &= 1 \\ (1+1)\bmod 2 &= 0 \\ (0-0)\bmod 2 &= 0 \\ (0-1)\bmod 2 &= 1 \\ (1-1)\bmod 2 &= 0 \end{aligned} \tag{4.45}$$

The notation

$$(x+y)\bmod 2 = x \oplus y \tag{4.46}$$

is shorthand for modulo 2 addition, also known as the "exclusive or" logical operation.

The difference matrix $\mathbf{\Delta}_{i,j}$

$$\Delta_{i,j} = \mathbf{A}_i \oplus \mathbf{A}_j \qquad (4.47)$$

is the analog of DE's traditional difference vector. For example, given the valid TSP matrices $\mathbf{A}_1$ and $\mathbf{A}_2$,

$$\mathbf{A}_1 = \begin{pmatrix} 0 & 1 & 0 & 0 & 1 \\ 1 & 0 & 0 & 1 & 0 \\ 0 & 0 & 0 & 1 & 1 \\ 0 & 1 & 1 & 0 & 0 \\ 1 & 0 & 1 & 0 & 0 \end{pmatrix}, \quad \mathbf{A}_2 = \begin{pmatrix} 0 & 1 & 0 & 0 & 1 \\ 1 & 0 & 1 & 0 & 0 \\ 0 & 1 & 0 & 1 & 0 \\ 0 & 0 & 1 & 0 & 1 \\ 1 & 0 & 0 & 1 & 0 \end{pmatrix}, \qquad (4.48)$$

their difference is

$$\Delta_{1,2} = \mathbf{A}_1 \oplus \mathbf{A}_2 = \begin{pmatrix} 0 & 0 & 0 & 0 & 0 \\ 0 & 0 & 1 & 1 & 0 \\ 0 & 1 & 0 & 0 & 1 \\ 0 & 1 & 0 & 0 & 1 \\ 0 & 0 & 1 & 1 & 0 \end{pmatrix}. \qquad (4.49)$$

From Eq. 4.49 it is apparent that $\Delta_{i,j}$ itself need not be a TSP matrix. For example, the first row and column of a valid TSP matrix cannot be all zeros.

Rotated but otherwise equal tours generate identical adjacency matrices. Consequently, the difference matrix between equal but rotated tours is always zero. Figure 4.35 shows that when two tours are not equal, the difference matrix drops connections that are common to both $\mathbf{A}_1$ and $\mathbf{A}_2$.

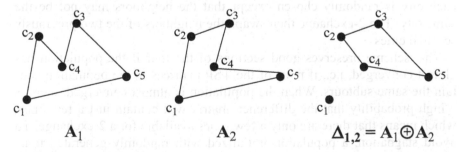

Fig. 4.35. Example for the difference matrix $\Delta_{i,j}$ and its graphical interpretation

The TSP's tight constraints make it unlikely that adding a difference matrix to a valid TSP base matrix will produce an adjacency matrix that satisfies the TSP's requirements. Consequently, invalid TSP matrices must be repaired to ensure that each city is connected to exactly two others. One possible repair mechanism is based on the bounce-back method of parameter constraint handling (Sect. 4.3.1). In particular, if adding the difference matrix to a randomly chosen TSP base matrix does not yield a valid TSP trial matrix, then this trial matrix is discarded, and a "2-exchange" (Syslo et al 1983) is performed on the TSP base matrix (Fig. 4.36) instead.

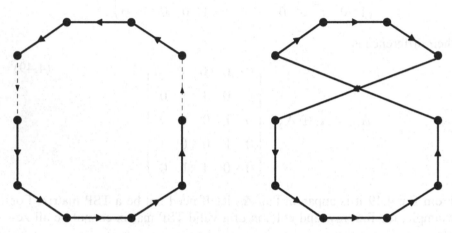

Fig. 4.36. The "2-exchange" heuristic mutation

The 2-exchange reconnects two cities in a different way. First, two cities are randomly selected except that both must be vertices of the difference matrix and not connected to each other. Next, an immediate neighbor of each city is randomly chosen except that the neighbors may not be the same city. The 2-exchange then swaps the neighbors of the two previously selected cities.

This scheme preserves good sections of the tour if the population has almost converged, i.e., if most of the TSP matrices in the population contain the same subtours. When the population is almost converged, there is a high probability that the difference matrix will contain just a few ones, which means that there are only a few cities available for a 2-exchange. To avoid stagnation, a population initialized with randomly generated tours must be large enough so that every city can be selected for a 2-exchange by a difference vector. This minimum population size depends on the probability that a city will be isolated in the difference matrix because two

adjacent city connections are the same. To reduce the probability of stagnation, entirely random 2-exchange moves are allowed with a probability 1
$- p$, $p \in [0, 1]$. Figure 4.37 shows that this method is competitive with
simulated annealing.

The dominant move in this scheme is the 2-exchange repair algorithm,
not the application of the difference matrix. This claim can easily be verified by running the algorithm without the 2-exchange repair algorithm
which quickly leads to stagnation. As such, selection is the only aspect of
DE in this technique that makes a significant contribution to resolving the
TSP. This is to be expected because the TSP is so heavily constrained that
a general-purpose mutation scheme is very unlikely to generate valid tours.

4.4.6 Summary

Although DE has performed well on wide-sense combinatorial problems, its suitability as a combinatorial optimizer is still a topic of considerable debate and a definitive judgment cannot be given at this time. Although the DE mutation concept extends to other groups, like the
permutation group, there is no empirical evidence that such operators are
particularly effective. Similarly, tagging what is fundamentally symbolic
data with numerals makes it possible to implement DE-style mutation operations, but the resulting differentials reflect arbitrary labeling choices,
not inherent metrical relationships or the population's correlation with the
objective function surface. Most of the gains seen when DE-style operators
are invoked in these circumstances can be traced to repair mechanisms and
DE's elitist selection scheme. More generally, the particular nature of
strict-sense combinatorial problems – their constraints and use of symbols
or numerical values – is more important in determining DE's success than
is the fact that they are strict-sense problems.

Certainly in the case of the TSP, the most successful strategies for the
TSP continue to be those that rely on special heuristics (Onwubolu 2003;
Michalewicz and Fogel 2000; Freisleben and Merz 1996; Dueck 1993).

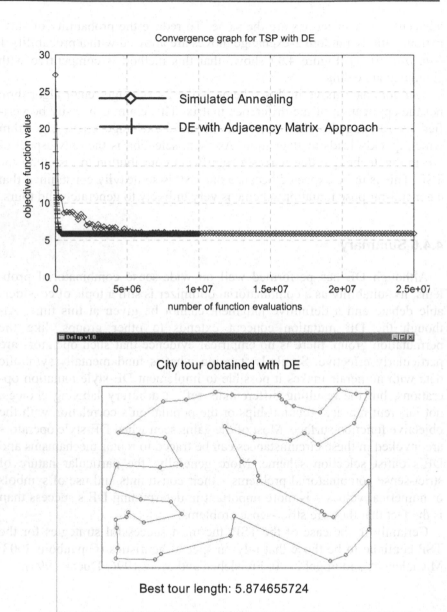

Fig. 4.37. Convergence graph (above) for a relatively small, 50-city TSP (below) that was solved with simulated annealing (Press et al. 1992) and with the adjacency matrix approach. The final tour length for SA was 5.8835 and 5.875 for DE. Control parameters for DE were $Np = 800$ and $p = 0.6$.

4.5 Design Centering

Design centering is a constraint satisfaction problem whose goal is to find the solution, which, when perturbed, remains feasible more often than any other solution. Such problems are common in manufacturing where imperfections in production processes are inevitable. For example, analog electronic circuit design relies on components like resistors and capacitors whose actual values inevitably differ from their stated values (Zhang and Styblinski 1995). Similarly, mechanical designers must contend with shape variations due to machine imprecision or tempering processes. Design centering also plays an important role in silicon chip manufacturing where processes cannot be perfectly controlled. Design centering maximizes production yield by finding the point least likely to produce an infeasible vector when manufacturing variations are taken into account.

When perturbed by the probability density function (PDF) that characterizes the production process, the vector that yields the most designs satisfying all constraints is the *design center*, x_0. Mathematically, design centering is the task of finding $x = x_0$ such that

$$\int_{ROA} PDF(x_0)dx_0 = \text{maximum}. \tag{4.50}$$

Loosely speaking, design centering tries to find the most interior point, for example, by finding the point that maximizes the minimum distance to the ROA's rim. This problem is non-trivial since the ROA is generally unknown. In addition, the PDF that models vector deviations plays an important role in defining the shape whose design center is sought. For example, the largest hypersphere that can be inscribed inside the ROA is not necessarily the best shape to represent the effects of perturbations.

4.5.1 Divergence, Self-Steering and Pooling

Populations driven by DE exhibit three properties that facilitate design centering: *divergence*, *self-steering* and *pooling*. Divergence is the tendency of the population's variance to increase over time when there is no selective pressure. *Self-steering* is the ability to adapt step sizes and orientations in response to the ROA's shape. *Pooling* describes the population's persistence in areas within the ROA where trial vectors have the greatest chance of survival.

Populations evolve within the ROA the same way that they do on objective function plateaus. Just as all vectors situated on a plateau have the same objective function value, those that inhabit the ROA are all considered equally feasible. In both cases, there is no selective pressure other than that exerted by trial vectors replacing their equally feasible targets. Since differential mutation inevitably places some trial vectors within the ROA but beyond the hull that encloses the current generation, the population expands if F is not too small (see Fig. 4.38). If, however, F is a normally distributed random variable that is below Zaharie's limit, the population will converge even if feasible trial vectors always replace feasible target vectors (see Sect. 2.5.1) (Zaharie 2002). Otherwise, vectors diverge, their differentials become longer and the expansion accelerates. Only the ROA's boundary halts the expansion because infeasible vectors cannot replace feasible ones. Figure 4.39 shows how divergence quickly disperses a highly localized initial population.

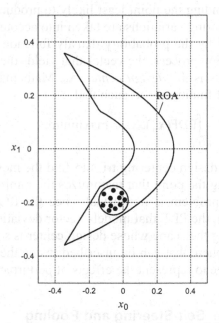

Fig. 4.38. Some of the feasible trial vectors that replace feasible target vectors will lie beyond the current population's boundary causing vectors to diverge over time if F is above Zaharie's limit.

While divergence helps DE quickly explore the ROA's full extent, self-steering adapts the rate of divergence to the ROA's shape. The second picture in Fig. 4.39 shows that as vectors conform to the "racket handle" shape, the differentials they generate reinforce the population's tendency to spread horizontally. Coupled with divergence, self-steering allows vectors to quickly escape from the racket handle. Once the population enters the

ROA's open area, steps begin to grow vertically while they continue to expand horizontally.

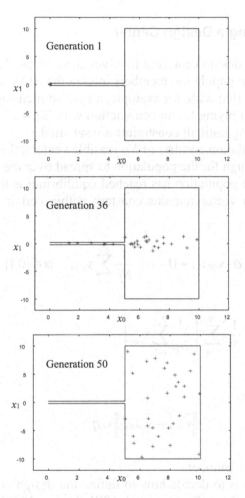

Fig. 4.39. Divergence, self-guiding and pooling of the vector population. Results were generated using DE/rand/1/bin with $Np = 30$, $F = 0.9$ and $Cr = 0.5$.

After the population expands into the ROA's open area, pooling keeps it there. Because vectors in the ROA's most expansive region are the points most likely to create a feasible trial vector, there is a natural tendency for vectors to pool there. As the final picture in Fig. 4.39 shows, tightly constrained areas like the racket handle are abandoned as vectors pool in a less restricted environment. The program, `racket.exe`, on the accompanying CD allows experimenting with the divergence, self-guiding and pooling effects.

4.5.2 Computing a Design Center

Applying DE to design centering involves three steps. The first step is to ensure that all Np population members inhabit the ROA, either by perturbing a known solution with, for example, a multi-dimensional Gaussian distribution, or by applying DE in conjunction with Eq. 4.21 (Lampinen's criterion) and waiting until all constraints are satisfied.

Once the population consists of Np feasible vectors, the second step is to run DE long enough for the population to spread over the entire ROA. One indicator that the population has reached equilibrium is that the time average of the mean vector remains constant within certain bounds, i.e., the time average

$$\overline{\mathbf{x}}_{av,k} = \alpha \cdot \overline{\mathbf{x}}_{av,k-1} + (1-\alpha) \cdot \frac{1}{Np} \sum_{i=1}^{Np} \mathbf{x}_{i,g}; \quad \alpha \in [0,1] \tag{4.51}$$

or

$$\overline{\mathbf{x}}_{av,k} = \frac{1}{N} \sum_{g=k}^{k+N-1} \left(\frac{1}{Np} \sum_{i=1}^{Np} \mathbf{x}_{i,g} \right)$$

should satisfy

$$\left\| \overline{\mathbf{x}}_{av,k} - \overline{\mathbf{x}}_{av,k-1} \right\| < \eta \tag{4.52}$$

where η is a small number.

The third step is to decide how to define the design center. For convex ROAs, several authors (Brayton et al. 1981; Lueder 1990; Sapatnekar et al. 1994) argue that it makes sense to use the average value itself as the design center. If, however, the ROA is non-convex, the mean vector can fall into an infeasible zone. Figure 4.39 illustrates a non-convex ROA.

Another measure that provides a rough estimate of the design center is the point that maximizes the *center index*:

$$c_j = \sum_{i=1}^{Np} \exp(-d_{ij}), \quad j = 1,2,...,Np, \quad d_{ij} = |x_i - x_j|. \tag{4.53}$$

Equation 4.52 assumes that all vectors are feasible. The center index is based on the idea that if feasible vectors are uniformly distributed, the ROA's most interior points also have the nearest neighbors. Figure 4.40 shows the center index for the problem defined in Eqs. 4.16–4.19 except that bounds have been changed from 1.04, 0.4, 0.8 to 1.1, 0.5, 0.4, respectively. The small circle shows that the design center is well placed even though the ROA is non-convex. It has to be admitted, however, that this method for estimating the design center is still in its infancy. Program code for this example can be found on this book's accompanying CD.

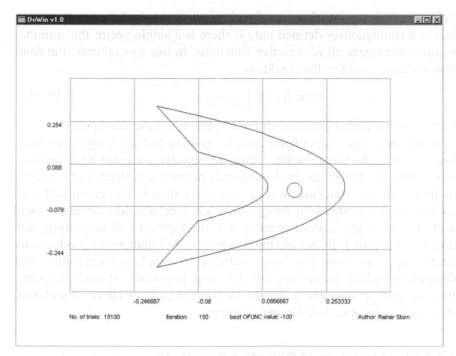

Fig. 4.40. The design center for ROA described by Eqs. 4.16–4.19 but with different constraints (1.1, 0.5, 0.4). Here, the design center is based on maximizing the center index (Eq. 4.53). Results were generated using DE/rand/1/bin with $Np = 100$, $Cr = 1$ and $F = 0.75$ with jitter: $F_j = 0.75 + \text{rand}_j((0,1) - 0.5) \cdot 0.0001$.

Design centers are less than optimal to the extent that neither the mean vector nor the center index takes into account the PDF to which parameters are subject (except to assume that it is uniform). While DE holds promise as a design centering method, more research is needed to develop an effective design centering implementation.

4.6 Multi-Objective Optimization

Multi-objective optimization attempts to simultaneously minimize K individual objective functions. The goal, therefore, is to

$$\text{Find} \tag{4.54a}$$

$$\mathbf{x} = (x_0, x_1, \ldots, x_{D-1})^T, \quad \mathbf{x} \in \mathfrak{R}^D$$

to minimize

$$f_k(\mathbf{x}), \quad k = 1, \ldots, K, \quad K \geq 2.$$

$\mathfrak{R}^D$ is the D-dimensional space of real numbers. The solution to this problem is unambiguously defined only if there is a single vector that simultaneously minimizes all K objective functions. In this special case, the solution vector, $\mathbf{x}$, satisfies the condition

$$\forall k \in \{1, \ldots, K\}: \mathbf{x} = \mathbf{x}_k^* \tag{4.54b}$$

where $\mathbf{x}_k^*$ is a global optimum of the kth objective function, $f_k(\mathbf{x})$.

   In practice, objectives often conflict, meaning that all K objective function extremes do not coincide. When all objectives cannot be simultaneously minimized, a single, best solution is not easily defined. For example, quality control technicians must inspect every item to be certain that each meets design tolerances. In many cases, however, absolute certainty about a product's quality is not necessary and the high cost of inspecting each unit is not justified. In this example, low cost and defect-free products are conflicting objectives. The "best" solution is usually a compromise that depends on which objectives are the most important. If each objective function can be assigned a weight that indicates its relative importance, then the best solution will be unambiguous.

4.6.1 Weighted Sum of Objective Functions

Minimizing a *weighted sum* of objective functions transforms a multi-objective optimization problem into one with a single objective to which DE or any other suitable optimizer can be readily applied. Summing weighted objective functions reduces the multi-objective goal to

Find   (4.55)

$$\mathbf{x} = (x_0, x_1, \ldots, x_{D-1})^\mathrm{T}, \quad \mathbf{x} \in \mathfrak{R}^D$$

to minimize

$$f'(\mathbf{x}) = \sum_{k=1}^{K} w_k f_k(\mathbf{x}), \quad K \geq 2.$$

The symbol, w_k, denotes the kth objective function weight.

Multi-objective optimization methods can be classified based on how they assign weights, i.e., how they articulate preferences (Hwang and Abu Syed 1979; Miettinen 1998; Deb 2001; Coello Coello et al. 2002).

- *A priori*. Weights are assigned prior to optimization based on expert knowledge.
- Progressive. An expert changes weights during optimization based on feedback from an updated set of solutions.
- *A posteriori*. Once a set of candidate solutions has been found, an expert selects one result and by so doing, implicitly specifies a set of weights.

A priori preference articulation assumes that objective preferences can be ordered and that weights do not change during optimization. If, however, there are many conflicting objectives, ordering them may be difficult. So that weights have the desired impact, objective functions may first have to be normalized to compensate for their different dynamic ranges. Determining the appropriate normalization scale factor can be difficult because it requires knowing the range of function extremes – knowledge that the optimization itself is supposed to provide. Without additional information about other potentially effective solutions, the intricacies of weight selection and normalization often make the *a priori* approach impractical.

Progressive preference articulation is more flexible than *a priori* weight selection because it exploits knowledge gleaned while the optimization is in progress. By periodically monitoring the optimization, an expert can change weights and affect corrections. Although progressive preference articulation represents a level of refinement over *a priori* weight selection, the expert's biases can inadvertently steer the population to what is ultimately seen as an undesirable compromise.

Searching for a single result that minimizes a weighted sum of objective functions excludes from consideration the many compromise solutions that may also be viable. Instead of imposing a set of weights, experts often want to find a set of competing solutions without biasing the result by either *a priori* or progressive preference articulation. In *a posteriori* preference articulation, experts implicitly apply weights by selecting one solution from a set of equally compelling final possibilities. Algorithms that employ *a posteriori* weight preference are typically based on Pareto-optimality.

4.6.2 Pareto Optimality

The concept of a *Pareto-optimum* was first introduced by the engineer/economist Vilfredo Pareto (Pareto 1886). The Pareto-optimization approach to multi-objective optimization can be characterized as follows: "the term 'optimize' in a multi-objective decision making problem refers to a solution around which there is no way of improving any objective without worsening at least one other objective" (Palli et al. 1998).

The central concept of Pareto-optimization is *Pareto-dominance*. A vector of objective function values *dominates* another if none of its objective values are higher and at least one is lower. More specifically, let vectors $\mathbf{y}$ = $\{f_1(\mathbf{x}), f_2(\mathbf{x}), ..., f_K(\mathbf{x})\}$ and $\mathbf{y}^* = \{f_1(\mathbf{x}^*), f_2(\mathbf{x}^*), ..., f_K(\mathbf{x}^*)\}$ be points in the objective function space $\mathbf{y} \in O \subset \Re^K$. Each vector's components are the objective function values of an associated point in the parameter or decision space, $\mathbf{x} \in S \subset \Re^D$. Mathematically, vector $\mathbf{y}^*$ dominates vector $\mathbf{y}$ *iff* $\mathbf{y}^*$ is *partially less than* $\mathbf{y}$, i.e., if

$$\forall k \in \{1,...,K\}: f_k(\mathbf{y}^*) \leq f_k(\mathbf{y}) \land \exists k \in \{1,...,K\}: f_k(\mathbf{y}^*) < f_k(\mathbf{y}). \qquad (4.56)$$

A solution that is not dominated by any other feasible solution is called *Pareto-optimal*, or *strongly efficient*. More precisely, a solution $\mathbf{x}^* \in S$ is Pareto-optimal if there is no other vector, $\mathbf{x} \in S$, whose objective function vector dominates that of $\mathbf{x}^*$. Thus, Pareto-dominance is a relationship between vectors in the objective function space, O, *not* the parameter space, S.

The *Pareto-front* is the hypersurface within the objective function space, O, that is defined by the set of all strongly efficient solutions. As such, the Pareto-front is a set of "best compromise" solutions that cannot be dominated – no objective can be improved without making some other objective worse. Armed with such a set of solutions, an expert can learn how much improving one objective worsens the others before picking one solution from the non-dominated set.

4.6.3 The Pareto-Front: Two Examples

Equation 4.57 outlines a two-dimensional, multi-objective optimization problem with two conflicting objectives:

Find   (4.57)

$$\mathbf{x} = (x_0, x_1)^{\mathrm{T}}, \quad \mathbf{x} \in \mathfrak{R}^2$$

to minimize

$$f_1(\mathbf{x}) = x_0^2 + x_1$$

$$f_2(\mathbf{x}) = x_0 + x_1^2$$

subject to bounds

$$-10 \leq x_0, x_1 \leq 10$$

Figure 4.41 plots 200 *non-dominated* solutions to Eq. 4.57 and one *dominated* solution that is outperformed by many other points with respect to both objectives. The distribution of non-dominated points in the objective function space approximates the true Pareto-front (Fig. 4.41, right).

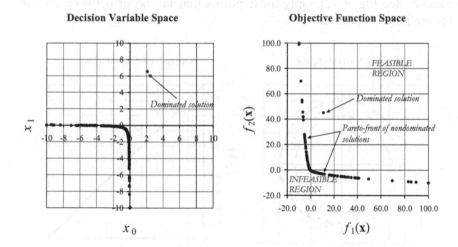

Fig. 4.41. Plots of 200 non-dominated solutions for the bi-objective problem in Eq. 4.57 in both the decision variable and objective function spaces. The Pareto-front separates feasible solutions from infeasible ones. Non-dominated points were generated with the modified version of DE that is described in the next section.

In this example, solutions also form a front when they are plotted in the decision variable space (Fig. 4.41, left). The distribution of Pareto-optimal points within the parameter space is a consequence of the particular way in which Eq. 4.57 associates points in the parameter space with their counterparts in the objective function space. Depending on the mapping, Pareto-optimal points may not form a front at all. In the following two-variable, two-objective optimization problem, solutions in the decision space are not distributed on a front:

$$\text{Find} \qquad\qquad\qquad\qquad\qquad\qquad (4.58)$$

$$\mathbf{x} = \left(x_0, x_1\right)^\mathrm{T}, \quad \mathbf{x} \in \mathfrak{R}^2$$

to minimize

$$f_1(\mathbf{x}) = \left| 6 - \sqrt{x_0^2 + x_1^2} \right|$$

$$f_2(\mathbf{x}) = \left| 8 - \sqrt{x_0^2 + x_1^2} \right|$$

subject to bounds

$$-10 \le x_0, x_1 \le 10$$

The dual objective in Eq. 4.58 is to simultaneously minimize the distance to each of two concentric circles, one with radius 6, the other with radius 8 (see Fig. 4.42). Only those points that fall on or between the circles are Pareto-optimal.

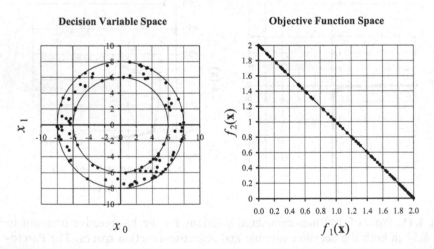

Fig. 4.42. A set of 100 different Pareto-optimal solutions for the bi-objective example problem Eq. 4.58. Only points on or between the circles are Pareto-optimal. The modified version of DE described in the next section generated the non-dominated set.

For any point that lies either inside or outside both circles, there will always be a solution that improves one objective without worsening the other. For example, (0, 5.9) is not a Pareto-optimal solution because it is dominated by the Pareto-optimal point (0, 6.1). Inspection shows that the point (0, 6.1) is Pareto-optimal because there is no point having a shorter distance to one circle, that does not have greater distance to the other.

In this example (Fig. 4.42), the non-dominated points approximate the Pareto-front in the objective function space, but not in the decision variable space where they form an annulus instead. In summary:

- The Pareto-front contains the Pareto-optimal solutions. The Pareto-front divides the objective function space into two parts: one that contains non-optimal solutions and one that contains infeasible solutions. There is no valid mapping from the decision variable space to the infeasible part of the objective function space because by definition there are no valid points beyond the Pareto-front.
- The Pareto-front is not always continuous. For example, constraints, quantized parameters or quantized non-continuous objective functions can produce a discontinuous Pareto-front.
- The Pareto-front is not always convex. It can be concave, e.g., serpentine, or consist of disjoint sections, each of which may be either concave or convex.
- The Pareto-front may coalesce to a single point if objectives do not conflict or if constraints so restrict it.
- The Pareto-front may extend toward infinity even if parameters are bound constrained. Consequently, it may be necessary to set bounds on objective function values.
- Depending on the problem, neighboring points in the Pareto-front are not always neighbors in the parameter space.

A more detailed discussion of Pareto-optimization and related concepts can be found in Miettinen (1998), Deb (2001) and Coello Coello et al. (2002).

For most nonlinear, multi-objective optimization problems, determining the entire continuous Pareto-optimal surface is practically impossible, but finding a discrete set of Pareto-optimal points that approximates the true Pareto-front (e.g., Fig. 4.42) is a realistic expectation. While even a simple random search can locate Pareto-optimal points, EAs can find multiple non-dominated solutions within a single run.

4.6.4 Adapting DE for Multi-Objective Optimization

Problems with multiple objectives resemble those with multiple constraints. In both cases, the goal is to find a set of parameters that minimizes a set of functions. Because problems in these two domains share this common structure, the methods used to solve them are often similar. For example, both constrained and multi-objective optimization may benefit from a judicious assignment of weights to raw function values. Pareto-optimal solutions, however, must be determined without weights so that an expert will have a set of unbiased options from which to choose. By using Pareto-dominance as a selection criterion, a population can be driven toward the Pareto-front in the same way that Lampinen's dominance-based criterion for constrained optimization (Eq. 4.21) pressures vectors toward feasible regions.

Incorporating dominance-based selection into DE involves nothing more than comparing the trial and target vectors to determine which one is dominant.

$$\mathbf{x}_{i,g+1} = \begin{cases} \mathbf{u}_{i,g} & \text{if } \forall k \in \{1,...,K\}: f_k\big(\mathbf{u}_{i,g}\big) \le f_k\big(\mathbf{x}_{i,g}\big), \\ \mathbf{x}_{i,g} & \text{otherwise.} \end{cases} \qquad (4.59)$$

According to Eq. 4.59, the trial vector $\mathbf{u}_{i,g}$ is selected if the target vector, $\mathbf{x}_{i,g}$, does not dominate it. Except for containing objective function values instead of constraint function violations, Eq. 4.59 is identical to Lampinen's criterion for comparing two infeasible vectors (Eq. 4.21).

After many generations, some population vectors will be dominated while others will not. As a final step, all dominated points in the last generation should be removed so that the remaining population approximates the Pareto-front. More formally, a member, $\mathbf{x}_b$, of the final generation's population should be removed from the final population if there exists another vector, $\mathbf{x}_a$, that satisfies the condition

$$\forall k \in \{1,...,K\}, \forall a,b \in \{1,2,...,Np\}: \qquad (4.60)$$

$$f_k(\mathbf{x}_a) \le f_k(\mathbf{x}_b) \wedge \exists k \in \{1,...,K\}: f_k(\mathbf{x}_a) < f_k(\mathbf{x}_b), \text{ where } a \ne b.$$

In many cases, the trial vector $\mathbf{u}_{i,g}$ can be rejected before all K objective functions have been evaluated. If this shortcut is exploited, DE will execute more quickly. The flowchart in Fig. 4.43 describes a Pareto-dominance selection criterion that minimizes the number of objective function evaluations.

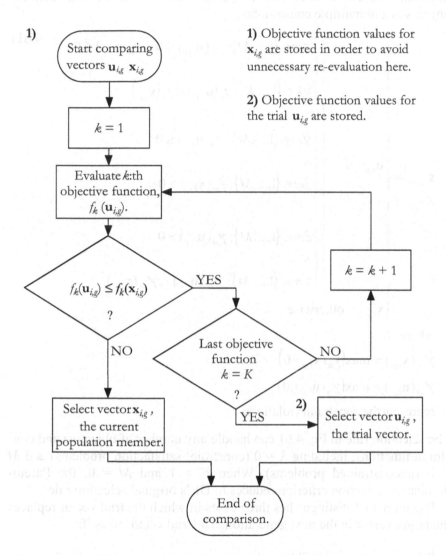

1) Objective function values for $\mathbf{x}_{i,g}$ are stored in order to avoid unnecessary re-evaluation here.

2) Objective function values for the trial $\mathbf{u}_{i,g}$ are stored.

Fig. 4.43. This implementation of the Pareto-dominance-based selection rule in Eq. 4.59 rejects a dominated trial vector at the earliest possible phase to avoid performing any unnecessary objective function evaluations.

Both the selection rule described above and Lampinen's criterion for direct constraint handling are based on Pareto-dominance. This similarity makes it easy to implement dominance-based selection for multi-objective problems that are also constrained. Equation 4.61 outlines a Pareto-dominance-based selection rule that is designed to handle both multiple objectives and multiple constraints:

$$
\mathbf{x}_{i,g+1} = \begin{cases} \mathbf{u}_{i,g} & \text{if} \quad \begin{cases} \begin{cases} \forall m \in \{1,...,M\}: \gamma_m(\mathbf{u}_{i,g}) \leq 0 \wedge \gamma_m(\mathbf{x}_{i,g}) \leq 0 \\ \wedge \\ \forall k \in \{1,...,K\}: f_k(\mathbf{u}_{i,g}) \leq f_k(\mathbf{x}_{i,g}) \end{cases} \\ \vee \\ \begin{cases} \forall m \in \{1,...,M\}: \gamma_m(\mathbf{u}_{i,g}) \leq 0 \\ \wedge \\ \exists m \in \{1,...,M\}: \gamma_m(\mathbf{x}_{i,g}) > 0 \end{cases} \\ \vee \\ \begin{cases} \exists m \in \{1,...,M\}: \gamma_m(\mathbf{u}_{i,g}) > 0 \\ \wedge \\ \forall m \in \{1,...,M\}: \gamma'_m(\mathbf{u}_{i,g}) \leq \gamma'_m(\mathbf{x}_{i,g}) \end{cases} \end{cases} \\ \mathbf{x}_{i,g} & \text{otherwise} \end{cases}
\tag{4.61}
$$

where

$$\gamma'_m(\mathbf{x}_{i,g}) = \max(\gamma_m(\mathbf{x}_{i,g}),0)$$
$$\gamma'_m(\mathbf{u}_{i,g}) = \max(\gamma_m(\mathbf{u}_{i,g}),0)$$

represent the constraint violations.

The selection rule in Eq. 4.61 can handle any number of objective and constraint functions, including $K = 0$ (constraint satisfaction problems) and $M = 0$ (unconstrained problems). When $K = 1$ and $M = 0$, the Pareto-dominance selection criterion reduces to DE's original selection rule.

Equation 4.61 distinguishes three cases in which the trial vector replaces the target vector in the next generation. The trial vector wins if:

1. Both vectors are feasible and the target vector does not dominate the trial vector in the objective function space.
2. The trial is feasible and the target vector is infeasible.
   Both vectors are infeasible and the trial vector's constraint violations are all less than or equal to those of the target vector.

Unless both the trial and target vectors are feasible, objective function values are not compared. Ignoring objective function values when one or both vectors are infeasible is justified in the same sense that it does not matter if a lens can be designed to give perfect images if it must be made of glass that does not exist. The flowcharts in Fig. 4.44 show the selection scheme outlined in Eq. 4.61 that minimizes the number of objective and constraint function evaluations. According to Fig. 4.44, the most computationally expensive constraint and objective functions should be evaluated last so that early detection of an inferior trial will save the most effort.

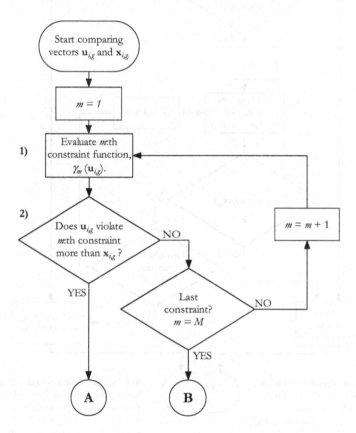

1) Objective and constraint function values for $x_{i,g}$ are stored in order to avoid unnecessary re-evaluation here.

2) After finding $u_{i,g}$ violating any of the constraints more than x_g, $u_{i,g}$ can be rejected immediately, without evaluating the remaining constraint function values at all.

Fig. 4.44a. Pareto-dominance-based selection with direct constraint handling (Eq. 4.60). Inferior trial vectors are rejected at the earliest possible phase to avoid performing unnecessary objective or constraint function evaluations.

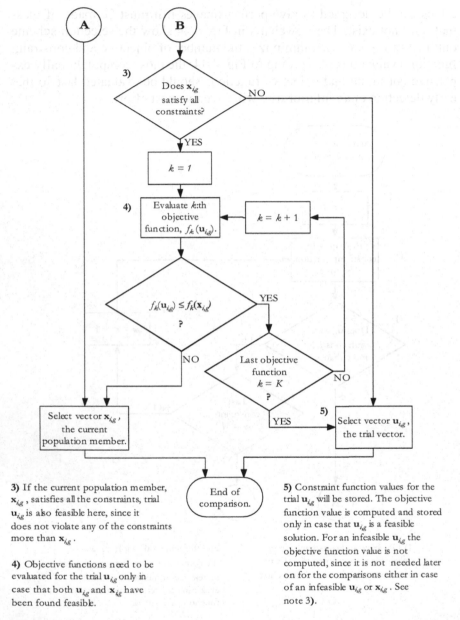

Fig. 4.44b. Continuation of Fig. 4.44a

3) If the current population member, $\mathbf{x}_{i,g}$, satisfies all the constraints, trial $\mathbf{u}_{i,g}$ is also feasible here, since it does not violate any of the constraints more than $\mathbf{x}_{i,g}$.

4) Objective functions need to be evaluated for the trial $\mathbf{u}_{i,g}$ only in case that both $\mathbf{u}_{i,g}$ and $\mathbf{x}_{i,g}$ have been found feasible.

5) Constraint function values for the trial $\mathbf{u}_{i,g}$ will be stored. The objective function value is computed and stored only in case that $\mathbf{u}_{i,g}$ is a feasible solution. For an infeasible $\mathbf{u}_{i,g}$ the objective function value is not computed, since it is not needed later on for the comparisons either in case of an infeasible $\mathbf{u}_{i,g}$ or $\mathbf{x}_{i,g}$. See note 3**)**.

The Pareto–DE approach described here is relatively easy to implement and should be effective on a wide range of problems. In some cases, however, this approach may suffer from the same problems that plague other multi-objective EAs, including:

- The approximated Pareto-front is too far from the true Pareto-front.
- Not enough non-dominated points are found.
- The non-dominated set does not cover the entire Pareto-front.
- The non-dominated set is distributed too non-uniformly along the Pareto-front.
- It is difficult to determine when the search is over. Compared to problems with a single objective, developing a stopping criterion for multi-objective problems is more difficult because the population cannot be expected to converge (see Figs. 4.41 and 4.42).

Examples of other ways in which DE has been adapted for multi-objective optimization can be found in Chang et al. (1999), Wang and Sheu (2000), Abbass et al. (2001), Abbass (2002a, 2002b, 2002c) and Madavan (2002). In addition, readers interested in other multi-objective EAs should refer to Miettinen (1998), Deb (2001) and Coello Coello et al. (2002).

4.7 Dynamic Objective Functions

Previous sections have assumed that a vector's objective function value is static (Fig. 4.45a). This section explores how DE handles *dynamic objective functions*, i.e., functions that do not always yield the same result each time a vector is evaluated. The source of an objective function's dynamism can be parameters (Fig. 4.45b), the objective function evaluation (Fig. 4.45c), or both (Fig. 4.45d).

In some cases, both the optimal vector and its objective function value can be reliably estimated by a time average; in others, the optimum vector drifts during the optimization process and must be tracked. Before outlining how DE handles optimum tracking, the next subsection considers the case in which the influence of objective function fluctuations can be averaged out.

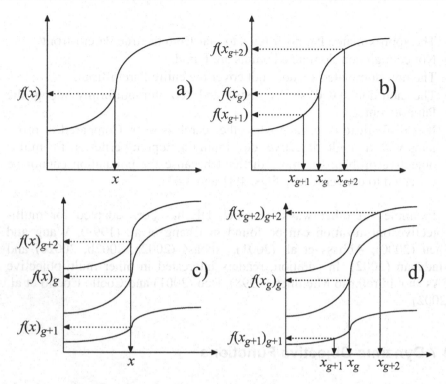

Fig. 4.45. a A static objective function; **b** parameter noise; **c** objective function evaluation noise; **d** both parameter and objective function evaluation noise. The subscript, g, is a discrete time index that indicates the population's current generation.

4.7.1 Stationary Optima

Probability distribution functions whose moments (i.e., expected value, variance, etc.) are time invariant are known as *stationary* distributions (Yaglom 1962). If the distribution of parameter or objective function evaluation fluctuations is stationary and its expectation is finite, then both the optimal vector and its objective function value can be reliably estimated. The remainder of this section assumes that objective function value noise is stationary and that the expectation of the underlying distribution is finite. The next subsection considers the effect that stationary parameter noise has on DE's ability to estimate the location of the optimal vector, $\mathbf{x}^*$.

Parameter Noise

Parameters may fluctuate because they depend on a manufacturing process that is subject to statistical variations. For example, manufacturing irregularities might limit the tolerance for an electronic component to 5% of its nominal value. For constraint satisfaction problems, the presence of parameter noise naturally leads to design centering (see Sect. 4.5). Noisy parameters can also occur when the precision of readings or adjustments of control variables in a physical experiment is limited.

Although parameter noise is not a factor during optimization's early stages, it interferes with convergence once fluctuations in the parameter space become comparable in size to vector differences in the population. One way to locate the optimum vector $\mathbf{x}^*$ is to estimate its position with the time-averaged mean value of the population's current best performing vector

$$\mathbf{x}^* \cong \overline{\mathbf{x}}_{best} = \frac{1}{g_2 - g_1 + 1} \sum_{g=g1}^{g2} \mathbf{x}_{best,g}, \qquad g_1, g_2 \text{ integer.} \qquad (4.62)$$

The integers g_1 and g_2 are the beginning and ending generations, respectively, over which the time average is taken. DE operates as usual, with both population and trial vectors competing in one-to-one, winner-takes-all competitions, except that at the end of each generation, $g_1 \leq g \leq g_2$, the population's current best vector, $\mathbf{x}_{best,g}$, is sampled so that the time-averaged best vector (Eq. 4.62) can finally be computed.

Figure 4.46 illustrates the effect that adding a uniformly distributed random variable to parameters has on DE's ability to optimize the ten-dimensional sphere objective function. The mean best vector (Eq. 4.62) outperforms the best vector in the final generation, but not the best vector obtained when noise is absent. Since there is no objective function evaluation noise in this example, the function value plotted for the mean best vector is accurate. The accuracy with which the minimum can be located can be improved by increasing the number of generations over which the position of the best performing vector is averaged. This, of course, only holds if the average is taken over those generations where the function value is basically constant. In Fig. 4.46 this region roughly starts at generation 2500.

Even though parameter noise makes an objective function value dynamic, it does not affect the location of the optimum. Like measurement error, parameter noise simply makes the optimum harder to locate. For the optimum to actually shift, the objective function evaluation itself must be noisy. The next subsection explores the effect of stationary function evaluation noise on DE's ability to both locate the optimal vector and determine its objective function value.

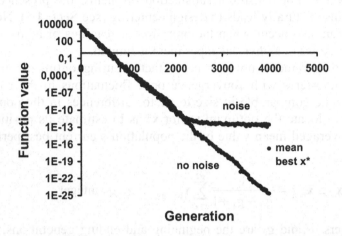

Fig. 4.46. The effect of stationary parameter noise when optimizing the ten-dimensional sphere objective function. To create the performance plot with noise, a random increment equal to $10^{-6} \cdot (\text{rand}_j(0,1) - 0.5)$ was added to each parameter prior to evaluation. The population's best value is plotted every 20 generations. The algorithm was DE/rand/1/bin, with $Np = 50$, $F = 0.9$, $Cr = 1$, no bound constraints. The mean best vector (Eq. 4.62) was averaged over the last 1000 generations.

Objective Function Evaluation Noise

Noise can also enter the optimization process during the objective function evaluation. For example, the on-line optimization of control parameters for an industrial process (controlling a chemical reactor, power plant, etc.) is a real-world scenario in which physical measurements of the objective function value (yield from the process, quality of product, etc.) have limited accuracy. In addition, random phenomena affecting the industrial process (incomplete mixing, thermal and chemical non-homogeneity, etc.) also generate some level of noise. Objective function noise can also arise from sensors like those used by autonomous robots seeking to optimize their response to their environment (Salomon 1997). Furthermore, any objective function that relies on a random number generator (e.g., simulations, game playing, etc.) also exhibits evaluation noise.

Figure 4.47 shows how DE responds when noise is added to the ten-dimensional sphere objective function. Results suggest that the time-averaged population's best vector is a good estimate for the optimal vector. In addition, averaging the current best vector's objective function value over the same span of generations will improve the estimate of the (nominal) minimum objective function value.

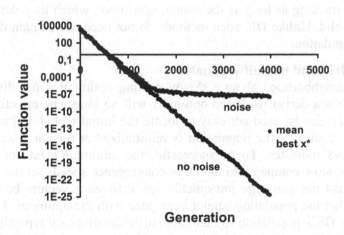

Fig. 4.47. The effect of adding stationary noise to the sphere objective function. A random increment equal to $10^{-6} \cdot \text{rand}_i(0,1)$ was added to each vector's function value. An instance of the best vector's objective function value is plotted as a function of the number of generations. The algorithm was DE/rand/1/bin with $Np = 50$, $F = 0.9$ and $Cr = 1$. The mean best value was averaged over the last 1000 generations.

4.7.2 Non-Stationary Optima

When an objective function's minimum is non-stationary, its (running) average location drifts and the optimization goal shifts to tracking the optimal vector at close range. Problems of this type arise in adaptive systems subject to environmental unpredictability. Very little experience has been amassed with DE on this type of application, but three cases can be distinguished based on the differentiability of the objective function and the optimum's speed relative to the population's rate of convergence. Each category poses problems that require modifying or supplementing classic DE.

- **Slow drift and differentiable**
  Problems where the objective function is multi-modal and subject to a slow drift may occur in real-world problems where, for example, temperature or an aging-process "slowly" alters the objective function. Here, the term "slowly" refers to the rate of change in the position of the minimum compared to the time required to find it. DE can be used for locating the (current) global minimum, but adaptive stochastic algorithms like the least-mean-square (LMS), the recursive-least-square (RLS) (Haykin 1991), or other deterministic optimizer are probably better for tracking as long as the assumptions upon which they depend remain valid. Unlike DE, such methods do not need to maintain diversity in a population.

- **Slow drift and non-differentiable**
  If the neighborhood about a slowly drifting optimum is not differentiable, then a derivative-based optimizer will no longer be effective. DE, however, can be used not only to locate the initial global optimum, but also to track it if the population is reinitialized at regular intervals, as Fig. 4.48 indicates. To be successful, the minimum's rate of change must be slow compared to the DE's convergence speed, yet the population must not converge too quickly lest difference vectors become so small that the population cannot keep pace with the optimum. Thus, requiring DE's population to maintain sufficient diversity typically limits not only the precision with which the minimum can be found, but also the precision with which it can be tracked. In addition, *parent vectors must be re-evaluated every generation* so that their objective function values remain current and the population is not constantly being dragged back to an earlier minimum that no longer exists.

- **Rapid drift and non-differentiable**
  An optimum that drifts rapidly and whose neighborhood is non-differentiable can immobilize a converged population. One strategy for keeping the search responsive is to operate in parallel several populations that have been initialized in a time-staggered fashion (Fig. 4.49). Each population has a different degree of convergence and each is reinitialized after a specified time. The current lowest objective function value taken over all populations is the point of choice for the application. Again, the parent population must be re-evaluated at each generation. The computational expense for this approach is potentially high, so it should only be considered when multi-modal, time-dependent objective functions are otherwise intractable.

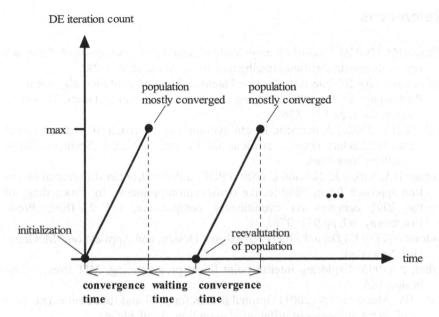

Fig. 4.48. Applying DE to a gradually changing objective function

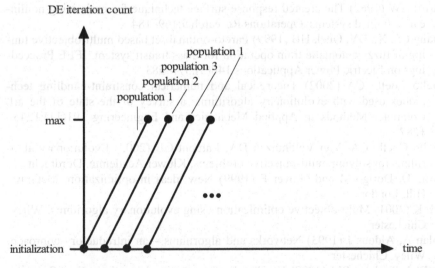

Fig. 4.49. An example of three independent populations, operating in parallel, whose initializations were time staggered

References

Abbass HA (2002a) An evolutionary artificial neural networks approach for breast cancer diagnosis. Artificial Intelligence in Medicine 25:265–281

Abbass HA (2002b) The self-adaptive Pareto differential evolution algorithm. In: Proceedings of the 2002 congress on evolutionary computation, Honolulu, Hawaii, vol 1, pp 831–836

Abbass HA (2002c) A memetic Pareto evolutionary approach to artificial neural networks. Lecture notes in artificial intelligence, vol 2256. Springer, Berlin Heidelberg New York

Abbass HA, Sarker R, Newton C (2001) PDE: a Pareto-frontier differential evolution approach for multi-objective optimization problems. In: Proceedings of the 2001 congress on evolutionary computation, vol 2, IEEE Press, Piscataway, NJ, pp 971–978

Antoniou A (1993) Digital Filters – Analysis, Design, and Applications. McGraw-Hill, New York

Arbel, A (1993) Exploring interior-point linear programming. MIT Press, Cambridge, MA

Babu BV, Munawar SA (2001) Optimal design for shell-and-tube heat exchangers by different Strategies of differential evolution. Available at: http://www.bvbabu.50megs.com/about.html

Brayton R, Hachtel G, Sangiovanni-Vincentelli A (1981) A survey of optimization techniques for integrated circuit design. Proceedings of the IEEE 70:1334–1362

Carrol CW (1962) The created response surface technique for optimizing nonlinear restrained systems. Operations Research 9:169–184

Chang CS, Xu DY, Quek HB (1999) Pareto-optimal set based multi-objective tuning of fuzzy automatic train operation for mass transit system. IEEE Proceedings on Electric Power Applications 146(5):577–583

Coello Coello CA (2002) Theoretical and numerical constraint-handling techniques used with evolutionary algorithms: a survey of the state of the art. Computer Methods in Applied Mechanics and Engineering 191(11):1245–1287

Coello Coello CA, Van Veldhuizen DA, Lamont GB (2002) Evolutionary algorithms for solving multi-objective problems. Kluwer Academic, Dordrecht

Corne D, Dorigo M and Glover F (1999) New ideas in optimization. McGraw-Hill, London

Deb K (2001) Multi-objective optimization using evolutionary algorithms. Wiley, Chichester

Dolan A, Aldous J (1993) Networks and algorithms – an introductory approach. Wiley, Chichester

Du D-Z, Pardalos PM (1998) Handbook of combinatorial optimization. Springer, Berlin Heidelberg New York

Dueck G (1993) New optimization heuristics – the great deluge algorithm and the record-to-record travel. Journal of Computational Physics 104:86–92

Freisleben B, Merz P (1996) Genetic local search algorithm for solving symmetric and asymmetric traveling salesman problems. In: Proceedings of the IEEE international conference on evolutionary computation (ICEC'96). IEEE Press, New York, pp 616–621

Frisch KR (1955) The logarithmic potential method of convex programming. Memorandum, University Institute of Economics, Oslo, May 13

Haykin S (1991) Adaptive filter theory. Prentice Hall, New Jersey

Hillier FS and Lieberman GJ (1997) Introduction to operations research, 6th ed. McGraw-Hill, New York

Hock W, Schittkowski K (1981) Test examples for nonlinear programming codes. Lecture notes in economics and mathematical systems, vol 187. Springer, Berlin Heidelberg New York

Hoefer K-H (1987) Schnelle Algorithmen zur multiplizierfreien rekursiven digitalen Signalverarbeitung. Ph.D. thesis, University of Stuttgart

Hwang Ching-Lai, Abu Syed Md Masud (1979) Multiple objective decision making – methods and applications. Springer, Berlin Heidelberg New York

Karmarkar, N (1984) A new polynomial-time algorithm for linear programming. Combinatorica 4:373–395

Kjellström G, Taxen L (1981) Stochastic optimization in system design. IEEE Transactions on Circuits and Systems CAS-28(7):702–715

Kondoz AM (1994) Digital speech. Wiley, New York

Kreutzer H (1985) Design centering for the enhancement of yield and for improvement of circuit properties. Ph. D. thesis (in German), University of Stuttgart

Lampinen J (2001) Solving problems subject to multiple nonlinear constraints by differential evolution. Private communication

Lampinen J (2002) A constraint handling approach for the differential evolution algorithm. In: Proceedings of the 2002 congress on evolutionary computation (CEC'02), vol 2, pp 1468–1473

Lampinen J, Storn R (2004) Differential Evolution. In: Godfrey C. Onwubolu, B. V. Babu (eds) New optimization techniques in engineering. Studies in fuzziness and soft computing, vol. 141, pp 123–166. Springer, Berlin Heidelberg New York

Lampinen J, Zelinka I (1999) Mechanical engineering design optimization by differential evolution. In: Corne et al. 1999

Lichtblau D (2002a) Discrete optimization using Mathematica. In: N Callaos, T Ebisuzaki, B Starr, JM Abe, D. Lichtblau (eds) World multi-conference on systemics, cybernetics and informatics (SCI 2002), International Institute of Informatics and Systemics, vol 16, pp 169–174. A Mathematica notebook version is available at:
http://library.wolfram.com/infocenter/Conferences/4317/

Lichtblau D (2002b) Private e-mail communication

Lueder E (1990) Optimization of circuits with a large number of parameters. Archiv fuer Elektrotechnik und Uebertragungstechnik (AEÜ) 44(2):131–138

Madavan Nateri K (2002) Multi-objective optimization using a Pareto differential evolution approach. In: Proceedings of the 2002 congress on evolutionary computation (CEC'02), vol 2, pp 1145–1150

Michalewicz Z, Fogel DB (2000) How to solve it: modern heuristics. Springer, Berlin Heidelberg New York

Miettinen KM (1998) Nonlinear multi-objective optimization. Kluwer Academic, Dordrecht

Mitra SJ and Kaiser JF (1993) Handbook for digital signal processing. Wiley, New York

Onwubolu GC (2003) Private e-mail communication

Onwubolu GC, Babu BV (2004) New optimization techniques in engineering. Springer, Berlin Heidelberg New York

Pahl PJ, Damrath R (2000) Mathematische Grundlagen der Ingenieur-Informatik. Springer, Berlin Heidelberg New York

Palli N, Azarm S, McCluskey P, Sandarajan P (1998) An interactive ε-inequality constraint method for multiple objectives decision making. Transactions of the ASME, Journal of Mechanical Design 120(4):678–686

Pareto V. (1886) Cours D'Economie Politique, vols. 1 and 2. F. Rouge, Lausanne

Press WH et al. (1992) Numerical recipes in C. Cambridge University Press

Rabiner LR, Gold B (1975) Theory and application of digital signal processing. Prentice Hall, New Jersey

Reeves C (1993) Modern heuristic techniques for combinatorial problems. Wiley, New York

Ruettgers M (1997) Differential evolution: a method for optimization of real scheduling problems. Technical report at the International Computer Science Institute, TR-97-013, pp 1–8

Salomon R (1997) Scaling behavior of the evolution strategy when evolving neuronal control architectures for autonomous agents. In: Angeline PJ, Reynolds RG, McDonnell JR, Eberhart R (eds) Proceedings of the 6th international conference on evolutionary programming. Springer, Berlin Heidelberg New York

Sapatnekar SS, Vaidya PM, Kang SM (1994) Convexity-based algorithms for design centering. IEEE Transactions on Computer-Aided Design 13(12):1536–1549

Storn R (1996a) Differential evolution design of an IIR-filter with requirements for magnitude and group delay. In: IEEE international conference on evolutionary computation (ICEC'96). IEEE Press, New York, pp 268–273

Storn R (1996b) On the usage of differential evolution for function optimization. In: Smith MH, Lee MA, Keller J, Yen J (eds) Proceedings of the North American Fuzzy Information Processing Society. IEEE Press, New York, pp 519–523

Storn R (1999a) System design by constraint adaptation and differential evolution. IEEE Transactions on Evolutionary Computation 3(1):22–34

Storn R (1999b) Designing digital filters with differential evolution. In: Corne D et al., pp 109–125

Storn R (2000) Available at:
http://www.icsi.berkeley.edu/~storn/fiwiz.html

Storn R, Price KV (1997) Differential evolution – a simple and efficient heuristic for global optimization over continuous spaces. Journal of Global Optimization 11:341–359

Syslo MM, Deo N, Kowalik JS (1983) Discrete optimization algorithms with Pascal programs. Prentice Hall, New Jersey

Wang F-S, Sheu J-W (2000) Multi-objective parameter estimation problems of fermentation processes using a high ethanol tolerance yeast. Chemical Engineering Science 55(18):3685–3695

Yaglom AM (1962) An introduction to the theory of stationary random functions. Prentice Hall, New Jersey

Zaharie D (2002) Critical values for the control parameters of differential evolution algorithms. International conference on soft computing (MENDEL 2002), pp 62–67

Zhang JC and Styblinski (1995) Yield and variability optimization of integrated circuits. Kluwer Academic, Dordrecht

5 Architectural Aspects and Computing Environments

5.1 DE on Parallel Processors

5.1.1 Background

Compared to gradient-based optimizers, evolutionary algorithms (EAs) demand more processing capacity because they typically require more objective function evaluations. Even so, DE's floating-point encoded selection and reproduction operations, which rely less on random number generation, are not as computationally expensive as are their binary-coded GA counterparts (Alander and Lampinen 1997). More often, the time required to generate child vectors is small compared to the time needed to evaluate the objective function. In real-world applications, it is not uncommon for the objective function evaluation to consume more than 95% of the total CPU time.

The need for faster processing is particularly acute when optimizing models based on simulations, since an acceptable solution may require tens of thousands of objective function evaluations. An efficient parallel approach to such problems is crucial, since a serial processor may take hundreds of hours to optimize models based on simulations.

5.1.2 Related Work

Research on parallel EAs gathered momentum in the mid-1980s. Jarmo Alander's comprehensive bibliography of distributed GAs (Alander 1997) indicates that a lot of work has been done since. The bibliography currently contains over 700 references, so this section provides only a very short and general overview. Udo Kohlmorgen surveyed implementation strategies for parallel GAs (Kohlmorgen 1995), classifying parallel GAs into three main types: the *farming model*, the *migration model* and the *diffusion model*.

Farming Model

In the *farming model*, the whole population is kept in a master processor that selects individuals for mating and sends them to slave processors who perform all other operations, including crossover, mutation and objective function evaluation.

STANDARD MODEL (a farming model)

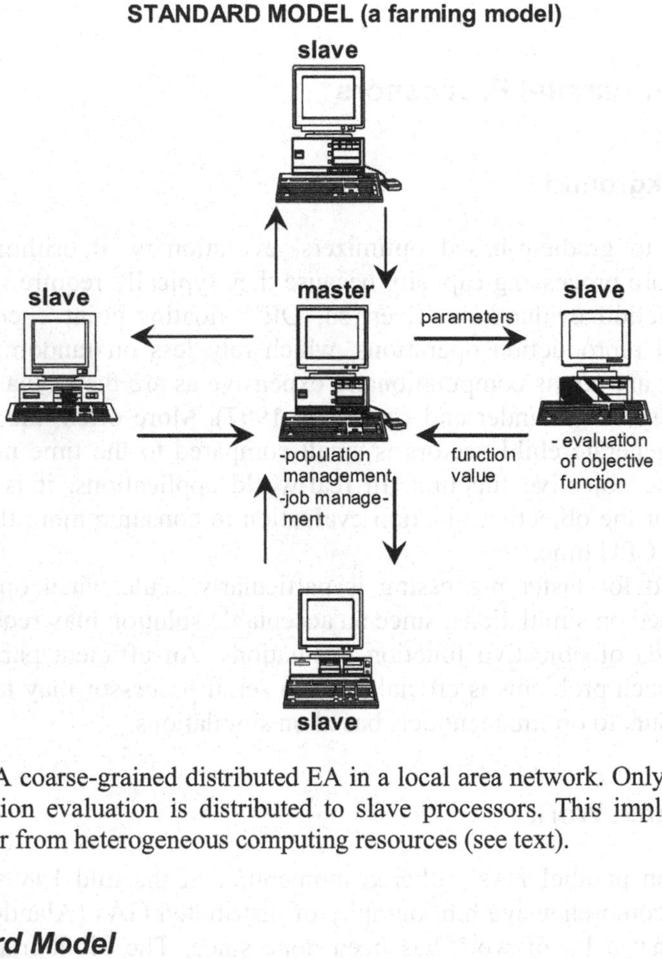

Fig. 5.1. A coarse-grained distributed EA in a local area network. Only the objective function evaluation is distributed to slave processors. This implementation may suffer from heterogeneous computing resources (see text).

Standard Model

The *standard model* is a variation of the farming model that distributes the objective function evaluation between slave processors (Fig. 5.1). This model has been used, for example, to implement distributed EAs on a local area network in Toivanen et al. (1995) and Alander et al. (1994, 1995). The *migration* version of the standard model divides the slave processors into separate populations (Fig. 5.2). To allow sub-populations to communicate with each other, EAs introduce an "immigration" operator that allows individuals to move between sub-populations.

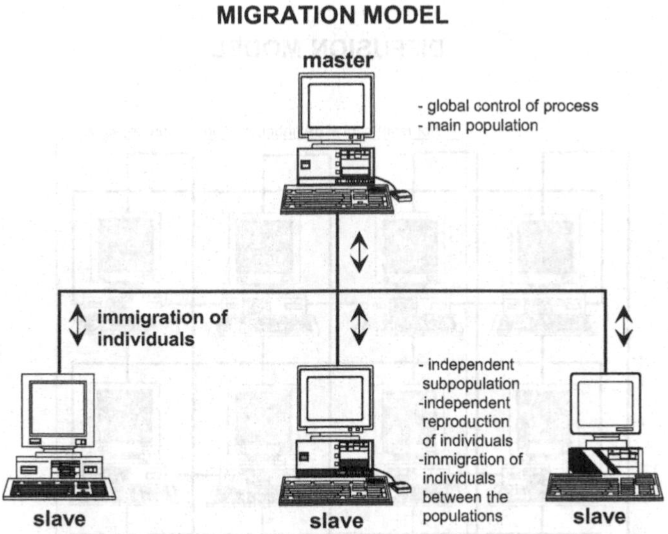

MIGRATION MODEL

Fig. 5.2. A coarse-grained, migration type, distributed EA in a local area network. The population is divided into independently operating subpopulations. The migration operator promotes interaction between the otherwise isolated populations by allowing individuals to immigrate into subpopulations.

Diffusion Model

The *diffusion* or *neighborhood model* distributes the population by mapping each individual to a single processor (Fig. 5.3). Processors are connected by a topology that defines how information can be exchanged between neighboring processors. Heinz Mühlenbein and colleagues (Mühlenbein et al. 1991) distinguished between the *coarse-grained* migration model and the *fine-grained* diffusion model based on the number of individuals assigned to a slave processor.

To ensure fast communication between processors, a parallel EA may need to reformulate its genetic operators (Salomon et al. 2000) as it becomes increasingly fine-grained and the demand for efficient, high-speed communication becomes crucial (Kohlmorgen 1995). If a traditional EA can be implemented as a coarse-grained model, then slower communication channels, like those available via a Local Area Network, may be adequate. In favorable cases, run-times scale almost linearly with the number of processors when the objective function evaluation is distributed across the network. If the objective function is computationally intensive, this model is often a good choice.

DIFFUSION MODEL

Fig. 5.3. A fine-grained, diffusion-type, distributed EA in a local area network. Each individual is mapped to a dedicated processor. In this example, the processors are connected in a 4 × 4 grid. Each individual is surrounded by a local neighborhood, in this case, a 3 × 3 grid around each processor (the gray-shaded local neighborhood of the processor marked X). All genetic operations for an individual are performed within this local neighborhood. In this example, the population is divided into multiple local neighborhoods (16 local neighborhoods here) that are partially overlapping. A variety of topologies and neighborhood definitions have been reported in the literature.

5.1.3 Drawbacks of the Standard Model

Jouni Lampinen's investigations into distributing DE across a cluster of PCs (Lampinen 1999) were inspired by the encouraging results reported in Toivanen et al. (1995), Alander et al. (1994, 1995), Alander and Lampinen (1997) and Lampinen and Alander (1998). In the latter three investigations, the objective function evaluation was distributed to PCs connected by a local area network, whereas Toivanen et al. employed a cluster of workstations. Both studies relied on the standard model (Fig. 5.1) because it is both straightforward and simple. If the slave processes run as background processes or with a low priority, then processors can also be used for other purposes during the optimization. If the network is not dedicated to DE, this approach allows the system to take advantage of processor capacity that is unused by the network without disrupting a processor's normally assigned tasks.

Despite its advantages, the standard model also has shortcomings. For example, the generational GA creates the next generation all at once. Individuals are then passed to the slave processes to be evaluated. The problem with this approach is that the next generation cannot be created until the previous generation has been evaluated, i.e., until the *slowest* slave process has returned an objective function value. Meanwhile, the faster slave processors that have already finished run idle. The total time spent idling can be high if the computer network is heterogeneous, the computing capacity of slave processors varies much, or if the time required for objective function evaluation varies a lot depending on the values of the parameters. Especially when objective functions are simulation models, evaluation time can be highly input dependent. For example, the size of the model could be a parameter, with large models requiring more time to evaluate than small ones. Alternatively, iterative solvers used in the simulator may execute a different number of iterations for different inputs. Furthermore, the computational path the simulation takes may depend on its input values.

In a heterogeneous computer network, the efficiency of a parallel system (*efficiency = speedup/number of processors*) may be less than 50% or even as low as 10%. In Kohlmorgen (1995), where the standard model was used on a homogeneous network, the efficiency of the standard model dropped to 85% when the number of processors was raised to four and to 70% when the number of processors was raised to eight. The situation will become even worse for a heterogeneous network because the slowest slave determines the overall speed of the optimization process.

Another drawback associated with generational reproduction in conjunction with the standard model distributed EA is that it is impossible to use a greater number of slave processors than the number of individuals in the population. Although the standard model cannot take advantage of more processors, it can be modified for doing so, and also to improve its efficiency. This will be described in the following sections.

5.1.4 Modifying the Standard Model

This section describes a simple, yet flexible, distributed DE algorithm method developed by Lampinen (1999). The method is generally based on the standard, coarse-grained model with computers distributed across a local area network (Fig. 5.1), but modifications avoid the drawbacks of the usual implementation. Since DE's selection operator works differently than those found in most GAs, earlier approaches (Alander and Lampinen 1997; Lampinen and Alander 1998) required some major changes.

Both the standard (Fig. 5.1) and migration (Fig. 5.2) models are relatively easy to implement and each offers its own advantages. The migration model, for example, is able to efficiently use all the computing resources available via local area network. Its network communication rate is low since only immigrating individuals need to be transferred. Ultimately, however, the migration model proves less adaptable to DE than the standard model.

First, the fact that the migration model can use all the processing capacity available via a local area network does not mean that it can solve a problem in the shortest time. The migration model is logically different from the sequential DE algorithm, so it is difficult to apply knowledge gathered with sequential DE to the migration model. There might be difficulties, for example, when selecting values for DE's control parameters. Furthermore, designing the migration operation entails deciding which individual migrates, where it migrates to and what the migration rate should be. Currently, no totally satisfying solutions appear to be available for these subproblems.

In its favor, the standard model is better suited to evaluating computationally expensive objective functions than the migration model because it can provide a quasi-linear speedup with respect to the available processor capacity. In the migration model, the same optimization program runs on all computers and the effectiveness of the parallelization depends on how efficient the migration operation is. For example, there is a high risk that all the subpopulations will converge to the same local optimum if the migration rate is too high. If the migration rate is too low, the benefit of using more than one processor will be lost. The optimum migration scheme depends on the problem to be solved, the number of slave processes and the control variables of slave processes and other factors. Currently, there is no general migration operation that can efficiently solve an arbitrary optimization problem. Since the efficiency of the migration model is highly dependent on the design of the migration scheme and because designing a good scheme is a non-trivial task, the standard model is the preferable way to implement a distributed DE algorithm.

The standard model's higher communication rate is only a minor drawback. Assuming that the objective function is computationally expensive, the communication rate will be comparatively low. Also, the time latency to establish a contact, as well as the time required to communicate, are virtually meaningless when compared to the total execution time of the optimization process. In most cases, the communication speed of a 10 Mbit/s Ethernet is high enough from both a latency and throughput point of view, although this situation can change if the network is also being used for other communications. Experience has shown that the modified standard model easily achieves over 95% efficiency and even 99% or more seems possible (Lampinen 1998).

5.1.5 The Master Process

Figure 5.4 shows a general overview of the modified standard model for a distributed EA. The master process performs all operations needed to create trial vectors. In addition, the master process selects trial vectors based on their objective function values and either inserts them into the population or rejects them. In addition, the master process evaluates trial objective function values and manages the slave processes. The slave processes are only responsible for computing the objective function values of the individuals that the master process sends them.

The master and slave processes communicate via two, shared disk files. Any computing resources with access to those two direct access files, e.g., via a local area network, can be exploited. The shared interface files can be physically located anywhere on the network, i.e., at any hard disk, as long as the master process and all slave processes have access to them. Table 5.1 shows the structure of the shared interface files.

The first interface file (Table 5.1) is a stack of unevaluated individuals. A stack of evaluated individuals is kept in a different interface file. The master process passes unevaluated trial vectors to the shared disk file of unevaluated individuals. A slave process then picks one individual from this file, evaluates it and returns the associated objective function value to the shared disk file of evaluated individuals. The slave then immediately picks a new individual, evaluates it and returns the corresponding objective function value, and so on.

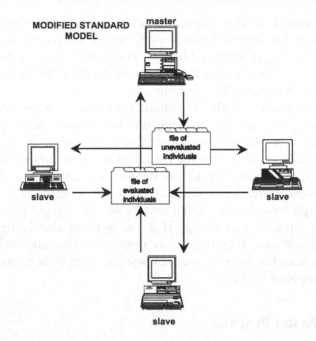

Fig. 5.4. A coarse-grained distribution of DE in a local area network based on the standard model except that shared interface files serve as a buffer for objective function evaluation tasks. The master process and the slave processes are only loosely coupled via these files and not synchronized to each other.

Table 5.1. The structure of shared interface files. The first file is for unevaluated individuals and the other is for finished objective function evaluation tasks. Each record of these files represents one trial individual.

SHARED FILE #1, unevaluated individuals

RECORD	read by slave	individual	cost	variable 1	variable 2	variable 3	variable 4
1	1	2	0.00	0.72	0.71	0.56	0.47
2	1	11	0.00	0.77	0.58	0.03	0.04
3	0	1	0.00	0.07	0.12	0.95	0.90
4	0	15	0.00	0.84	0.46	0.73	0.78
5	0	9	0.00	0.57	0.15	0.01	0.18

SHARED FILE #2, evaluated individuals

RECORD	read by master	individual	cost	variable 1	variable 2	variable 3	variable 4
1	1	19	1.99	0.53	0.58	0.72	0.17
2	1	4	2.30	0.29	0.78	0.88	0.36
3	1	2	1.48	0.75	0.36	0.02	0.35
4	0	16	1.95	0.81	0.04	0.80	0.29
5	0	7	2.12	0.24	0.66	0.28	0.94

The master process works like ordinary DE, but with some exceptions. Before evaluating a new trial individual, the master process checks the shared file of unevaluated tasks. If there is space for a new task, the master process moves an objective function evaluation task to the file of un-evaluated individuals. If there is currently no space in the file of unevaluated tasks, the master process evaluates the current trial individual itself. After that, the master process proceeds to the next trial individual and the process is repeated.

At the end of each generation, the master process reads the shared file of evaluated individuals. A value of "1" in the first field of a record indicates that the task is finished because the master process reads the file before the end of the previous generation. To preserve DE's one-to-one selection scheme, each trial is assigned a unique index that pairs it with a vector in the current population (called "individual" in Table 5.1). When the evaluated trial vector returns to the master process, this index indicates the member of the current population with which the trial is compared during survivor selection (Sect. 2.7). Any trial with a lower objective function value than that of the corresponding target in the current population is transferred from the shared file to the current population where it over-writes its competitor. Otherwise the trial from the shared file is simply ignored.

In Table 5.1, the first field of records contains either zero or one. If a one appears in the file for unevaluated tasks, the record has been read by a slave and the task has either already been evaluated, or is currently being evaluated. Because this record has already been read, the master process is allowed to overwrite it with a new task. A zero indicates that a task is still unevaluated and has not yet been assigned to a slave.

The size of files (the number of records) should be the same as the population size of the master process, or higher. This ensures that the file of unevaluated individuals always contains work for slave processes and that there is always space for evaluated individuals in the shared file of evaluated tasks. In Lampinen (1999), the number of records was set to twice the population size, i.e., $2Np$.

Because the master process uses shared files as buffers, it does not need to be synchronized with the slave processes. For example, it is possible that a child does not return back to the master process during the same generation in which it was created. Preliminary tests suggest that this does not diminish DE's effectiveness, efficiency or robustness. Furthermore, it seems possible that this delay slightly improves DE's robustness by maintaining a higher level of diversity in the population. When populations are small, this added diversity helps DE to avoid premature convergence. This effect, however, still needs to be investigated and supported by experimental data.

Synchronization is necessary only if two processes (master or slave) try to access the same file simultaneously, in which case the process attempting to access the file must wait a few milliseconds and try again. Access to shared files, by either the master process or a slave process, should be set to *deny-read/write* mode; no other process may access a file that is already open. Alternatively, some programming environments allow a single record to be locked against writing to minimize file access conflicts between processors. The waiting times due to file access collisions, however, are typically insignificant compared to the time required to evaluate a computationally expensive objective function. Note also that, due to fact that master and slave processes are not synchronized on a generation level, the number of slave processes is not limited by the population size, as is the case for the classic standard model. For further details on master–slave synchronization, see Lampinen (1999).

5.2 DE on Limited Resource Devices

Limited resource devices are special-purpose processors operating in environments in which one or more of the following three categories is subject to non-negligible limitations: available program memory, data memory and processing capacity. Many consumer products like wireless phones, palmtop computers, toys, etc., as well as controller units in cars, home appliances, etc., face one or more of these limitations. Often, simple processors must operate with fixed-point arithmetic so the underlying goal is to save program memory space and processing time by avoiding divisions and high-precision multiplication. This section gives some hints on how to write efficient code for limited resource environments. Some of the proposed, "tricks" may alter DE's convergence behavior, so adequate testing is important.

5.2.1 Random Numbers

Random number generators seldom receive much attention because they are usually functions integrated into a high-level language. Their design must be rethought, however, when random number generators are implemented on devices that lack a floating-point unit or where memory space is a limited resource. Emulating floating-point operations, especially division, is costly in terms of both program memory and processing time.

Since generating floating-point values in the interval [0, 1] requires division, random number generators for limited resource devices need to be carefully designed.

Random Numbers for Initialization

Populations can be initialized in a variety of ways. Initialization can specify predefined points, a collection of points around a nominal point, etc., but most of the time, parameters are initialized with values that are randomly chosen from within the allowed range according to:

$$x_j = x_{j,L} + \mathrm{rand}_j(0,1) \cdot (x_{j,U} - x_{j,L}), \tag{5.1}$$

where $\mathrm{rand}_j(0,1)$ is a uniformly distributed random number from the interval $(0,1)$ that is chosen anew for each parameter. Since initialization requires only $Np \cdot D$ random numbers, the generator's sequence length is not of primary importance. For this reason, and also because DE is fairly insensitive to initial conditions, pseudo-random numbers only need be reasonably random. Efficient code is more important on limited resource devices than accurate random simulation. The three random number generators described in Figs. 5.5–5.7 are both small and efficient. All three random number generators can be found in the file simplerand.cpp on the accompanying CD.

```
#define PI   3.14159265

float rand01(float *fp_seed)
//Generates random numbers from the interval [0,1].
//Initialize *fp_seed with 4.0 (for example)
{
    float f_x;
    f_x = *fp_seed + PI;
    f_x = f_x*f_x*f_x*f_x*f_x; //f_x^5
    *fp_seed = f_x - floor(f_x);
    return(*fp_seed);
}
```

Fig. 5.5. A simple floating-point random number generator from the HP-35 application programs (Miller 1981).

```
unsigned int rand15(unsigned long *ulp_seed)
//Generates random numbers from the interval [0,2^15-1],
//so it is suited as a fixed point random number
//in 1.15 format.
//Initialize *ulp_seed with 1 (for example)
{
    *ulp_seed = *ulp_seed*1103515245 + 12345;
    return(unsigned int)((*ulp_seed>>16)& 32767);//modulo 2^15
}
```

Fig. 5.6. A simple fixed-point random number generator provided by the ANSI C committee (Press et al. 1992). The computation modulo N is a logical AND with $N - 1$, provided that N is a power of two. In this case, $N = 2^{15}$.

```
unsigned int rand15a(unsigned long *ulp_seed)
//Generates random numbers from the interval [0,2^15-1],
//so it is suited as a fixed point random number
//in 1.15 format.
//Initialize *ulp_seed with 1 (for example)
{
    *ulp_seed = *ulp_seed*1664525 + 1013904223;
    return(unsigned int)((*ulp_seed)& 32767);//modulo 2^15
}
```

Fig. 5.7. A simple random number generator according to Knuth (1981) and Press et al. (1992). This generator can operate with modulo 2^{32}, in which case the return value must be "unsigned long".

Random Numbers for Generating Trial Vectors

To index the population vectors that will be combined into a mutant, most DE variants only need random numbers that are integers from the range [0, $Np - 1$]. The conceptually simplest method for generating random population indices is to compute floor($r \cdot (Np - 1)$), where $r \in [0, 1]$ is the floating-point output of a uniform random number generator. As Fig. 5.8 shows, the multiplication, $r \cdot (Np - 1)$, can be avoided if $Np = 2^k$, k integer.

If the population size does not equal a power of 2, then ui_N can be set to the first power of 2 greater than Np. Returned values should then be checked to see that they belong to the allowed range, [0, $Np - 1$]. If a returned value is out of range, then new values are generated until one falls within the permitted range. Since all numbers are equally likely, elements of the subset [0, $Np - 1$] will also be generated with equal probability. A disadvantage of this method is that generating invalid numbers steals valu-

able processor time. C code for trial vector selection appears in the file
rndselect.cpp on the accompanying CD.

```
unsigned int randNa(unsigned long *ulp_seed, unsigned int ui_N)
//Generates random numbers from the interval [0,ui_N-1] with
//ui_N being a power of two.
{
    *ulp_seed = *ulp_seed*1103515245 + 12345;
    return(unsigned int)((*ulp_seed>>16)&(ui_N-1));//modulo ui_N
}
```

Fig. 5.8. This code modifies the random number generator in Fig. 5.6 to yield
numbers in [0, ui_N − 1].

```
unsigned int randNb(unsigned long *ulp_seed, unsigned int ui_N)
//Generates random numbers from the interval [0,ui_N-1] with
//ui_N being a power of two.
{
    *ulp_seed = *ulp_seed*1664525 + 1013904223;
    return(unsigned int)((*ulp_seed)&(ui_N-1));//modulo ui_N
}
```

Fig. 5.9. This code modifies the random number generator in Fig. 5.7 to yield
numbers between [0,ui_N − 1].

5.2.2 Permutation Generators

One disadvantage of generating population indices with a random number
generator is that successive indices may not be distinct. If successive indices are equal, the generator must be run until a distinct index occurs. The
random permutation generator avoids the computational expense caused
by repeated calls for a distinct index, because successive indices are always distinct. Figure 5.10 illustrates the "urn algorithm" permutation generator that was developed by C. L. Robinson and published in Herrmann
(1992). As its name implies, the process can be modeled using two urns.
The first urn contains marbles marked with the numbers from 0 to $Np - 1$
and the second urn is initially empty. Marbles are randomly selected from
urn number 1 and placed into urn number 2. The sequence of numbers of
marbles placed in urn number 2 defines the permutation. The operation of
the algorithm is depicted in Figs. 5.10 and 5.11.

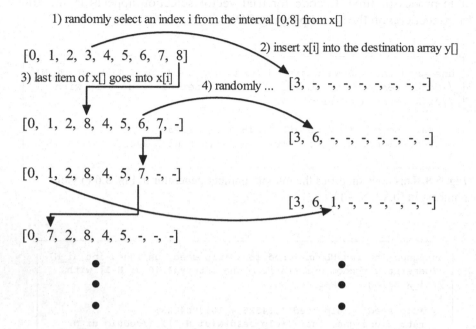

1) randomly select an index i from the interval [0,8] from x[]

[0, 1, 2, 3, 4, 5, 6, 7, 8]

2) insert x[i] into the destination array y[]

3) last item of x[] goes into x[i]

4) randomly ...   [3, -, -, -, -, -, -, -, -]

[0, 1, 2, 8, 4, 5, 6, 7, -]

[3, 6, -, -, -, -, -, -, -]

[0, 1, 2, 8, 4, 5, 7, -, -]

[3, 6, 1, -, -, -, -, -, -]

[0, 7, 2, 8, 4, 5, -, -, -]

Fig. 5.10. Visualizing the "urn algorithm" for permuting an array

```
...
k      = NP;
i_urn1 = 0;
i_urn2 = 0;
for (i=0; i<NP; i++) ia_urn1[i] = i; //initialize urn1

while (k >= NP-M+1)//M is the amount of indices wanted (must be <= NP)
{
    i_urn1 = (rand15a(&gul_seed)*k)>>15;      //choose a random index
    ia_urn2[i_urn2] = ia_urn1[i_urn1];        //move it into urn2
    ia_urn1[i_urn1] = ia_urn1[k-1]; //move highest index to fill gap
    k = k-1;               //reduce number of accessible indices
    i_urn2 = i_urn2 + 1; //next position in urn2
}
...
```

Fig. 5.11. C code snippet that illustrates the "urn algorithm"

The more efficient algorithm shown in Figs. 5.12 and 5.13 implements the urn idea with only one array. The trick is to use a bit-swap pointer that moves upward one index with every swap. Numbers that are randomly drawn from the upper portion of the array are recorded, then placed in the lower part of the array so that they will not be chosen again until the next generation.

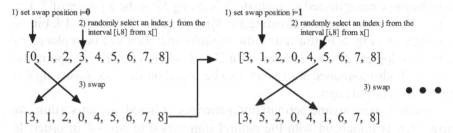

Fig. 5.12. The modified urn algorithm with just one array

```
...
for (i=0; i<NP; i++) ia_urn1[i] = i; //initialize urn1

for (k=0; k < M; k++)//M is the amount of indices wanted (must be <= NP)
{
    i_urn1 = ((rand15a(&gul_seed)*(NP-k))>>15) + k;  //choose a random index
    i_swap = ia_urn1[i_urn1];           //element to be swapped
    ia_urn1[i_urn1] = ia_urn1[k];       //swap element
    ia_urn1[k]      = i_swap;
}
...
```

Fig. 5.13. C code snippet for the modified urn algorithm with just one array

```
...
NP = 2<<k; //NP = 2^k
i_alpha = (rand15a(&gul_seed)*(NP-1))>>15;  //choose a random multiplier
if ((i_alpha%2) == 0 || (i_alpha == 0)) //make sure i_alpha is odd
{
    i_alpha = i_alpha+1;
}
...
for (i=0; i < NP; i++)//generate permuted sequence
{
    i_perm = (i*i_alpha)&(NP-1);  //bitwise AND makes mod NP if NP=2^k
}
...
```

Fig. 5.14. A permutation generator that needs no array memory

If memory is a serious issue, then permutations can be generated without arrays. One method is based on the number-theoretic result that the sequence $i = (\alpha \cdot j)\mathrm{mod}(Np)$, $j = 0, 1, 2, \ldots, Np - 1$, is a permutation of $j = 0$,

1, 2, ..., $Np - 1$ if α and Np are relatively prime (McClellan and Rader 1979). For example, if Np is itself prime then α can be any number. Making Np prime can be a drawback, however, because the modulo function will be more complicated to evaluate. Choosing Np to be a power of 2, i.e., $Np = 2^k$, is faster because dividing by Np just shifts the dividend k bits to the right. As Fig. 5.14 illustrates, the modulo operation can be replaced by a bitwise logical AND operation with $2^k - 1$ when $Np = 2^k$. The permutation methods mentioned above can also be found on the accompanying CD in the file permute.cpp.

Another, more arcane permutation generator is based on "un-sorting" an array that is initialized with the natural numbers 0 to $Np - 1$ in order. In un-sorting, a random decision, i.e. a "coin-flip", decides whether to swap two elements. Repeated random selections produce an unsorted array. The "coin-flip" can be done by rounding all output from a [0, 1] random number generator less than 0.5 to "0" and numbers above or equal to 0.5 to "1". This method may prove useful if existing source code contains a sorting routine that can be modified to either sort or un-sort.

5.2.3 Efficient Sorting

As has been shown in previous chapters, $(\mu + \lambda)$ is a viable alternative to DE selection that sorts both the current and trial populations according to their objective function value. Once the population has been initially sorted (ranked), it never again needs to be sorted in its entirety. When a trial vector is accepted, it is merged into the population array, while the member with the worst objective function value is eliminated. A simple sort algorithm, like the bubble sort (Standish 1995), will save program memory, but if memory is an issue, then many copy actions may be required. If, however, speed is an issue, a linked list is the most efficient approach, although the linked list itself requires additional memory. Figure 5.15 illustrates both approaches. Note how the highest value has to vanish because Np must stay constant.

5.2.4 Memory-Saving DE Variants

If memory in a device is limited, it may be necessary to implement DE with only $Np{\cdot}D$ array locations instead of $2{\cdot}Np{\cdot}D$. In addition, a single vector holds the current trial vector, which, if it wins, immediately replaces the target vector. With the one-array implementation, the concept of a generation becomes meaningless because there is no effective separation between current and trial vectors. A new generation only means that the loop through the population has begun again with the first individual. This form of evolution is similar to the way in which a steady-state GA operates. Figure 5.16 shows how to implement the DE/rand/1/bin scheme with a single population.

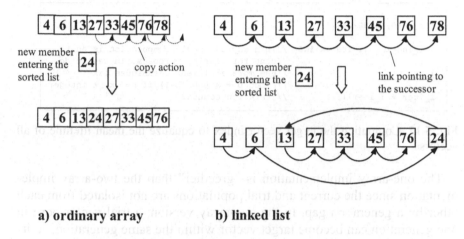

a) ordinary array   b) linked list

Fig. 5.15. Two different approaches for organizing a sorted number sequence. The array approach requires the least memory, while the linked list approach allows new members to be quickly merged.

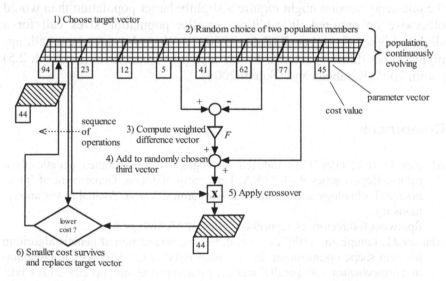

Fig. 5.16. A DE variant with only one population array of *Np* individuals

If target vectors are always chosen in the order 0, 1, 2, ..., $Np - 1$, then the vector selection scheme becomes biased because vectors with lower indices will be replaced by better solutions before vectors with high indices. Computing the target vector index according to the algorithm shown in Fig. 5.17 mitigates this bias.

```
...
for (k=0; k<NP; k++)
{
  i_index = (i_iter+k)%NP; //This way of modulo computation is just
  ...                      //done for clarity purposes. In real code
}                          //use more efficient techniques like
                           //a mod b = a AND (b-1) if b=2^k, k integer.
i_iter = i_iter+1;         //iteration counter
```

Fig. 5.17. Computing the target vector index to equalize the mean lifetime of all vectors

The one-array implementation is "greedier" than the two-array implementation since the current and trial populations are not isolated from each other by a generation gap. In the one-array version, a trial vector born in one generation can become target vector within the same generation, so its influence on the population's subsequent evolution is immediate. Similarly, writing a new vector into a single array of Np individuals improves the population more quickly than delaying a surviving trial vector's participation until the next generation. To compensate for this "greediness", the one-array version might require a slightly larger population than would otherwise be required. In addition, smaller population sizes call for a slightly higher Cr and/or F (Zaharie 2002). When Np must be small, applying either dither or jitter to F may also prove beneficial (see Sect. 2.5) (Storn 2005; Karaboga and Ökdem 2004).

References

Alander JT (ed.) (1997) An indexed bibliography of distributed genetic algorithms. Report series 94-1-PARA, University of Vaasa, Department of Information Technology and Production Economics, Vaasa. Available via anonymous ftp:
ftp.uwasa.fi directory cs/report94-1 (file gaPARAbib.ps.Z)
Alander JT, Lampinen J (1997) A distributed implementation of genetic algorithm for cam shape optimization. In: Topping BHV (ed.): Advances in computational mechanics with parallel and distributed processing, pp 209–217, Civil-Comp Press, Edinburgh

Alander JT, Ylinen J, Tyni T (1994) Optimizing elevator control parameters. In: Alander, JT (ed.), Proceedings of the second Finnish workshop on genetic algorithms and their applications (2FWGA), Vaasa (Finland), March 16–18, 1994. Report series 94-2. University of Vaasa, Department of Information Technology and Production Economics, Vaasa, pp 105–114

Alander JT, Ylinen J, Tyni T (1995) Optimizing elevator group control parameters using distributed genetic algorithms. In: Pearson DW, Steele NC, Albrecht RF (eds) Artificial neural nets and genetic algorithms, Alés (France), April 19–21, 1995. Springer-Verlag, Vienna, pp 400–403

Herrmann D (1992) Algorithmen-Arbeitsbuch. Addison-Wesley, Bonn

Karaboga D, Ökdem S (2004) A simple and global optimization algorithm for engineering problems: differential evolution algorithm. Turkish Journal of Electrical Engineering & Computer Sciences 12(1): 53–60

Knuth DE (1981) The art of computer programming, vol 2, semi-numerical algorithms. Addison-Wesley, Reading, MA

Kohlmorgen U (1995) Overview of parallel genetic algorithms. In: [13] of Alander et al. (1994), pp 135–143

Lampinen J (1999) Differential evolution – new naturally parallel approach for engineering design optimization. In: Topping BHV (ed.), Developments in computational mechanics with high performance computing. Civil-Comp Press, Edinburgh, pp 217–228

Lampinen J, Alander JT (1998) Shape design and optimization by genetic algorithm. In: Topping BHV (ed.), Advances in computational mechanics with high performance computing. Civil-Comp Press, Edinburgh, pp 187–196

McClellan JH, Rader CM (1979) Number theory in digital signal processing. Prentice Hall, Englewood Cliffs, NJ

Miller AR (1981) Pascal programs for scientists and engineers. Sybex, Berkeley, CA

Mühlenbein H, Schomisch M, Born J (1991) The parallel genetic algorithm as function optimizer. Parallel Computing 17:619–632

Press WH et al. (1992) Numerical recipes in C. Cambridge University Press

Salomon M, Perrin G-R, Heitz F (2000) Parallelizing differential evolution for 3-D medical image registration. Rapport de Recherche 00-06, September. Available via Internet at:
http://icps.u-strasbg.fr/~salomon/

Standish TA (1995) Data structures, algorithms & software principles in C. Addison-Wesley, Reading, MA

Storn R (2005) FIWIZ – a versatile program for the design of digital filters using differential evolution. Chapter 7.8 in this book

Toivanen J, Mäkinen RAE, Lahdelma R (1995) The reconstruction of an airfoil in 2D potential flow using a genetic algorithm on a parallel computer. In: [13] of Alander et al. (1994), pp 229–240

Zaharie D (2002) Critical values for the control parameters of differential evolution algorithms. In: 8th international conference on soft computing MENDEL 2002, Brno (Czech Republic), June 5–7, pp 62–67

6 Computer Code

6.1 DeMat – Differential Evolution for MATLAB®

This chapter describes DeMat – a collection of MATLAB® m-files that
provides a framework for solving function optimization problems with differential evolution (DE). The accompanying CD contains examples with
code that includes easily modified graphics support. The following subsections detail DeMat's architecture.

6.1.1 General Structure of DeMat

DeMat was developed to make it easy to apply DE to an arbitrary optimization problem. DeMat also provides graphics support, since visually
monitoring an ongoing optimization is often very helpful. DeMat consists
of the files shown in Fig. 6.1 and requires no special installation routine to
run.

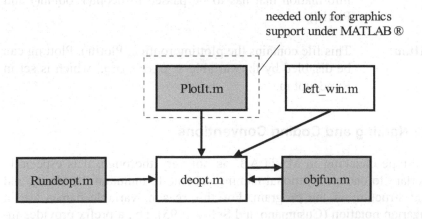

Fig. 6.1. The DeMat files. Gray-shaded files must be adapted to each new problem. The left_win.m file may also need to be modified if the selection criterion
changes.

Here is a brief description of the DeMat files:

deopt.m: This file contains the main DE engine, deopt(), along with several different DE mutation and selection operators. It also handles crossover and implements boundary constraints. This file controls the output of results.

objfun.m: This file contains the objective function, objfun(), which computes objective function (cost) values and evaluates constraints. The results of the objective function evaluation are returned in the structure S_MSE.

left_win.m: left_win() is a function that defines under what circumstances the trial (child) vector (corresponding to left input argument) wins against the target vector (corresponding to right input argument). This function takes both constraints and objective function values into consideration when deciding which vector survives.

Rundeopt.m: This is the main script file for configuring the optimization and for experimenting with DE strategies, population sizes, etc. All control variables are listed here. To simplify parameter passing, control variables are handed over to the structure variable, S_struct, that acts as a container for information that has to be passed to deopt(), objun() and PlotIt().

PlotIt.m: This file contains the plotting routine, PlotIt(). Plotting can be disabled by the variable I_plotting, which is set in Rundeopt.m.

6.1.2 Naming and Coding Conventions

Since type checking in MATLAB® is not very thorough, it is especially important to choose a rational naming scheme to maintain the clarity and sound structure of the program. For this reason, variable names are in Hungarian notation (Cusumano and Selby 1995), i.e., a prefix provides information about a variable's type. Hungarian notation also helps to identify where assignments may lead to problems.

The prefix, which can consist of several characters, is followed by an underscore, which is then followed by a descriptive name. The highest precedence in prefix construction is assigned to the characters "I", "F" and "S", which denote integer, float and structure, respectively. The next highest precedence goes to "Vr", "Vc" and "M" for row vectors, column vectors and matrices, respectively. For example, a row vector of floating-point variables would have the prefix FVr_, while a matrix of integers would have the prefix IM_. For the sake of simplicity, simple loop counter variables do not need a prefix and may be named k, l, m, etc. MATLAB®, however, reserves the characters i and j for the imaginary constant, so these names should not be used for counter variables (even though MATLAB® does not forbid this).

Table 6.1 shows the prefixes used in the code. Additional prefixes may be defined as needed.

Table 6.1. Variable naming conventions using Hungarian Notation

Variable	Prefix	Example	Remark
Integer	I	I_refresh	Even though typing is not strong in MATLAB® it is good to know whether a variable is intended as an integer
Float	F	F_cost_tol	–
Structure	S	S_x	Can have an arbitrary number of attached variables, e.g., S_x.I_nc S_x.I_no S_x.FVr_ca S_x.FVr_oa
Row vector	Vr	FVr_lim_up	Length of vector is not indicated in the name
Column vector	Vc	FVc_test	Length of vector is not indicated in the name
Matrix	M	FM_pop	Row and column dimensions are not shown in the name

Most functions also have a comment header that provides information about its arguments and return values.

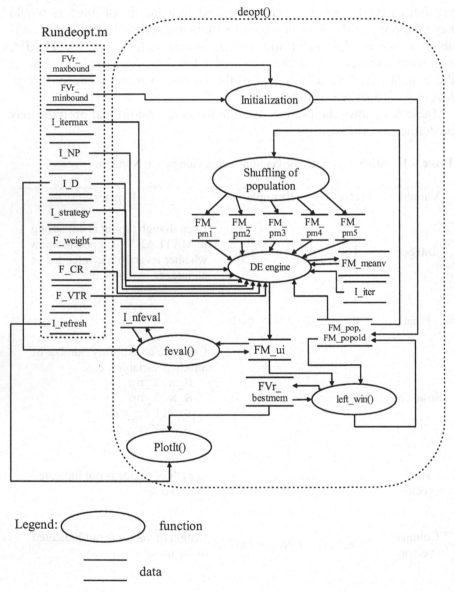

Fig. 6.2. A data flow diagram of DeMat's most important parts

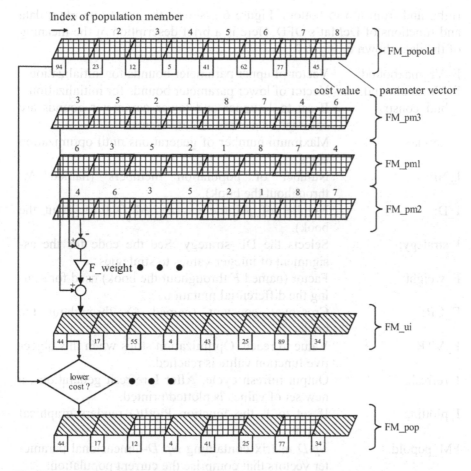

Fig. 6.3. A flowchart for DE/rand/1/bin (classic DE) in deopt.m. For simplicity, the crossover operation is not shown ($Cr = 1$).

6.1.3 Data Flow Diagram

The data flow diagram (DFD) (Yourdon 1989) provides an overview of a program's functionality. In contrast to flow charts that emphasize algorithmic design, a DFD shows which data is used and which functions or processes change the data. By convention, data is indicated by two parallel horizontal bars while functions are denoted by ellipses. Function names followed by () are explicitly named in the program. If parentheses are missing, then the functionality is embedded in a sequence of statements, not an explicit function. Arrows indicate the data flow and contain additional but limited time information. In general, time increases from left to right, and from top to bottom. Figure 6.2 shows the most important data and functions of DeMat's DFD. Here is a brief description of the meaning of the data shown in Fig. 6.2:

F_Vr_maxbound:	Vector of upper parameter bounds for initialization.
F_Vr_minbound:	Vector of lower parameter bounds for initialization.
I_bnd_constr:	If set to 1, upper and lower parameter bounds are enforced as parameter constraints.
I_itermax:	Maximum number of generations until optimization stops.
I_NP:	Number of population members (named Np throughout the book).
I_D:	Number of parameters (named D throughout the book).
I_strategy:	Selects the DE strategy. See the code for the assignment of integer values to strategies.
F_weight:	Factor (named F throughout the book) used for scaling the differential mutation.
F_CR:	Crossover constant (named Cr throughout the book).
F_VTR:	Value to reach. Optimization stops when this objective function value is reached.
I_refresh:	Output refresh cycle. After I_refresh generations, a new set of values is plotted/printed.
I_plotting:	If set to 1, the function PlotIt() renders graphical output.
FM_popold:	$Np·D$ matrix containing Np D-dimensional parameter vectors that comprise the current population.
FM_pop:	$Np·D$ matrix containing Np D-dimensional parameter vectors that comprise the next population.
FM_pm1:	Same as FM_popold, but with shuffled rows. FM_pm2, 3, 4, 5 are similar, but contain a different shuffling.
FM_meanv:	Contains Np instantiations of FM_popold's mean vector. Note: Keeping an array of Np instantiations of the same vector is for coding convenience and clarity, but it is not very memory efficient.
FVr_bestmem:	Best-so-far population member.
I_iter:	Generation (iteration) counter.
I_nfeval:	Counter for function evaluations.

Figure 6.3 illustrates how deopt.m implements DE. All random vectors are taken from the shuffled versions FM_pm1, FM_pm2 and FM_pm3 of

FM_popold to exploit MATLAB®'s built-in matrix and vector manipulation routines.

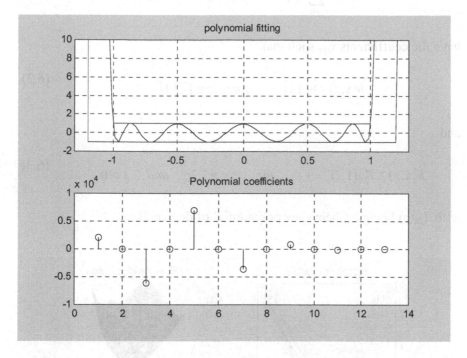

Fig. 6.4. An example of the graphics output for the polynomial fitting problem

6.1.4 How to Use the Graphics

The graphics routines in DeMat have been kept fairly independent of the objective function to give the user flexibility in deciding what to monitor. MATLAB®'s powerful and versatile graphics capabilities are well suited for this purpose. Setting the variable, I_plotting in the rundeopt.m file to "1" enables graphical monitoring via the plot function PlotIt() in PlotIt.m. Any other value for I_plotting disables plotting. It is good practice to disable plotting while trying to get a new optimization to work. Later, the plot function, PlotIt(), can be adapted to the new problem to provide a meaningful graphical output. Figure 6.4 shows an example of DeMat's graphical output. The example plotted is the Chebyshev polynomial fitting problem by Storn and Price (1997), which is repeated here for convenience.

Let

$$h(\mathbf{x},z) = \sum_{j=0}^{2k} x_{j+1} \cdot z^j , \quad k \text{ integer and } >0, \tag{6.1}$$

have the coefficients x_{j+1} such that

$$h(\mathbf{x},z) \in [-1,1] \qquad for \quad z \in [-1,1] \tag{6.2}$$

and

$$h(\mathbf{x},z) \geq T_{2k}(1.2) \quad +\varepsilon \quad for \quad z = \pm 1.2; \quad and \quad \varepsilon > 0 \tag{6.3}$$

with $T_{2k}(z)$ being a Chebyshev polynomial of degree $2k$.

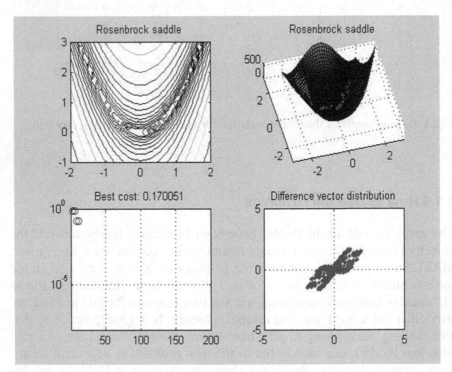

Fig 6.5. An example of a DeMat plot of the "Rosenbrock Saddle" that shows the results after generation 27.

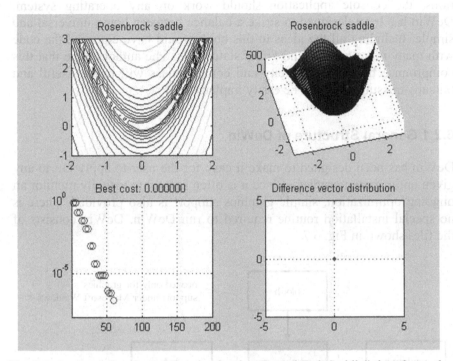

Fig. 6.6. An example of a DeMat plot for the "Rosenbrock Saddle" that shows the results after generation 75

The Chebyshev polynomials are recursively defined according to the difference equation $T_{n+1}(z) = 2z \cdot T_n(z) - T_{n-1}(z)$, with $n > 0$, integer and with the initial conditions $T_0(z) = 1$ and $T_1(z) = z$. The solution to the polynomial fitting problem is $h(\mathbf{x},z) = T_{2k}(z)$, a polynomial that oscillates between -1 and 1 when its argument, z, is between -1 and 1. Outside this "tube", the polynomial rises steeply in direction of high positive ordinate values. Figures 6.4, 6.5 and 6.6 illustrate the versatility of DeMat's graphics with an example that can be generated with the software on the accompanying CD.

6.2 DeWin – DE for MS Windows®: An Application in C

This section contains a brief overview of DeWin, a C-based application framework for solving function optimization problems with DE. The code provides easily modified graphics support for the MS Windows® operating system. By setting a compiler switch, the code can also be compiled as a console application devoid of any graphics support. After minor modifications, the console application should work on any operating system. DeWin has been designed to strike a balance between being universal and simple. Including all the ideas in this book would have cluttered the code with many more #define and #ifdef statements. The authors hope that this compromise between simplicity and completeness will prove useful and that any missing ideas can be easily implemented.

6.2.1 General Structure of DeWin

DeWin has been designed to make it easy for the user to apply DE to any given optimization problem. Since it is often helpful to visually monitor an ongoing optimization, simple graphics support is also provided. There is no special installation routine required to run DeWin. DeWin consists of the files shown in Fig. 6.7.

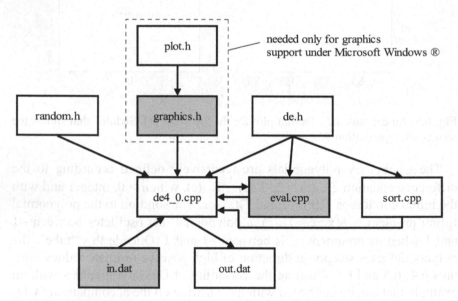

Fig. 6.7. The files required to compile, link, and run DeWin. The gray-shaded files must be modified to the demands of each new problem.

Here is a brief description of the DeWin files:

de4_0.cpp: This file contains the main DE engine, devol(). It also contains several different mutation and selection operators and handles both crossover and boundary constraints. In addition, this file controls the output of results. Defining the compiler switch DO_PLOTTING enables graphics support for programs compiled under Microsoft Windows® as Microsoft Windows® applications (see Fig. 6.8). If DO_PLOTTING is undefined, then the program must be compiled as a console application (see Fig. 6.9). This console version should work on other operating system, although tests have only been done for Microsoft Windows®. Another compiler switch, BOUND_CONSTR, enforces the parameter bounds given in fa_minbound[] and fa_maxbound[] (for definitions see below) throughout the optimization when BOUND_CONSTR is defined. If BOUND_CONSTR is undefined the parameter bounds are enforced during initialization only.

de.h: This file contains some general constants and, more importantly, the definition of a population member. The structure below defines a population member:

```
typedef struct
          //************************************
          //** Definition of population member
          //************************************
          {
              //parameter vector
              float fa_vector[MAXDIM];
              //vector of objectives (costs)
              float fa_cost[MAXCOST];
              //vector of constraints
              float fa_constraint[MAXCONST];
          } t_pop;
```

This structure contains not only the vector components (parameters), but also the objective function values (costs) and constraint arrays needed for multi-objective and/or constrained optimizations.

eval.cpp: This file contains the objective function, evaluate(), and the left_vector_wins() function that defines the circumstances under which the trial vector wins against the target vector. The left_vector_wins() function is the routine that takes constraints and objective function (cost) values into consideration when deciding which vector survives.

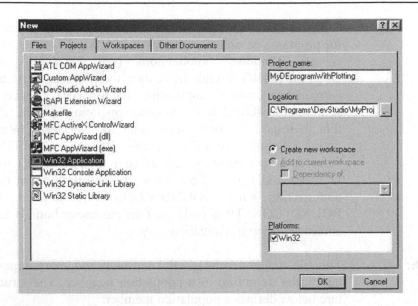

Fig. 6.8. An example of how to start building DeWin for Microsoft Visual C++ ®
(version 5.0) when DO_PLOTTING is defined. The Win32 application allows the
plotting routines to be used.

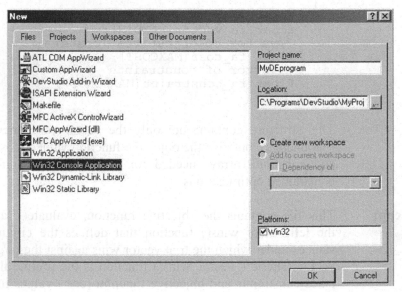

Fig. 6.9. An example of how to start building DeWin for Microsoft Visual C++®
(version 5.0) when DO_PLOTTING in undefined. In this case, only console output will be active.

random.h: This file contains the "Mersenne twister" random number generator created by Matsumoto and Nishimura (1998).

sort.cpp: This file contains the sorting function, sort(), that ranks the population array in ascending order where cost[0] is the sorting criterion (even if more than one cost exists). The sort() function is needed for the "best of parent and child" selection, i.e., $(\mu + \lambda)$ selection.

in.dat: This file holds the control variables that allow the user to experiment with various DE strategies, the number of population members, etc., without recompiling the entire program. An example of the values in in.dat is

```
3                choice of method i_strategy
10000            maximum number of iterations i_genmax
10               output refresh cycle i_refresh
19               number of parameters gi_D
200              number of parents i_NP
0.85             weighting factor f_weight
1.               crossover constant f_cross
1345             random seed i_seed
1                selection flag i_bs_flag
-1000            lower parameter bound
                 fa_minbound[0]
-1000            lower parameter bound
                 fa_minbound[1]
...
-1000            lower parameter bound
                 fa_minbound[18]
+1000            lower parameter bound
                 fa_maxbound[0]
...
+1000            lower parameter bound
                 fa_maxbound[18]
```

The source code contains further details on the purpose of these variables.

out.dat: At the end of the optimization, this file will contain information showing the convergence behavior, i.e., the improvement of the best objective function value (cost) over time. It also contains the final parameter values.

plot.h: This is the public domain graphics library written by Eric Brasseur and enhanced by the authors.

graphics.h: This file contains the graphic functions graphics_init(), draw_graph() and update_graphics(), all of which must be adapted to the demands of each problem.

6.2.2 Naming and Coding Conventions

DeWin's coding conventions follow a similar line of thought as those for DeMat. Constants defined with #define are in uppercase letters to distinguish them from variables. Except for the graphics library in plot.h, variable names are in Hungarian notation (Cusumano and Selby 1995), i.e., a prefix gives information about a variable's type. Hungarian notation helps to identify where typecasting is necessary and where assignments may lead to problems that the compiler does not report. Hungarian notation also warns of the possibility of side effects when accessing a global variable.

The prefix, which can consist of several characters, is followed by an underscore that is then followed by a descriptive name. The highest precedence in prefix construction is assigned to the character "g", which denotes a global variable. The next highest precedence goes to "p" for pointer variables. The next highest precedence goes to data representations like "f" for float, "i" for integer, etc. The lowest precedence has "a" for arrays. For example, a global array of floating-point variables would have the prefix gfa_, while a global pointer to a floating-point array would have the prefix gpfa_. For the sake of simplicity, simple loop counter variables may be named i, j, k ,etc., and do not need a prefix.

Table 6.2 shows the prefixes that this code uses. If needed, one may define additional prefixes, e.g., d_ for double-precision variables.

Most functions contain a comment header that gives information about both global variables and those that are passed as function arguments. An (I) indicates an input variable, whereas (I/O) stands for input/output and means that the function will change the value of the corresponding variable.

6.2.3 Data Flow Diagram

As in DeMat, the DFD indicates data by two parallel horizontal bars and functions by ellipses. Function names followed by () are explicitly named in the program. If () is missing, then the functionality is embedded in a sequence of statements, not an explicit function. Arrows indicate the data flow and contain limited additional time information. In general, time increases from left to right, or from top to bottom. Images of three-dimensional cylinders indicate files. Figure 6.10 shows the DFD of DeWin. Here are the meanings of the data in Fig. 6.10:

i_seed:	Seed value for the random number generator, which should be positive.
fa_minbound[]:	Array of upper parameter bounds.
fa_maxbound[]:	Array of lower parameter bounds. All parameters of a population vector are randomly initialized within the limits defined by fa_minbound[] and fa_maxbound[]. If the compiler switch BOUND_CONSTR is defined, then these bounds also serve as bound constraints.
i_genmax:	Maximum number of generations until optimization stops.
gi_NP:	Number of population members (named Np throughout the book).
gi_D:	Number of parameters (named D throughout the book).
i_strategy:	Selects the DE strategy. See code for the assignment of integer values to strategies.
f_weight:	Factor (named F throughout the book) used for scaling the differential mutation.
f_cross:	Crossover constant (named Cr throughout the book).
i_bs_flag:	Flag indicating which selection method is used. i_bs_flag = 1 (TRUE), enables "best of parent and child selection", i.e., $(\mu + \lambda)$ selection, whereas i_bs_flag = 0 (FALSE) enables DE's standard "trial vs. target" selection.
i_refresh:	Output refresh cycle. After i_refresh generations, a new set of values is plotted/printed.
gla_mt[]:	State array for the random number generator.
gl_mti:	State variable for the random number generator.
i_r1, …, i_r4:	Random variables $\in [0, i\_NP]$ that are mutually exclusive.
t_mean:	Contains the mean population vector of the current population.
gi_gen:	Generation (iteration) counter.
t_temp:	Temporary structure that holds the trial vector.
gt_best:	Current best population member.
ta_pop:	The population array that contains the current (old) and new populations side by side. Two pointers,

pta_old and pta_new, indicate the start of the corresponding array (Fig. 6.11).

Table 6.2. Naming variables in Hungarian Notation.

Variable	Prefix	Example	Remark
typedef struct	t	t_pop	Structure defining a population member
char	c	c_dummy	–
int	i	i_strategy	Integer variable where bit-width is not a major concern
long	l	l_iter	long is defined as 32 bits
float	f	f_x	–
global	g	gl_nfeval	A long variable which can be accessed globally
pointer	p	pt_pold	A pointer to a structure
array	a	ta_pop[2*MAXPOP]	An array of structures
File pointer	Fp	Fp_out	–

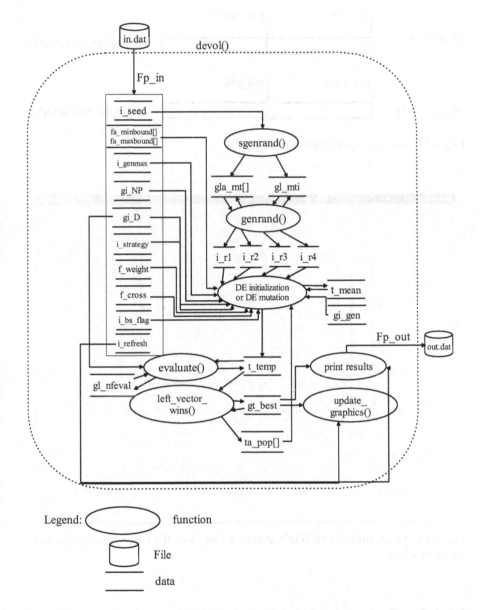

Fig. 6.10. A data flow graph of DeWin's most important parts

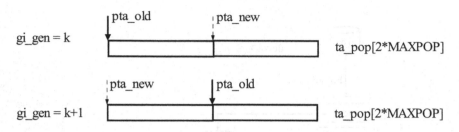

Fig. 6.11. Assigning populations to ta_pop[ ]

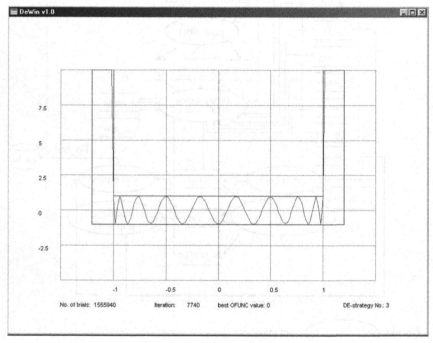

Fig. 6.12. An example of DeWin's graphics that plots the Chebyshev polynomial fitting problem.

6.2.4 How To Use the Graphics

The graphics in DeWin are only loosely coupled to the objective function so that the user can decide what the graphics should monitor. To keep code simple, DeWin's graphics capabilities are basic, yet zooming in and out is possible by drawing a rectangle around the region of interest with the left mouse button clicked (zoom in). A right-click of the mouse on the graph zooms out to the original scale. A right-click in the picture also clears out any artifacts that may occur.

Figures 6.12 and 6.13 plot the Chebyshev polynomial fitting problem described earlier.

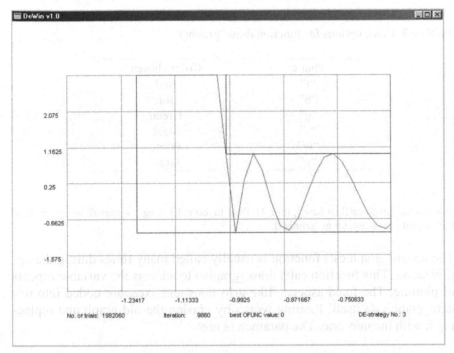

Fig. 6.13. Zoomed-in graph of the Chebyshev polynomial fitting problem

The basic graphics routines are located in plot.h, which has been written by Eric Brasseur and extended by the authors. Ordinarily, the plot.h file should not be changed. All functions needed to modify the graphics display are located in graphics.h.

6.2.5 Functions of graphics.h

void graphics_init(void):

This function defines the plotting ranges for the x– and y–axes. See the code for further details.

void draw_graph(float fa_params[], int i_D, char color):

This function defines what should be drawn on the screen. Usually fa_params[] holds the current best parameter vector and the variable i_D defines its dimension. The variable "color" defines the color in which the graph is drawn. The currently available options are shown in Table 6.3:

Table 6.3. Color options for function draw_graph()

char c	Color chosen
"r"	Red
"b"	Blue
"g"	Green
"s"	Black
"w"	White
"y"	Gray

void update_graphics(float best[], int i_D, float fa_bound[], long   l_nfeval, int i_gen, float f_emin, int i_strategy, int gi_genmax):

The update_graphics() function is usually called many times during the optimization. This function calls draw_graph() to address the variable aspects of plotting. The fixed aspects, like grid lines and axes, are coded into update_graphics() itself. Plotting works by erasing the old graph and replacing it with the new one. The parameters are:

```
**  Parameters    :best[]       (I)   Parameter vector
**                 i_D          (I)   Dimension of the parameter
**                                    vector
**                 fa_bound[]   (I)   Array defining a tolerance
**                                    scheme for the
**                                    current example
**                 l_nfeval     (I)   Current number of accumu-
**                                    lated function evaluations
**                 i_gen        (I)   Current number of accumu-
**                                    lated generations
**                 f_emin       (I)   Current best objective
**                                    function value
**                 i_strategy   (I)   DE strategy (coded as
**                                    a number)
**                 gi_genmax    (I)   Maximum number of genera-
**                                    tions
```

Some Basic Functions from plot.h

Below are the basic drawing functions in plot.h that will be of greatest help when writing a customized plot with draw_graph() and update_graphics().

void   fline(float x0, float y0, float x1, float y1, char c):

Draws a line from (x0,y0) to (x1,y1) with the color defined by the character variable c (see above).

void   fcircle(float xm, float ym, float radius, char c):

Draws a circle with radius "radius" around the point (xm,ym) with the color defined by the character variable c (see above).

void   frect(float xlu, float ylu, float xrl, float yrl, char c):

Draws a rectangle from the upper left upper corner (xlu,ylu) to the lower right corner (xrl,yrl) with the color defined by the character variable c (see above).

void   box(char c):

Draws a box around the plot with the color defined by the character variable c (see above).

void   grid(char c, int ix, int iy):

Draws ix and iy grid lines in the vertical and horizontal direction, respectively, and prints the tic labels. Drawing is done with the color defined by the character variable c (see above).

void   myprint(float x, float y, char *s):

Prints the string pointed to by *s starting at the location (x,y).

6.3 Software on the Accompanying CD

Figure 6.14 shows how the software on the accompanying CD is organized. There is a directory with code that has been exclusively written for this book. This code is in either MATLAB® or C, i.e., it is contained in the DeMat and DeWin directories. There is also code that has been contributed by various authors, which in many instances has been written before work on this book even started. Although most of this code does not contain the latest versions of DE, it has been included to offer a wider variety of languages. A third section contains demo programs that may be part of commercial packages that use DE. To use this code, refer to the relevant readme files.

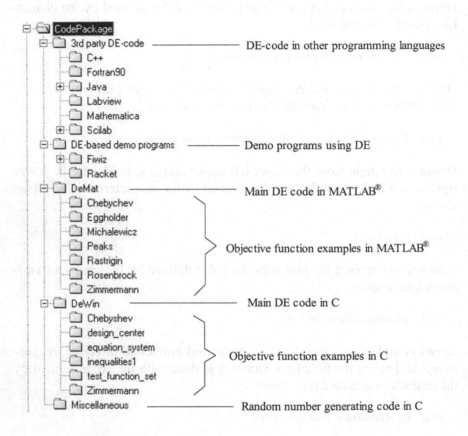

Fig. 6.14. The directory structure of the accompanying CD

Disclaimer of Warranty

THIS SOFTWARE AND THE ACCOMPANYING FILES ARE PROVIDED "AS IS" AND WITHOUT WARRANTIES AS TO PERFORMANCE OR MERCHANTABILITY OR ANY OTHER WARRANTIES WHETHER EXPRESSED OR IMPLIED. NO WARRANTY OF FITNESS FOR A PARTICULAR PURPOSE IS OFFERED.

Good data processing procedure dictates that any program should be thoroughly tested with non-critical data before relying on it. The user must assume the entire risk of using the programs.

MATLAB® is a registered trademark of The Math Works, Inc.

WINDOWS® is a registered trademark of Microsoft Corporation.

References

Cusumano MA, Selby RW (1995) Microsoft secrets. The Free Press, Simon & Schuster, New York

Matsumoto M, Nishimura T (1998) Mersenne Twister: A 623-dimensionally equidistributed uniform pseudorandom number generator. ACM Transactions on Modeling and Computer Simulation 8(1):3–30

Storn R, Price KV (1997) Differential evolution – A simple and efficient heuristic for global optimization over continuous spaces. Journal of Global Optimization 11:341–359

Yourdon E (1989) Modern structured analysis. Prentice Hall, Englewood Cliffs, NJ

7 Applications

The test bed of multi-modal functions is a valuable tool for benchmarking algorithmic performance. Test functions also make it easy to study how dimension, epistasis, the number of local optima and other variables affect an optimizer's performance. Even so, separable or symmetric test functions are typically easier to solve than epistatic, real-world problems of similar size, yet epistatic test functions with millions of local optima are much more difficult to minimize than the majority of real-world optimization tasks. Furthermore, real-world problems involve the art of objective function design, something that is absent when it comes to predefined test functions. A clever design of the objective function may considerably decrease the size of the search space, or convert the problem into one that is easier to solve. As an outcome, often a better solution may be obtained with less computation, and with a lower probability of failure. Therefore, the design of the objective function is as important a success factor as is the design of the optimization algorithm. Sometimes a good objective function design can even be crucial for the solution of the entire problem. This is also something to be learned from real-world examples.

Those who have real-world problems to solve are often the best judges of an optimizer's practical value. Their experiences more closely reflect both the true effort that an optimizer requires and the utility of the results that it provides. In addition, independent research with DE helps to ensure that a balanced view of the algorithm's abilities and potential emerges without any bias that DE's creators may inadvertently display in their analysis.

Taken together, test functions and practical examples give a more complete picture of DE's abilities than test functions alone can provide. This chapter is, therefore, devoted to presenting the experiences of scientists and engineers who have successfully applied DE to the optimization tasks that are central to their research. In part, these applications serve as "worked examples" that can help answer questions about DE that are not addressed in this book.

Several applications highlight DE's performance on highly epistatic, multi-modal optimization problems like the many-body problem in Sect. 7.1 and the inverse problems in Sects. 7.5, 7.10 and 7.11. In addition, several applications compare DE's performance to that of more traditional techniques like Powell's method (Sect. 7.5), the downhill simplex method (Sect. 7.5), simulated annealing (Sect. 7.1) and the simple genetic algorithm (Sects. 7.1 and 7.4). Sometimes DE is faster, sometimes not, but its success is less dependent on initial conditions than competing algorithms and it almost always yields an equal or better result that is more precise, especially when handing data with a wide dynamic range.

Like all random search algorithms, DE sometimes suggests novel or unconventional designs. Such was the case when DE designed an air compressor system with commercially available components (Sect. 7.3). On occasion, DE arrives at a solution that was not just unexpected, but also unintended. For example, when DE optimized a reflector's performance, its solution was to enclose the light source (Sect. 7.2)! DE has also found algebraic codes that not only perform near the theoretical limit, but also execute faster than their classically derived counterparts because they are simpler (Sect. 7.7). In fields like digital filter design, DE has produced designs that traditionally would require considerable expert knowledge (Sect. 7.8). DE's improvements over existing designs are often significant. When used in conjunction with finite element modeling, DE improved the performance of a radial active magnetic bearing by 8% (Sect. 7.9).

A few applications adapt DE for special purposes. For example, in Sect. 7.6, DE is implemented in parallel to perform image registration. Performance scaled almost linearly as the number of processors increased. Section 7.12 outlines how DE optimized parameters on-line to control plasma for semiconductor processing.

As these and other researchers have discovered, DE is robust, versatile, accurate and reliable. There is every reason to believe that other applications that adopt DE can reap benefits on a par with those reported in this chapter.

7.1 Genetic Algorithms and Related Techniques for Optimizing Si–H Clusters: A Merit Analysis for Differential Evolution

Nirupam Chakraborti

Abstract. Differential evolution (DE) has been successfully utilized to predict the ground state structures of a number of Si–H clusters. The computations are based upon a non-orthogonal tight-binding model developed for the Si–H system. The energy functional constructed from this model includes contributions of both electronic and pairwise interaction between the atoms and the energy minimization has been carried out using DE. The results are compared with the previous studies of these clusters where other tight-binding models, simple genetic algorithms and simulated annealing were used. Also, some specific advantages of DE over the other techniques have been identified and highlighted in this study.

7.1.1 Introduction

The need for newer and often exotic materials now seems to be ever increasing. The challenge now is not just to analyze the structure and properties of some existing materials, but to design, on the computer screen, some hitherto unseen clusters and molecules, tailor-made for some specific purpose. This has rendered the atomistic structure calculations of clusters and molecules of immense practical importance. Since the ultimate objective of most of the simulations is to identify a *ground state* corresponding to a global minimum of the energy functional, the problem essentially requires a fast and efficient optimizer, for which techniques like simulated annealing (SA) have been traditionally used, and genetic algorithms (GAs) are actually late entrants. However, some of the recent success stories of calculating the structures of large carbon molecules (Wang and Ho 1997), large polymeric materials (Keser and Stupp 2000) and various silicon (Morris et al. 1999) and Si–H clusters (Chakraborti et al. 1999, 2002) simply indicate that GAs are actually going to stay in this area. This indeed has a far-reaching consequence in the area of computational material design, as the GAs have actually rendered the search for the ground state energy minimum a thorough and efficient process. They are already showing an edge over a number of existing techniques and are definitely worthy of further exploration.

Some of the inherent shortcomings of the simple genetic algorithm (SGA), however, quite regularly show up in such complex computing processes. In the case of a large number of variables, typical of many molecules and clusters, mapping all of them in a binary format to acceptable accuracy requires manipulating lengthy arrays containing ones and zeros, and this often slows down the computation process to an unacceptable level. Furthermore, binary arithmetic has the implicit disadvantage of sometimes being stranded in a *Hamming cliff*. In this situation, as is well known, any small change in the real space would require a very large change in the corresponding binary. This halts progress of the solution and often makes fine convergence impossible in a near-optimal scenario. Furthermore, in SGA, the rate of mutation is not self-adjusting in nature. The requirements of mutation at the beginning and end of the computational process are, however, not necessarily the same. The user can, and sometimes does, induce an ad hoc adjustment of the mutation probability as the solution progresses, but a realistic estimate of the required temporal changes in the mutation probability still remains quite cumbersome. Finally, SGA is not geared for searching beyond the prescribed variable bounds, which, at least in few cases, may not be precisely known.

All such problems can be easily tackled in the computing environment prescribed by DE (Price and Storn 1997), where all the genetic operations are performed on the real-coded variables themselves, thereby totally eliminating the risks of ending up in a Hamming cliff. The concept of *mutating with vector differentials* that DE proposes automatically makes mutation a self-adjusting phenomenon, because the vector differential remains quite large for the initial random population and, analogous to a noise term, progressively decays as the population gradually converges. So far, it appears that DE has been tested for just one materials-related problem where it has successfully calculated the ground state structures of a number of Si–H clusters (Chakraborti et al. 2001). Since this particular system has also been studied by the *ab initio* method (Balamurugan and Prasad 2002; Prasad 2002), SA (Gupte and Prasad 1998; Gupte 1998) and SGA with and without *Niching* (Chakraborti et al. 1999, 2002), it is taken here as the paradigm case to establish the efficacy of DE for such types of studies. We begin with a description of the system model.

7.1.2 The System Model

The Basics and the Background

A study of Si–H clusters is quite important from a practical point of view because hydrogenated amorphous silicon is a highly promising opto-electronic material and during its formation through the glow discharge of silane (SiH_4) gas, various assemblages of silicon and hydrogen are known to play very important roles (Gupte 1998). As indicated above, in order to locate the ground state energy levels of the clusters, the computational task of the DE algorithm is to locate the atomic coordinates corresponding to the minimum energy values. The computing difficulties are enormous, since it is virtually possible to construct an infinite number of candidate solutions in a messy multi-dimensional *fitness landscape*, and the task of identifying the ground state solution there is often worse than locating the proverbial needle in a haystack! This is essentially an energy minimization problem; therefore, the fitness value can be taken as the negative of the energy functional. One needs a rigorous description of the energy, considering the interactions between the electrons, as well as the ionic cores of the atoms. This can be done from first principles devoid of any empirical or adjustable parameters, as attempted in the Car–Parrinello approach (Car and Parrinello 1985). Although known for their accuracy, such *ab initio* schemes are often unmanageably computing intensive. For the covalently bonded materials an excellent alternate is the *tight-binding* approach, which has been used in conjunction with DE (Chakraborti et al. 2001). Further details are provided below.

The Tight-Binding Model

The tight-binding approximation treats the system as consisting of ionic cores and an electron gas and is now very widely used for studying covalently bonded materials (Wang and Ho 1997). It attempts to calculate the total energy functional for the entire system $\left(E^{total} \right)$ by adding up the one - particle eigenvalues and the individual pair potential terms, such that

$$E^{total} = U_0 + E^{el} + E^{pair} \tag{7.1}$$

where the constant U_0 shifts the cohesive energy as prescribed by the user. E^{el} denotes the energy associated with the occupied eigenvalues of the electronic system and E^{pair} is the sum of pair potential terms caused by the repulsion between the ionic cores. Denoting the occupancy of the kth eigenstate as g_k, and N^{occ} as the number of occupied orbitals, the electronic contribution to the total energy is expressed as

$$E^{el} = \sum_{k=1}^{N^{occ}} g_k \varepsilon_k. \tag{7.2}$$

Furthermore, summing the pair potential terms related to repulsion between the ionic cores, E^{pair} is obtained as

$$E^{pair} = \sum_{i<j} \chi(r_{ij}). \tag{7.3}$$

Utilizing this basic definition for total energy, the wave functions of these eigenstates are given in terms of the non-orthogonal basis as

$$|\psi_n\rangle = \sum_i C_i^n \cdot |\phi^i\rangle \tag{7.4}$$

where $|\phi^i\rangle$ are the basis functions. In the non-orthogonal tight-binding theory employed here the basis functions are localized on each atom resembling its atomic orbital, and spherical harmonic functions (Y_{im}) are used to describe the angular parts. The characteristic equation is then expressed as

$$\sum_i (H_{ij} - \varepsilon_{ij} S_{ij}) C_j^{ij} = 0 \tag{7.5}$$

where H_{ij} denotes the Hamiltonian matrix elements between the ith and jth orbitals, such that

$$H_{ij} = \langle \phi^i | H | \phi^j \rangle. \tag{7.6}$$

The overlap matrix elements between them are expressed as

$$S_{ij} = \langle \phi^i \| \phi^j \rangle. \tag{7.7}$$

Further details of calculating the Hamiltonian and the overlap elements are provided elsewhere (Gupte 1998).

7.1.3 Computational Details

In order to apply DE to the present problem (Chakraborti et al. 2001), a Cartesian coordinate system was adopted in a cubic search space of 5 Å each side and a search was conducted for the atomic coordinates corresponding to the ground state structure. Out of the infinite atomic arrangements possible in this solution domain, the task of DE was to locate the unique configuration leading to the ground state with minimum energy. DE could perform this task quite satisfactorily. A DE code was tailor-made for this particular work, following the guidelines available in the literature (Price and Storn 1997). A population size of ten times the number of variables appeared to be adequate in most cases and a scheme for adjusting the mutation constant and crossover probability was evolved through systematic trial and error. For a number of clusters DE reached the near-optimal range rather quickly compared to the SGA-based studies performed previously (Chakraborti et al. 1999, 2002). The effect was more pronounced in some of the larger clusters: Si_6H, for example, was computed within just 350 generations. In general, however, a few hundred to over a thousand generations were necessary to resolve a structure, and often the computation for larger assemblages took less time compared to some of the smaller ones.

All the calculations were performed in a local area network of a number of Silicon Graphics workstations in the SG 200 Origin series. The calculations were very computationally intensive; a few even took several months to converge when submitted as background jobs in a multiple-users environment, and apparently there was a nonlinear increase in the problem stiffness with increasing problem dimension. Even then, the use of DE resulted in a considerable amount of savings in terms of CPU time, as the recent *ab initio* calculations for the same system (Balamurugan and Prasad 2002; Prasad 2002) took considerably longer than the evolutionary approach. In fact, DE could locate the near-optimal range rather quickly compared to some other techniques – SA, for example (Gupte and Prasad 1998; Gupte 1998). An *elitist* feature was introduced into the DE algorithm used in this problem, which turned out to be an absolute necessity for this sort of calculation. The movement of the elite was closely monitored and the decision for convergence was taken on the basis of the performance of the best individual, rather than an average member of the population.

7.1.4 Results and Discussion

Evolution of the Three-Dimensional Clusters

Some of the hydrogenated silicon clusters were previously studied using an *empirical tight-binding* (ETB) approach (Katircioğlu and Erkoç 1993). The three-dimensional nature of most of the clusters was not adequately highlighted in that work, as the cluster geometries could not be optimized. It has been possible to overcome this problem by coupling the DE with a tight-binding formulation. A typical case is shown in Fig. 7.1 for Si_2H_2. This is a symmetrical structure where the two silicon atoms are bound to each other and both the hydrogen atoms are bonded strongly to a silicon atom as the mirror image of each other.

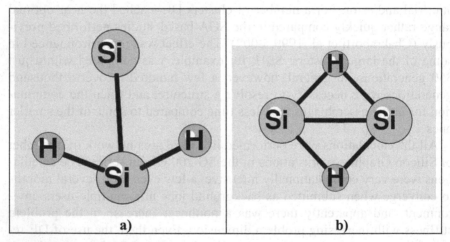

a)   b)

Fig. 7.1. The ground state structure of the Si_2H_2 cluster: **a** calculation by DE and SGA; **b** prediction from ETB model

As evident from Fig. 7.1, these structural features could not be fully revealed using the ETB formulation alone and, therefore, the optimization through DE has a very special role to play. The Si–H bond length is calculated as 1.54 Å. The calculated value of cohesive energy is 10.14 eV, which is in excellent agreement with the earlier computations using SGA and SA (Chakraborti et al. 1999, 2002; Gupte 1998). DE was able to resolve this structure approximately within 500 generations and, in fact, a near-optimal situation was obtained much sooner. A variable mutation constant between 0.02 and 0.3 and variable crossover probability between 0.5 and 0.9 were judiciously used over a population size of 120. The mechanisms of probability variation were evolved through systematic trial and error.

Several Si–H clusters have been calculated fairly recently using DE (Chakraborti et al. 2001). Here we will analyze a few selected ones, showing some characteristic trends.

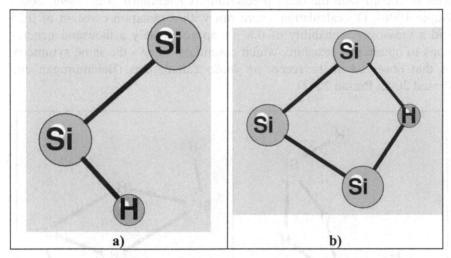

Fig. 7.2. Ground state structures of **a** Si_2H and **b** Si_3H calculated by DE and SGA.

The Symmetric and Asymmetric Clusters

Many of the clusters calculated in a recent study (Chakraborti et al. 2001) were highly symmetric, and a few were not. DE was successful in resolving both types, as shown in Fig. 7.2. Like other studies using SA (Gupte 1998) and SGA (Chakraborti et al. 1999, 2002), the calculations using DE have also determined the Si_2H structure as an asymmetric planar structure with the hydrogen atom located closer to one silicon atom than the other. The ground state cohesive energy is determined as -6.56 eV which is very similar to the values obtained in earlier calculations. Although the cohesive energy calculated through DE is actually identical to what was obtained through SGA (Chakraborti et al. 1999, 2002), the bond lengths are now slightly altered. Furthermore, some *ab initio* calculations (Balamurugan and Prasad 2002; Prasad 2002) reported a symmetric structure for this cluster. This perhaps suggests a significant multi-modality of the solution space, an important issue that warrants further analyses. Because of the structure's high asymmetry, calculation of this structure was a little cumbersome. It was necessary to run the DE code for about 400 generations to obtain complete convergence.

In the Si_3H structure two silicon atoms are situated symmetrically with respect to the third. The lone hydrogen atom is situated at the axis of symmetry. The structure is planar with an Si–H bond distance of the order of 1.8 Å, which is greater than that existing in the SiH cluster. DE determined the cohesive energy of the ground state structure as −10.61 eV, which is well in accord with the other predictions (Chakraborti et al. 1999, 2002; Gupte 1998). DE calculations were run with a mutation constant of 0.25 and a crossover probability of 0.8 for approximately a thousand generations to obtain this structure, which essentially shows the same symmetry as that observed in the recent *ab initio* calculations (Balamurugan and Prasad 2002; Prasad 2002).

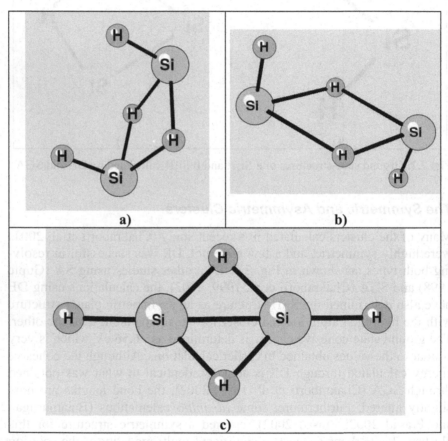

Fig. 7.3. The ground state structure of Si_2H_4 cluster: **a** DE and SGA calculation; **b** prediction by SA; **c** prediction from ETB model.

The Consequences of a Complex and Multi-modal Fitness Landscape

The energy functional that has been constructed here gives rise to a strong multi-modality in a complicated fitness landscape, containing closely spaced energy peaks. Negotiating such a rough terrain would be a Herculean task for any optimizer, and in this situation DE seems to have done a very commendable job. For example, in case of Si_2H_4 (Fig. 7.3), DE produced a structure energetically superior to those obtained by both SA and SGA (Gupte 1998; Chakraborti et al. 1999), fully resolving the symmetry predicted by ETB calculations (Katircioğlu and Erkoç 1993) in three dimensions. The structure shown in Fig. 7.3 corresponds to a cohesive energy of -17.02 eV, where the values computed through SA and SGA were -16.93 and -17.01 eV, respectively. This structure looks similar to what has been obtained through SGA (Chakraborti et al. 1999, 2002) but in variance with the structure determined through SA (Gupte 1998). Although DE converges to a lower minimum, energetically, the minima predicted by DE and SA are just 0.09 eV apart. The existence of two different minima at such a close proximity is indicative of a strong multi-modality, and the success of DE in locating the better of the set, without resorting to any *niching* strategy (Goldberg 1998), as was done in the case of SGA, speaks volumes for the excellent searching ability of this scheme. To ensure stability of the structure, the calculations were run for about 1300 generations. A mutation constant of 0.2 was mostly used along with a crossover probability of 0.8.

The Behavior of Hydrogen

The nature of the hydrogen bonding plays a key role in the stability of Si–H clusters. In amorphous hydrogenated silicon, hydrogen often forms some weak *dangling bonds* with silicon, which tend to deteriorate on absorption of some photon energy, rendering the material unsuitable for most practical applications. Evolutionary computing, through both DE and GAs, have revealed that hydrogen can be bonded to a single silicon atom or can be made to form a *bridge bond* with a number of silicon atoms. Two typical clusters, Si_4H and Si_2H_3, are shown in Fig. 7.4. In Si_4H the lone hydrogen is bonded to an silicon atom, while in Si_2H_3, two hydrogen atoms form bridge bonds with a pair of silicon atoms, and the remaining hydrogen is bonded to a single silicon atom.

The ground state structure of Si_4H is quite similar to that of Si_4 discussed earlier (Gupte 1998). The presence of hydrogen causes some distortion in the structure, but retains the essential geometric features of Si_4. To resolve this fully, the DE calculations were run for about 2000 generations with a mutation constant of 0.3 and a crossover probability of 0.8. The cohesive energy for the ground state structure was calculated as -14.81 eV, which is quite comparable to the values calculated by the other techniques (Gupte 1998).

The structure of Si_2H_3 obtained by DE (Fig. 7.4) qualitatively shows the same atomic arrangements as what was obtained earlier by both SGA and SA (Chakraborti et al. 1999, 2002; Gupte 1998). However, by some quirks of computing DE predicted the cohesive energy as -7.04 eV, as opposed to a value of -12.87 eV computed earlier. In fact, this is the only case where DE failed to reach the correct convergence and performed in an inferior way compared to both SGA and SA. The calculations were continued for about 900 generations, varying the mutation constant between 0.0001 and 0.2 and the crossover probability between 0.5 and 0.99, and an apparently premature convergence was obtained. The reasons for the poor performance of DE in this case are not clearly understood. Due to its unique reproduction strategy, DE essentially remains a *greedy scheme*, which converges really fast, but sometimes may lead to problems. However, this result can be taken as a very rare exception rather than a rule, as this problem was not encountered for any other clusters in the study.

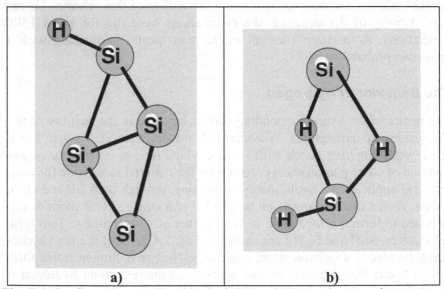

Fig. 7.4. Configurations predicted by both DE and SGA: **a** the ground state structure of the Si_4H cluster; **b** the ground state structure of the Si_2H_3 cluster

Hydrogen in Relatively Large Clusters

In all the clusters studied here by DE the hydrogen atoms were observed to occupy positions either outside or at the surface of the clusters. In no case was hydrogen found to be situated inside the clusters. Hydrogen also seems to form multi-centered bonds with two or more silicon atoms, where it is not strongly attached to any particular one of them. Also, it appears that even a single hydrogen atom can very significantly alter the geometry of a silicon cluster. Such findings are quite prominent, particularly in relatively large clusters like Si_5H (Fig. 7.5) and Si_6H (Fig. 7.6).

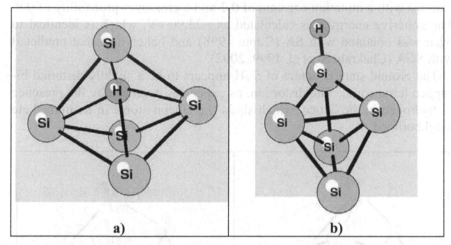

Fig. 7.5. Si_5H cluster: **a** the ground state structure predicted by DE; **b** a slightly higher energy structure obtained by *ab initio* calculations.

In the ground state configuration of Si_5H revealed through DE (Fig. 7.5), three silicon atoms are situated on the same plane as the lone hydrogen. The structure is symmetric along this plane with two silicon atoms symmetrically placed on two sides of it. This is quite deviant from the reported structure of Si_5, and in fact it is more like the structure of Si_6 (Gupte 1998). The hydrogen therefore causes a very large distortion in the Si_5 structure, which was also observed in the earlier studies. The cohesive energy was calculated as -23.90 eV, and with variable mutation constants between 0.1 and 0.3 and the corresponding crossover probabilities between 0.5 and 0.85, DE was able to obtain this relatively large structure within just 600 generations.

An alternate structure with slightly higher cohesive energy obtained in a recent *ab initio* calculation (Balamurugan and Prasad 2002; Prasad 2002) is also shown in Fig. 7.5. The hydrogen in this configuration is completely outside the silicon lattice and is attached to just one silicon atom. The multi-centered bonds in larger clusters therefore break with very little energy input, and this phenomenon may have some significant implications in selecting materials for opto-electronic devices like light sensors, thin film transistors, light-emitting diodes, etc., where hydrogenated amorphous silicon is currently emerging as a strong candidate.

Similar trends were also observed in Si_6H, and both the DE and SGA calculations have placed the hydrogen atom at the surface of cluster (Fig. 7.6). DE quite efficiently resolved this relatively large configuration. The ground state structure shown in Fig. 7.6 was obtained within just 350 generations with a mutation constant of 0.2 and a crossover probability of 0.9. The cohesive energy was calculated as −23.90 eV, which is identical to what was obtained with SA (Gupte 1998) and better than that predicted with SGA (Chakraborti et al. 1999, 2002).

The ground state structure of Si_6H appears to be a slightly distorted bicapped tetrahedron. The distortion, as expected, is caused by the presence of hydrogen in the lattice, which shifts the silicon atoms in its immediate neighborhood.

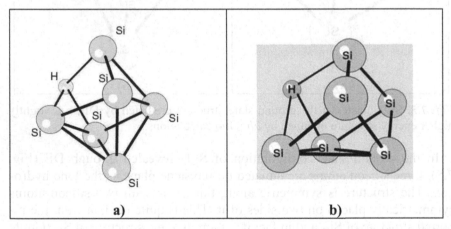

Fig. 7.6. The ground state structure of the Si_6H cluster: **a** DE calculation; **b** SGA calculation

An Evaluation of the Performance of DE

From the experience of studying Si–H (Chakraborti et al. 2001), a clear picture emerges regarding the efficacy of DE. A few salient points are highlighted below.

- Real coding in DE requires far less data storage, and the absence of any mapping between binary and real numbers makes it execute faster. These features make DE an ideal candidate for the calculation of large clusters, where the CPU time is crucial.
- Except for one single exception, DE in the present context always produced results either comparable to or better than those obtained by SGA and SA.
- DE effectively is a greedy scheme, preferring either the better child or the better parent. This leads to a speedy convergence in most cases, but may backfire in some.
- The self-adjusting mutation scheme in DE is a definite advantage.
- DE can negotiate non-smooth fitness landscapes, as encountered in the present study.
- The introduction of elitism in DE was found to be highly beneficial in the present case.
- At least during this investigation, DE resolved very closely spaced multi-modality without resorting to any niching strategies.

For large computations of this sort, the power of DE can perhaps be more efficiently realized by developing a parallel version of the current DE methodology. Future researchers in this area might closely explore such an option.

7.1.5 Concluding Remarks

The success of DE use demonstrated here could be of immense significance to the general area of computational materials science, because the methodology described here can be easily extended to materials like giant fullerenes and carbon nanotubes and can be effectively tried out for a large number of hitherto unsolved materials-related problems. Applications of GAs have opened up a new pathway in materials research, particularly in materials design. This can now be further enriched by the adoption of DE, and the time is quite ripe for its vigorous promotion by materials researchers at large.

References

Balamurugan D, Prasad R (2002). Ground state structures and properties of Si_2H_n Clusters. In: Chakraborti N, Chatterjee UK (eds.), International conference in advances in materials and materials processing (ICAMMP-2002), February 1–

3, 2002, Indian Institute of Technology, Kharagpur, India. Tata-McGraw-Hill, New Delhi, pp 609–613

Car R, Parrinello M (1985) Unified approach for molecular dynamics and density-functional theory. Physical Review Letters 55(22):2471–2474

Chakraborti N, De PS, Prasad R (1999) A study of Si-H system using genetic algorithms and a tight binding approach. Zeitschrift für Metallkund 9093:508–513

Chakraborti N, Misra K, Bhatt P, Barman N, Prasad R (2001). Tight-binding calculations of Si-H clusters using genetic algorithms and related techniques: studies using differential evolution. Journal of Phase Equilibria 22(5):525–530

Chakraborti N, De PS, Prasad R (2002) Genetic algorithms based structure calculations for hydrogenated silicon clusters. Materials Letters 55():20-26

Goldberg DE (1989) Genetic algorithms in search, optimization and machine learning. Addison-Wesley, Reading, MA

Gupte GR (1998) Molecular dynamics studies of small hydrogenated silicon clusters and hydrogenated amorphous silicon. Ph.D. dissertation, Indian Institute of Technology, Kanpur

Gupte GR, Prasad R (1998) Ground state geometries and vibrational spectra of small hydrogenated silicon clusters using nonorthogonal tight-binding molecular dynamics. International Journal of Modern Physics B 12(15):1607–1622

Katircioğlu S, Erkoç S (1993) Empirical tight-binding total energy calculation for Si_nH_{2m} (n = 1 to 6, m = 1 to 3) Clusters. Physica Status Solidi (b), 177(373):373–378

Keser M, Stupp SI (2000) Genetic algorithms in computational materials science and engineering: simulations and design of self-assembling materials. Computer Methods in Applied Mechanical Engineering 186:373–385

Morris JR, Deaven DM, Ho KM, Wang CZ, Pan BC, Wacker JG, Turner DE (1999) Genetic algorithm optimization of atomic clusters. In: Davis LD, DeJong K, Vose MD, Whitley LD (eds.), Evolutionary algorithms. IMA volumes in mathematics and its applications, vol 111. Springer, Berlin Heidelberg New York

Prasad R (2002) Ground state structures of small hydrogenated silicon clusters. In: Chakraborti N, Chatterjee UK (eds.), International conference on advances in materials and materials processing (ICAMMP-2002), February 1–3, 2002, Indian Institute of Technology, Kharagpur, India. Tata-McGraw-Hill, New Delhi, pp 741–748

Price K, Storn R (1997) Differential evolution: a simple evolution strategy for fast optimization. Dr. Dobb's Journal 22(April):18–24

Wang CZ, Ho KM (1997) Materials simulations with tight-binding molecular dynamics. Journal of Phase Equilibria 18(6):516–529

7.2 Non-Imaging Optical Design Using Differential Evolution

David Corcoran and Steven Doyle

Abstract. The application of differential evolution to non-imaging optical design is explored here. The objective is to create a mirror shape which reflects light from a source to produce a desired light distribution in some target region. Differential evolution uses a cost measure to numerically determine the quality of a proposed solution against a desired solution and various cost measures specific to non-imaging optical design are examined. A reverse engineering strategy is used to test the design methodology for a point light source, which lends insight into the differential evolution approach, and validates it for two geometric classes of problems. In these the target distribution comes from either a parabolic mirror shape or an elliptical mirror shape. The methodology is also validated for an extended light source.

7.2.1 Introduction

Luminaire reflectors can be found in car head-lamps, lighting fixtures, indeed anywhere there is some form of artificial lighting. It may be surprising but the design of such reflectors can be a time-consuming and costly exercise taking of the order of years to complete in the case of car head-lamps. The reason is the trial and error approach which is adopted by the designer. Software packages which aid in the design process exist, yet they still require an interactive procedure of varying the design and then testing it over many iterations, a process which is not just time consuming but must eventually fail as the complexity of a design increases. Observation of the design search space for such problems demonstrates the necessity for a global optimization process (Doyle et al. 1999a). Prompted by the genetic algorithm work of Ashdown (1994), we have explored the use of differential evolution (DE) to automate non-imaging optical design and found that the technique is not only valid but also feasible (Doyle et al. 1999a, 1999b, 2001).

7.2.2 Objective Function

The task of a non-imaging optical designer is that, given a light source, a reflector design is required which will cast a desired distribution of light on some target surface (henceforth referred to as the target distribution). The objective function must therefore provide a means of calculating a light distribution and measuring the difference between this and the target distribution. In this regard the key elements of the objective function are a ray tracer and a cost measurement.

Ray Tracer

The distribution from a light source reflector combination can be calculated using a ray-tracing approach as illustrated in Fig. 7.7.

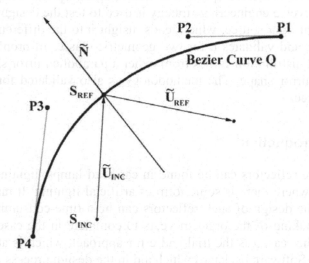

Fig. 7.7. Ray tracing a Bezier curve

In ray tracing one makes use of the geometric nature of light: that is, light travels in straight lines and upon reflection from a smooth surface the angle of incidence of a light ray is equal to its angle of reflection. A light source can then be considered a source of light rays with each ray being described by a parametric vector of the form

$$\mathbf{R} = \mathbf{S} + t_R \mathbf{U} \tag{7.8}$$

where $\mathbf{R}$ is the ray position, $\mathbf{S}$ is a vector locating the source of the ray, $\mathbf{U}$ is the ray direction and t_R is the ray parameter with $t_R \in [0, 1]$.

Mathematically a mirror can be represented by a Bezier curve (Bezier 1974), a continuous parametric curve determined by a small number of parameters called control points (see Fig. 7.7). In two dimensions a cubic Bezier curve can be written as

$$Q(t_B) = \begin{bmatrix} x(t_B) \\ y(t_B) \end{bmatrix} = (1-t_B)^3 \mathbf{P}_1 + 3t_B(1-t_B)^2 \mathbf{P}_2 + 3t_B^2(1-t_B)\mathbf{P}_3 + t_B^3 \mathbf{P}_4 \tag{7.9}$$

where $\mathbf{P}_i$ are the control points $\begin{bmatrix} P_{ix} \\ P_{iy} \end{bmatrix}$ and t_B is the Bezier parameter, $t_B \in [0, 1]$.

The parameters $\mathbf{P}_1$, $\mathbf{P}_4$ and $\mathbf{P}_2$, $\mathbf{P}_3$ determine the end position points of the curve/mirror and the magnitude and direction to the curve at these points. Changing the control parameters thus allows a continuum of potential mirror designs $\mathbf{P} = [\mathbf{P}_1, \mathbf{P}_2, \mathbf{P}_3, \mathbf{P}_4]$ to be generated. The objective function can then be written in general terms as $f(\mathbf{P}) = C$, which is a cost (C)-based function to be minimized using DE.

To calculate the resultant light distribution for a particular mirror design, $\mathbf{P}$, rays are launched discretely and are reflected from the mirror according to

$$\mathbf{U}_{ref} = \mathbf{U}_{inc} - 2\mathbf{N}\left(\frac{\mathbf{U}_{inc} \cdot \mathbf{N}}{|\mathbf{N}|^2}\right) \tag{7.10}$$

where $\mathbf{N}$ is the normal to the mirror at the point of intersection, the latter obtained from solving for the incident ray $\mathbf{R}_{inc} = Q(t_{inc})$. The reflected ray is then described by

$$\mathbf{R}_{ref} = \mathbf{S}_{ref} + t_R \mathbf{U}_{ref} \quad \text{where} \quad \mathbf{S}_{ref} = Q(t_{inc}). \tag{7.11}$$

Multiple reflections are also allowed by iteration in which $\mathbf{R}_{inc}$ becomes $\mathbf{R}_{ref}$ and the process continues to some desired limit imposed by computational constraints. The resultant distribution can then be determined by calculating the intersection of the reflected rays with bins or ray-collecting elements distributed over the target surface. The area for ray collection is called the intercept region (see Fig. 7.8 for an example). As the purpose here is mirror and not source design, rays that are cast directly from the source to the target surface without reflection are not considered in the calculation of the resultant distribution.

Cost Measurement

A figure of cost for a particular mirror design can be based in general terms on some numerate difference between the target and resultant distributions. We have investigated various cost measurements (Doyle et al. 2001) and it is instructive to examine a few examples to gain an insight into the operation of DE for design problems.

In preliminary work, problem-specific cost measurements were used (Doyle et al. 2001). For example, a focusing solution could be achieved by requiring DE to maximize the number of reflected rays in a single intercept bin, of narrow angular width with respect to the source, accomplished by applying the cost measurement

$$C = T - R \tag{7.12}$$

where T is the total number of rays and R is the number of intercepting rays. Interestingly, observation revealed that DE first attempted mirror designs that would result in any rays being directed toward the intercept region, then later attempted designs that maximized the reflected ray number. This points toward an inherent strength of DE: that is, it is capable of finding not only an optimum solution but also developing a solution strategy.

To generalize the approach to non-specific design problems, the angular width of the intercept region was extended and divided into a number of ray-collecting bins. The following simple cost measurement was applied where T_i is now the desired number of rays in a collecting bin, R_i the number reflected into that bin and N the number of bins:

$$C = \sum_{i=1}^{N} |T_i - R_i|. \tag{7.13}$$

There is, however, an underlying difficulty with this approach: that is, the cost biases the selection of mirror solutions toward those with divergent ray reflections. A mirror design proposed by DE which casts rays outside the intercept region can potentially have a lower cost than one which casts rays into the region but in the wrong bins. To circumvent this the number of these outer rays was added as a penalty to the cost:

$$C = \sum_{i=1}^{N} |T_i - R_i| + R_\infty. \tag{7.14}$$

The ray number R_∞ is determined by counting those reflected rays that intersect the target surface but not the finite region in which the intercept bins are located. This is equivalent to including an infinite intercept region, though note that individual mirror shape ultimately determines how much of this region is available.

Remarkably, with the new cost measurement implemented, it was found that when the target was in the far field (taken to be a distance several times the mirror dimension) the generated optimum mirror shapes often *enclosed* the light source preventing *any* rays from reaching the intercept region. To understand this, consider first that the initial population of trial mirrors used by DE is generated randomly. As the binning region, i.e., the region containing the N collecting elements, is small in comparison to the intercept region, one might expect the initial cost of the population members to be

$$C \sim \sum_{i=1}^{N} |T_i - \alpha R| + (1 - \alpha)R \quad \text{or} \quad C \sim \sum_{i=1}^{N} |T_i| + (1 - 2\alpha)R \tag{7.15}$$

where R is the total number of rays reflected and α is the fraction of these rays in the binning region. It is assumed here that $\alpha \le 0.5$ and R_i, the number of rays reflected into each bin, is less than T_i. A closed mirror solution will therefore have the lower cost

$$C \sim \sum_{i=1}^{N} |T_i| \tag{7.16}$$

particularly in the far field (as α decreases with distance) and DE simply selects the optimum solution. It must be stressed that DE is working correctly; the issue is that a simple cost measurement may have hidden unexpected behavior. The final version of the cost measurement and the one used in the results detailed below (unless otherwise stated) solves this problem by the inclusion of a component that penalizes any difference between the total desired ray number in the binning region and the total reflected ray number which reaches the entire intercept region:

$$C = \sum_{i=1}^{N} |T_i - R_i| + R_\infty + \left| \sum_{i=1}^{N} T_i - \sum_{i=1}^{N} R_i - R_\infty \right|. \tag{7.17}$$

7.2.3 A Reverse Engineering Approach to Testing

To investigate the DE approach to non-imaging optical design, it is useful to choose test problems for which the solution is already known. In this regard two classes of mirror design problems have been examined (Doyle et al. 1999a), one in which the output light distribution diverges/converges and one in which the light distribution is maintained. These correspond to elliptical and parabolic mirror designs respectively. Casting the light distribution from either of these mirrors on a distant surface produces a target distribution for a known mirror shape. Using this as the target for the objective function, DE should in principle be capable of reverse engineering the problem to determine the original mirror shape. A DE algorithm (Storn and Price 1997) was implemented initially with a population size of 40, a crossover constant of 0.2 and a noise scaling factor of 0.8. Target distributions were generated on planes from point light sources at near, middle and far fields defined by the reducing angular width of the distributions with respect to the source. Further details of the parameters for the parabolic and elliptical problems can be found in our earlier work (Doyle et al. 1999a).

Sample results are presented in Fig. 7.8 for a near-target distribution generated by a parabolic mirror with the point source located at its focus. In this figure the generating mirror is presented with the solution obtained by DE. Overall for this problem the average cost (from ten computational runs), after a maximum of 10^4 iterations and expressed as a percentage of rays launched, was 2%. The low average cost and visual inspection of Fig. 7.8 demonstrate that the DE approach has performed exceptionally well. Interestingly, while all mirror solutions presented produce close to, and in Fig. 7.8c exactly, the desired distribution, only in one of these cases is the mirror close to being the generating parabolic shape. This is a recurrent feature of the DE solutions. It arises because there is more than one solution to the design problem and the DE strategy is general enough to be able to find these solutions. From the point of view of a designer this offers the very beneficial property of flexibility. For instance, in addition to meeting a physical requirement one might choose from the collection of designs one which is aesthetically pleasing. As the target distribution is removed to further distances and its angular width decreases with respect to the source, the design problem becomes more difficult. It has been observed that at the benchmark 10^4 iterations the average cost increases to 18% and 38% in the middle and far field. Nevertheless by increasing the number of iterations by a factor of four it has been shown that the cost for a middle-field problem can be reduced to 5% (Doyle et al. 1999a).

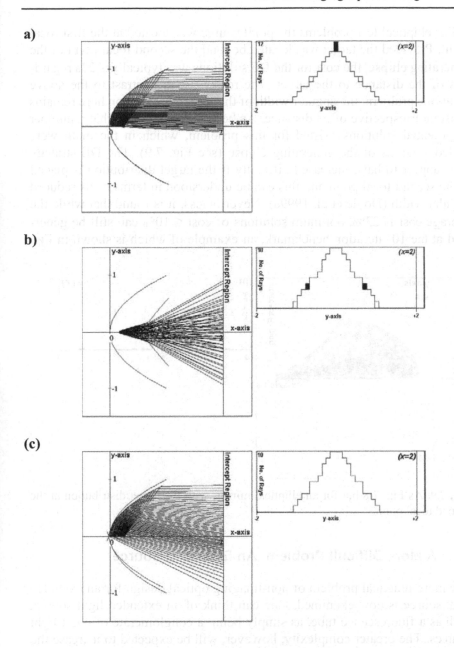

Fig. 7.8. Sample results from the DE approach to mirror design. On the left a selection of ray-traced optimum mirror designs proposed by the strategy. The parabolic mirror used to generate the target distribution is also shown. On the right the overlaid target distribution and resultant distributions are shown. Black and white boxes represent an excess and deficit respectively.

For elliptical test problems the point source was located at the first focal point. Provided the target was located beyond the second focal point of the generating ellipse, the cost for the DE solutions was typically ≤ 2% regardless of the distance to the target. The reason is, in contrast to the above parabolic problem, the angular width of the target distribution here remains constant irrespective of its distance. As before, it was found that a number of potential solutions existed for this problem, which in the main were scaled versions of the generating ellipse (see Fig. 7.9). The DE strategy does appear to have increased difficulty if the target distribution is placed at the second focal point and this can be understood in terms of its reduced angular width (Doyle et al. 1999a). Nevertheless, it is found that while the average cost is 23%, optimum solutions of cost ≤ 10% can still be generated at the 10^4 iteration benchmark, an example of which is shown in Fig. 7.9.

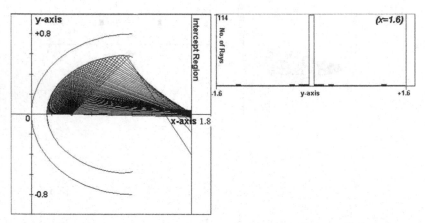

Fig. 7.9. As Fig. 7.8 but for an elliptical mirror, with the target distribution at the second focal point.

7.2.4 A More Difficult Problem: An Extended Source

The more practical problem of non-imaging optical design for an extended light source is now examined. One can think of an extended light source, such as a fluorescence tube, as simply being a conglomerate of point light sources. The greater complexity, however, will be expected to increase the number of computations necessary for ray tracing a mirror design.

The choice of DE strategy and corresponding crossover and scaling factors was considered with the objective of improving computational efficiency. To this end a parabolic middle-field distribution with point light source (see above) was selected as the desired or target distribution and a selection of eight DE strategies applied (Doyle et al. 1999b). For each strategy, three values, 0.5, 0.7 and 0.9, were used for both the crossover and noise scaling factor, and a cost was established based on an average of eight computational runs benchmarked at 5×10^3 iterations. The DE/best/2/bin strategy produced the lowest average cost of 3%, for a crossover of 0.9 and noise weighting of 0.7. This is the method used to generate the results presented below. Within statistical error the strategy DE/rand/1/bin would also have been acceptable giving an average cost of 3.4% for a crossover of 0.9 and noise weighting of 0.9. The large values for the crossover and noise in both cases indicate a bias toward the random exploration of the problem search space.

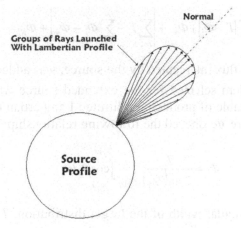

Fig. 7.10. A circular light source model

A circular light source model was selected as the extended source (see Fig. 7.10). In photometry the quantity of light power is the luminous flux Φ (Pedrotti and Pedrotti 1993). The circular source here was discretely sampled at a number of equally spaced points and each point allowed to generate a Lambertian distribution of rays at regular angular intervals, i.e., one following a luminous flux profile of $\Phi = \Phi_o \cos \theta$. The geometry selected for the target region was also circular, centered on the source, and consequently the angle subtended by the ray-collecting bins remained constant. Accordingly, the initial cost measurement used for the extended source was

$$C = \sum_{i=1}^{N} \left| T_i - \Phi_i \right| + \Phi_\infty + \left| \sum_{i=1}^{N} T_i - \sum_{i=1}^{N} \Phi_i - \Phi_\infty \right| \qquad (7.18)$$

$$\underbrace{}_{\text{Term 1}} \qquad \underbrace{}_{\text{Term 2}}$$

where in each element T_i is the target flux, Φ_i the reflected flux and Φ_∞ is the reflected flux outside the ray-collecting region.

An implication of an extended source is that by virtue of its size, it may obstruct reflected rays and in reality might prevent them from reaching the target region. To allow for this, any reflected rays found to intersect the circular source were excluded. In addition, it was recognized that a mirror design which reduced the ray number from reaching the target would be a poorer design than one that does not. The second term in the cost measurement above already penalizes such mirror designs implicitly, as a reduction in luminous flux reaching the overall intercept region would cause this term to increase. An explicit penalty scheme was also explored:

$$C = \sum_{i=1}^{N}|T_i - \Phi_i| + \Phi_\infty + \left|\sum_{i=1}^{N}T_i - \sum_{i=1}^{N}\Phi_i - \Phi_\infty\right| + \Phi_{RET} \tag{7.19}$$

where Φ_{RET}, the flux intercepted by the source, was added to the cost.

The test problem selected for the extended source was to determine a mirror shape capable of producing a limited Lambertian output defined by us to be one where Φ_i obeyed the following relationship:

$$\Phi_i = \frac{T}{2\sin\left(\alpha/2\right)} \int_{(i-1/N-1/2)\alpha}^{(i/N-1/2)\alpha}\cos\theta\,d\theta \tag{7.20}$$

where α is the angular width of the target distribution, T is the total luminous flux in the target distribution, θ is the angular position in the target (with respect to the source) and N is the number of ray-collecting elements. In addition, we have explored the behavior of DE strategy as the source size is increased, α is decreased and T is increased. Details can be found in our earlier work (Doyle et al. 1999b).

As in the case of the point source, over a wide range of the parameters explored DE performed exceptionally well with typical costs of ~6%. For 80% of the cases studied it was found that the costs of the mirror solutions proposed by the explicit scheme were lower than those from the implicit scheme. Moreover, the fraction of luminous flux returned to source was also always lowest in the explicit approach. While the cost was observed to increase with target size reduction and source size increase, this was understood in terms of the limitations imposed on the search space size. For example, if we restrict the size of the mirror to be comparable to a mirror size, clearly there is a physical limitation to the optimum solution DE can provide.

7.2.5 Conclusion

The use of DE has been explored as a design methodology for non-imaging optics, in which the objective is to generate a mirror shape, for a given light source, that produces a desired light distribution. Overall it appears that the approach is adept at generating mirror solutions for two broad classes of design problem, namely parabolic and elliptical target distributions produced for a point light source. The approach has also been validated for a more realistic extended light source.

The investigation has revealed some interesting features with regard to the general application of DE to design problems. It has been observed, for example, that in addition to generating a solution, design strategies for obtaining a solution may emerge. With regard to the cost measure, one must caution against naively choosing a measure that simply indicates a numerical difference between a desired design quantity and a solution-generated quantity. As has been seen here, this can lead to unexpected results which, although meeting the cost measure criteria, fall far short of the design criteria. The case in point is where DE generates a mirror design that simply encloses the light source to achieve a minimum cost. Lastly, one should be aware that for certain design problems there may be more than one potential solution in the search space. The distinct advantage of DE here is that its stochastic nature allows these to be discovered, presenting the designer with a possible choice of potential designs.

References

Ashdown I (1994) Non-imaging optics design using genetic algorithms. Journal of the Illumination Engineering Society 3(1):12–21

Bezier P (1974) Mathematical and practical possibilities of UNISURF. In: Barnhill RE, Riesenfeld RF (eds.), Computer aided design. Academic Press, New York, pp 127–152

Doyle S, Corcoran D, Connell J (1999a) Automated mirror design using an evolution strategy. Optical Engineering 38(2):323–333

Doyle S, Corcoran D, Connell J (1999b) Automated mirror design for an extended light source. Proceedings of the SPIE 3781:94–102

Doyle S, Corcoran D, Connell J (2001) A merit function for automated mirror design. Journal of the Illumination Engineering Society 30(2):3–11

Pedrotti FL, Pedrotti LS (1993) Introduction to optics. Prentice Hall International, London

Storn R, Price K (1997) Differential evolution – A simple evolution strategy for fast optimization. Dr. Dobb's Journal 22(4):18–24 and 78

7.3 Optimization of an Industrial Compressor Supply System

Evan P. Hancox and Robert W. Derksen

Abstract. This section demonstrates a modified version of differential evolution (MDE) that produces the n best solutions for the real-world problem of selecting the optimum combination of compressor supply system components. The selection is based on the plant's unique compressed-air requirements. The cost function considers the initial purchase price and the cost of operating over a user-specified number of years. The results demonstrate MDE's ability to produce non-intuitive solutions. Data representing 4060 unique industrial plants is presented to demonstrate the effect of various settings of F, K, and population size on convergence rates. The performance of MDE was verified using test cases with known solutions in order to obtain 100% certainty of the results presented here.

7.3.1 Introduction

Differential evolution (DE) was applied to the real-world engineering problem of selecting equipment for an industrial compressor system. Due to the fact that compressor system components are available in discrete sizes, databases of actual components available in the marketplace were created that represented each of the parameters in the population vectors. Databases were created using information retrieved from freely distributed catalogue software from *ZeksPro*^{TM} (ZEKSPRO 2002) and *Compressed Air Systems* (Compressed Air Systems 1999).

The industrial compressor system can be broken into two parts: the supply side and demand side. For the system studied here, the supply side (Fig. 7.11) consisted of air compressors, pre-filters to remove any compressor oil or other mists from the air stream that may damage the desiccant material in the air dryer, desiccant air dryers that remove moisture from the air stream down to a dew point of −40°C, after-filters to remove any desiccant dust that may carry over from the dryer, storage receivers (tanks), and a proportional–integral–derivative (PID) flow controller. The flow controller facilitates high-pressure storage in the receiver on the supply side and precise supply to the low-pressure demand side. The demand side represents the remaining system piping and end use equipment.

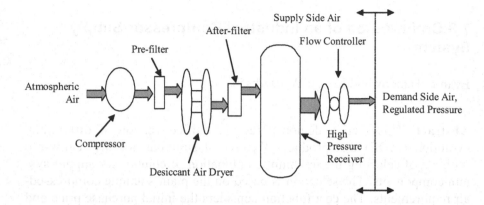

Fig. 7.11. A schematic layout of the components in a pneumatic supply system. Air enters the system from the atmosphere and passes through each component bank in the sequence indicated. The bank consists of an array of a sufficient number of identical components to meet demand. This problem has six degrees of freedom as we have five separate discrete values that we can select, namely the compressor, pre-filter, desiccant air dryer, after-filter and flow controller; and one continuous value, the receiver volume.

7.3.2 Background Information on the Test Problem

Typically, industrial compressors are sized according the largest expected flow conditions plus a safety factor to ensure adequate supply. Other air treatment equipment is then sized to allow maximum delivery rates of the compressors. Where downtime is seen as unacceptable, one or more additional compressors of full capacity are installed for backup purposes. Finally, the sizing of storage receivers is often based on a rule of 1 US gallon for each standard cubic foot per minute (*scfm*) of compressor capacity (Compressed Air Systems 1999). This approach simply does not afford the optimum setup.

7.3.3 System Optimization

Engineers are beginning to recognize the importance of improving conditions on the demand side, along with the need for flow controllers and increased supply side storage (Foss 2002). These improvements increase the efficiency of the system and reduce operating costs. However, the overall system is not as efficient as it could be: system designers still rely on rules of thumb and the advice of equipment suppliers when it comes to sizing receivers and other supply equipment. It is the authors' contention that in order to achieve a complete peak operating efficiency, the entire system must be optimized. The test case presented here considers optimization of the supply side of the system and assumes the demand side has been effectively designed or upgraded.

It is only necessary to consider one type of industrial compressor system, to show that DE can be easily applied to a real-world application such as this. Including other system types would require expansion of the objective function and mean little to the investigation. The test problem assumes the use of rotary screw compressors, mist/oil eliminator pre-filters, desiccant air dryers, after-filters, storage receivers and flow controllers. Further, the addition of a wider range of component types only serves to make the number of catalogue entries larger and make the search take longer. We elected to limit both the number of components within a catalogue and the types of components considered, as characteristics of the search should be the same.

While the test program optimizes the supply side, it does so according to the fluctuating compressed-air requirements of the demand side. We can construe the demand as a variable that can be broken into different constituents made up of profiles.

7.3.4 Demand Profiles

For purposes of the test program, a demand profile is described as the demand for compressed air over a given period and is rated in scfm (standard cubic feet per minute). The operational workweek is divided into different segments in an attempt to define the overall consumption rates into unique flow profiles: off hours, base and peak demand.

Off hours represents the smallest consumption rate and may occur overnight or on weekends. *Base* demand represents the overall average consumption rate that occurs during normal operation. *Peak* demand depicts the enduring highest consumption rate that can occur at peak times of operation. For example, peak rate can occur at the beginning or end of any shift or while a high-demand process runs.

After determining the demand rates for each of the above flow profiles, it is necessary to determine the accumulated hours per week over which each profile occurs. The hours are accumulated and are not necessarily contiguous. In other words, peak hours are a summation of timed events added up over the week and can occur during base and off hours flow profiles, or they can include long-duration events that may represent an entire work shift. Finally, the test program requires the total number of hours per year for each of the above profiles.

The test program requires three more flow types that must be identified in order to fully describe typical flow conditions. Maximum plant flow is the absolute maximum sustained demand that can be created at any given time. *Spike* demand is demand that may occur for a very short duration, on the order of a few minutes, that surpasses the output capacity of all compressors. Surge demand is a similarly sudden, short-duration increase in demand that surpasses *currently loaded* compressors.

7.3.5 Modified Differential Evolution; Extending the Generality of DE

The optimization method selected for this project was DE. The theory and details of DE are thoroughly discussed in Price (1999) and will not be expanded on here. This selection was based on previous experience of finding the optimum shape of turbo machinery cascade aerofoil profiles (see Rogalsky et al. 2000). The conclusion of this work was that DE was highly effective at finding the global minimum, as it did so without fail in a comparison with other optimization methods. This becomes more important as the topology of the solution space becomes more complex. As is the case when the number of degrees of freedom and nonlinearity increase, the number of minima and maxima also increase. This, as well as the ability to deal with integer and discrete parameters, stimulated the interest in this work.

Modified differential evolution (MDE) extends the generality of DE for engineering design, by seeking more than a single best solution. In the case of the test program, producing n best solutions allows an engineer to deal with unique circumstances after the fact simply by selecting the appropriate best solution that fits that set of circumstances.

The main advantage of finding the n best solutions is that the designer has more choices. For example, the test program often produced results that showed that different component combinations had minor differences in overall costs over the test period, leaving the final selection to the designers' preference. Preference cannot easily be modeled into an objective function.

Table 7.1 lists the ten best solutions from one of the test cases. Each solution is a unique combination of components that vary in model, size or number of specific components, and is capable of meeting operational requirements. The narrow range of total costs clearly demonstrates the value of the n-best approach.

Table 7.1. An example of the *n*-best solutions (*n* = 10) obtained using MDE

Configuration	Total costs	Component change from global best
Best	$207,053	Best system
2nd	$207,290	Larger after-filter, fewer replacement cartridges < service
3rd	$207,385	Single larger receiver (18 ft high) (best had two smaller units, 13.8 ft high)
4th	$207,386	Six smaller receivers (8 ft high)
.	.	
.	.	
.	.	
10th	$207,738	Single larger receiver (18 ft high), larger after-filter, fewer replacement cartridges

7.3.6 Component Selection from the Database

Separate databases were created for each of the components, based on supplier catalogues, with information such as brake horse power (*BHP*), pressure drops (*ΔP*), throughput capacities and list prices. Each parameter was represented by a floating-point value equal to its integer index in the catalogue. This allows the parameter to be treated as a continuous variable. Checking that the value is within the appropriate range for the parameter, and randomly selecting a value within the range if it is outside, ensures that the boundary constraints of the database are maintained. To avoid a downward bias and ensure that the final database entries can be selected, the floating-point value is incremented by 0.5 and converted to integer value when the database was referenced.

7.3.7 Crossover Approaches

Biased Toward the Best-So-Far Vector

Initial trials focused on simply tracking the *n best-so-far* candidates in an array, but still relied on the single *best-so-far* for mutation purposes. Tests began with the following crossover function (Storn and Price 2002) DE/rand-to-best/exp:

$$\text{child[n]} = \text{child[n]} + F \cdot (\text{Current\_best[n]} - \text{child[n]}) + \tag{7.21}$$

$$F \cdot (\text{Parent[r1][n]} - \text{Parent[r2][n]}).$$

While the single best solution was consistently located, this approach often missed some of the n best values. This behavior was due to the effect of the current best vector, which pulled the entire population in its direction very quickly. This of course was the intended purpose of DE's inventors Storn and Price. However, this influence proved too great to ensure finding all n local minima along with the global minimum.

Non-Biased Approach

Another test was devised using a non-biased approach, which used the following crossover function (Storn and Price 2002) DE/rand/1/exp:

$$\text{child[n]} = \text{child[n]} + F \cdot (\text{Parent[r1][n]} - \text{Parent[r2][n]}). \tag{7.22}$$

While a non-biased crossover method showed some success in preliminary investigations, it was very slow to converge to all n-best solutions and was soon abandoned. The question became, how could DE be forced to find all n minima, while doing so with the speed and accuracy the original inventors intended for the absolute minimum?

New Approach

To resolve the speed/accuracy convergence problem for n best solutions, a solution was devised based on one of Storn and Price's original approaches, namely "DE/rand-to-best/exp". The new approach uses a biased crossover method, but randomly biases the search toward one of the n best-so-far as opposed to the single best-so-far. This has the effect of speeding up the search while maintaining the integrity and spirit of the original scheme.

The new version, which we can denote "DE/rand-to-rand-n_best/exp" in keeping with the nomenclature set out by Storn and Price (Price 1999), is as follows:

$$\text{child[n]} = \text{child[n]} + K \cdot (\text{n\_best[m][n]} - \text{child[n]}) + \tag{7.23}$$

$$F \cdot (\text{Parent[r1][n]} - \text{Parent[r2][n]}).$$

Note that the [m] in the two-dimensional array represents the random selection from the n best list, which causes the population to converge toward all n best minima and not just the global minimum. This new approach proved to be successful and was used for the data collection and analysis of F and K. The n best list was maintained in a fashion depicted by the following pseudo-code:

$$\text{if(current < n\_best) insert\_candidate();} \tag{7.24}$$

where the insert_candidate() function places the candidate in the list according to cost.

Objective Function

Once a component is selected the number of components was computed to satisfy minimum capacities based on demand requirements. In this case, compressors are added until their combined output exceeds the peak output required during plant operation. Finally, the number of compressors is saved. A similar approach is repeated for each component except the receiver.

The volume of the receiver was calculated as the sum of the spike and surge volumes. The spike and surge volumes were calculated using

$$\tag{7.25}$$

$$\text{spike volume} = \frac{t_{spike} \cdot (\text{spike flow scfm}) \cdot P_{atm} \cdot SF}{\Delta P_{allowable}}$$

where the spike volume is the volume to be added to the receiver capacity in ft^3, t_{spike} is the duration of spike in minutes, spike flow scfm is the flow rate above total compressor capacity, P_{atm} is atmospheric pressure, SF is a user-defined safety factor, and $\Delta P_{allowable}$ is the allowable pressure drop in the receiver. A similar formula was defined for the surge volume, but *surge flow scfm* represents the flow rate above that of the currently loaded compressors.

Now that there are enough components to assemble a workable system, the cost of the system is determined across the three main flow regimes. The objective function determines the cost of the system by summing the initial capital requirements and the accumulated cost of operating the system for a given time period – five years for the test program. The operational costs include the cost of electricity and replacement filters.

The off hours flow profile performance and cost is computed as follows. First the total capacity of operating compressors must be equal to or greater than off hours demand. This approach allows us to tailor the number of operating compressors for each demand profile. Now, the portion of the trim compressor output that will contribute to repressurizing the receiver is given by

$$\text{Xscfm} = (\text{\# operating compressors}) \cdot (\text{compressor scfm}) - \tag{7.26}$$
$$(\text{off hours demand scfm})$$

The flow the receiver will contribute to the system demand while the trim compressor is in the unloaded state is given by

$$\text{Yscfm} = (\text{off hours demand scfm}) - ((\text{\# operating compressors}) - 1) \cdot \tag{7.27}$$
$$(\text{compressor scfm}).$$

We can find the time required to repressurize the receiver from

$$t_1 = (\text{receiver volume}) \cdot \Delta P_{allowable} / (\text{Xscfm} \cdot \text{Pa}), \tag{7.28}$$

and the time that the receiver will supply the trim portion of the air with the trim compressor in the unloaded state from

$$t_2 = (\text{receiver volume}) \cdot \Delta P_{allowable} / (\text{Yscfm} \cdot \text{Pa}) \tag{7.29}$$

The cycle time of the trim compressor is $t_1 + t_2$, and must be within the compressor manufacturer's specifications to prevent overloading of the motor.

We assume a proportional pressure drop occurs across units and calculate the drop as

$$\Delta P_{prefilters} = \left(\frac{1}{\text{\# prefilters}} \right) \left(\frac{\Delta P_{low} + \Delta P_{high}}{2} \right) \left(\frac{\text{off hours scfm}}{\text{prefilter scfm}} \right) \tag{7.30}$$

The pressure drops for the dryers and after-filters are found in the same way. Then the total pressure drop becomes

$$\Delta P_{total} = \Delta P_{prefilters} + \Delta P_{dryers} + \Delta P_{afterfilters}. \tag{7.31}$$

The cost of operating the compressors while refilling the receiver occurs during t_1, which was found earlier. The actual BHP load during this time must account for an extra 0.5% to overcome each 1 psig pressure drop (Compressed Air Systems 1999):

$$\text{BHP}_{actual\_1} = t_1 \cdot (\text{\# operating compressors}) \cdot \text{BHP}_{loaded} \cdot \Delta P_{total} \cdot \tag{7.32}$$
$$1.005$$

The units for BHP_{actual_1} are (BHP $\cdot$ minutes). The actual BHP load during t_2 must account for the unloaded trim compressor and the remaining fully loaded compressors. An unloaded compressor is running but not producing air. Hence

$$\text{BHP}_{actual\_2} = t_2 \cdot [(\text{\# operating compressors} - 1) \cdot \text{BHP}_{loaded} \cdot \Delta P_{total} \tag{7.33}$$
$$\cdot 1.005 + \text{BHP}_{unloaded}].$$

The cost of purged air from the heat-less desiccant dryers is based on the volume and cost of air purged for a given profile:

$$\text{purge volume} = \frac{\%\text{purged} \cdot \text{profile volume}}{1 - \%\text{purged}} \qquad (7.34)$$

$$\text{total purge volume} = \left(\text{purge volume}\,\frac{\text{ft}^3}{\text{min}}\right)\left(\frac{60\,\text{min}}{\text{hr}}\right)\left(\frac{\text{profile hrs}}{\text{yr}}\right)(\text{yrs}) \qquad (7.35)$$

$$\text{hrs to produce purged air} = \left(\text{total purge volume ft}^3\right)\left(\frac{\text{min}}{\text{compressor ft}^3}\right)\left(\frac{\text{hr}}{60\,\text{min}}\right) \qquad (7.36)$$

$$\text{cost of purged air} = \qquad\qquad\qquad\qquad\qquad\qquad (7.37)$$

$$(\text{hrs to produce}) \cdot \text{BHP}_{loaded} \cdot \left(\frac{0.7457\,\text{kW}}{\text{HP}}\right)\left(\frac{\$}{\text{kW} \cdot \text{hr}}\right)\left(\frac{1}{\eta_{motor}}\right)$$

where η_{motor} is motor efficiency, and BHP_{loaded} is the horse power draw of the compressor while the unit is producing air. The total cost of the profile becomes

$$\text{cost of off hours} = \qquad\qquad\qquad\qquad\qquad\qquad\qquad (7.38)$$

$$\left(\frac{\text{profile hrs}}{\text{yr}}\right)\left(\frac{\text{BHP}_{actual\_1} + \text{BHP}_{actual\_2}}{\text{cycle time}}\right)\left(\frac{\$}{\text{kW} \cdot \text{hr}}\right)\left(\frac{\text{yrs}}{\eta_{motor}}\right) + \text{cost of purged air.}$$

The calculations shown to this point for the off hours profile are repeated for the base and peak profiles. Finally, the total cost of operating the components (during each profile) over the study period is added to the total capital cost of all components along with the replacement cost of all filter elements.

The number of filter elements required for the pre-filters and after-filters is determined for the study period. The replacement interval will be the rated element life scaled by the ratio of total rated throughput to total actual profile flow rate. The total number of elements required over the study period is simply the time multiplied by the element life.

Thus, the study period has a direct effect on the selection of the components in that as the years of the study increase, the effect of the purchase price reduces and the effect of operating the components increases.

No consideration was given to interest rates and the cost of electricity was fixed. However, this is a minor point considering that the test software is meant simply to demonstrate that MDE will work on a real-world industrial problem.

7.3.8 Testing Procedures

Population sizes were tested at increments of 60, from 120 up to 3000 members. Each population size was tested on (up to) 4060 unique input profiles which were defined to cover the full range of the databases, and represented 4060 separate and unique industrial compressed-air demand profiles. In order to speed data collection, MDE was required to converge to all n best values within an arbitrary ceiling of 8,064,000 objective function evaluations. This number matched the complete comparison of all possible parameter combinations for the test program.

If the modified DE failed to converge (within the arbitrary ceiling) to the n best for any one of the 4060 input variations during the testing of a given population size, that population size was discarded. This is not to say that MDE would have failed to converge for the given population size without the arbitrary ceiling, but no further testing was conducted in this direction. Upon successful convergence for all 4060 input variations or failure for one, the population size was incremented and the inputs began again from the initial settings. Figure 7.12 shows the results of the data collection.

7.3.9 Obtaining 100% Certainty of the Results

To facilitate data collection when testing for multiple best solutions, it became necessary to know ahead of time, with 100% certainty, the n best solutions that the MDE software would hunt for. Therefore, a separate program was written using the same input parameters, settings and objective function that the MDE program would use, but that evaluated every possible combination of components. An arbitrary value of 10 was chosen for n, thus the ten best solutions for each input profile were saved in a uniquely named file that was subsequently loaded for the MDE trials with the corresponding input profile. This way the ten best solutions were known prior to testing the MDE approach. This approach also meant that each unique profile setting required only one analysis with the separate program that compared all combinations of components

Extensive testing was not conducted with other values of n. The known solution of the ten best vectors had no influence on the MDE process and was accessed simply to verify convergence and reduce the time it took to collect data. To that end, code was modified so that the MDE version of the program would load the correct answers and, after each generation, it would compare the list of values it had found with the correct values. When all values matched, the MDE program would cease to run. The number of objective function evaluations required to converge to the ten best solutions was tracked by updating the current convergence count at the time of each cost improvement. A standalone version of the test program limits the number of function evaluations to some arbitrary number greater than the expected convergence rate based on experimentation.

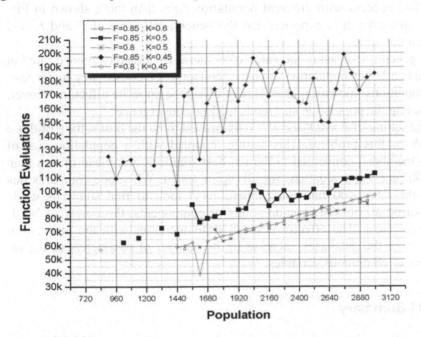

Fig. 7.12. Test results showing the average number of function evaluations required to find the ten best solutions as a function of population size, F and K.

7.3.10 Results

The results of this test program showed that the optimum system changes from favoring single compressor systems to dual compressor systems when longer time intervals were examined. This is not what an engineer would intuitively expect, and indicates the importance of DE as a design tool.

Each point on the graph in Fig. 7.12 represents the average number of function evaluations it took to converge to the ten best solutions in 4060 unique tests of a given population size.

The tested values of $F = 0.5$ with $K = 0.5$ failed to converge within the arbitrary ceiling for at least one of the 4060 input variations for each population size tested – hence the absence of representative data for that setting.

Similarly, the absence of data depicting a setting of $F = 0.8$ and $K = 0.6$ is due to the same cause.

The results seem to indicate some of the outer limits of the values of F and K that can ensure convergence under the test conditions enforced here. It is quite likely that exact limitations will be problem specific and others will find success with different population sizes than those shown in Fig. 7.12. However, it is expected that the general influence of F and K will remain the same.

In general, increasing the value of K increases the convergence rate but requires a larger population to ensure convergence consistency under testing conditions. Increasing the value of F has the opposite effect; moreover, increasing the population size tends to slow convergence.

The setting of $F = 0.8$ and $K = 0.45$ proved to be the best setting of those tested for this problem. This setting, coupled with a population size of 1620 members, converged to the ten best solutions at a mean of less than 40,000 function evaluations for all inputs tested. When one compares this with the 8,064,000 function evaluations it took to find solutions while comparing all possible combinations of components, the power of MDE becomes clear. The effectiveness of MDE increases with the increase in the size of the databases, increase in the number of parameters or the inclusion of continuous variables.

7.3.11 Summary

The results of this work clearly show the beneficial value of MDE for design optimization. The method maintains the speed and properties of DE, and reliably retrieves the n best solutions to the component selection. The n best multiple solutions allow the designer choice without having to code preference into the objective function. Ultimately, the designer can easily select the most effective combination of components without giving up the flexibility of preference.

The results demonstrated that DE and its offspring MDE can result in the natural discovery of non-intuitive solutions to a design optimization. This was demonstrated by the finding that the optimum number of compressors would change as the study period increased; using current design methods most engineers would assume that number to be constant. We have concluded that DE and MDE are the most appropriate choice for the construction of design optimization codes.

This section also demonstrated that increasing the value of K increases convergence rate but requires a larger population to ensure convergence consistency. Increasing the value of F has the opposite effect and increasing the population size tends to slow convergence. These findings should be valid for any version of DE but the precise values of F, K and population size that prove the most effective will be problem specific.

References

Compressed Air Systems (1999) 1999 Buying guide and technical handbook. Compressed Air Systems, Division of Hydra-Pak, Inc.

Foss S (2002) Compressed air systems workshop. Dissertation, Greenwood Inn, Winnipeg, Manitoba, Canada, May 1 & 2

Price KV (1999) An introduction to differential evolution. In: Corne D, Dorigo M, Glover F (eds.), New ideas in optimization. McGraw-Hill, London, pp 79–108

Rogalsky T, Kocabiyik S, Derksen R (2000) Differential evolution in aerodynamic optimization. Canadian Aeronautics and Space Journal 46(4):178–183

Storn R, Price KV (2002) C code for differential evolution. Available at: www.icsi.berkeley.edu

ZEKSPRO^{TM} (2002) Compressed air treatment specifier, Version 2.1, ZEKS COMPRESSED AIR SOLUTIONS^{TM}, West Chester, PA 19380, USA

7.4 Minimal Representation Multi-Sensor Fusion Using Differential Evolution

Rajive Joshi and Arthur C. Sanderson

Abstract. We present the application of differential evolution to solve a class of multi-sensor fusion problems commonly encountered in building intelligent robotic systems. The class of multi-sensor fusion problems is characterized by a set of sensors observing a common environment model. The observed data features are modeled as a projection or a "view" of an (unknown) underlying environment model into the sensor space, with unknown uncertainty and correspondence transformation injected in the observation process. The observed features may include outliers or may correspond to environment model features, via an unknown correspondence mapping.

The goal of multi-sensor fusion is to find the *best* environment model identity, transformation parameters and the correspondence mapping that map the model features to observed data features. We use the minimal representation size criterion to formulate the model search problem. The minimal representation approach is based on an information measure as a *universal yardstick* for fusion, and provides a framework for integrating information from a variety of sources. The minimal representation size criterion favors the selection of the simplest explanation that is the most likely explanation of the observed multi-sensor data.

We develop a differential evolution approach to the search for minimal representation multi-sensor fusion solutions. Laboratory experiments in robot manipulation using both tactile and visual sensing demonstrate that differential evolution is effective at finding useful and practical solutions to this problem for real systems. Comparison of this differential evolution algorithm to traditional genetic algorithms shows distinct advantages in both accuracy and efficiency.

7.4.1 Introduction

Multi-Sensor Fusion and Model Selection in Robotics

Multi-sensor fusion is a central problem in robotic systems, where interaction with the environment is critical to successful operation. It is a key component in systems capable of interacting with their environments and making semi-autonomous decisions based on sensory data to accomplish various manipulation, navigation and assembly tasks.

Figure 7.13 shows an example of multi-sensor fusion used to guide manipulation of an object, taken from our laboratory setup. The *Anthrobot* (Ali and Engler 1991) five-fingered hand grasps an object, and senses the contact points with the surface of the object using tactile sensors. The tactile sensors extract a touch position in the kinematic reference frame of the hand. In addition, a CCD camera views the position of the same object and extracts vertex/edge features of the object image. Both the tactile features and the visual features are related to the position and orientation ("pose") of the object, and in practice we wish to combine these two sources of information to improve our ability to accurately manipulate the object.

In this setup, the hand is visible in the field of view of the camera, and introduces extraneous vertex features in the image. Typically, about half of the image vertex features are due to the hand occluding parts of the object – thus the vision data alone may not be sufficient to estimate the three-dimensional pose of the object. We used a five-fingered hand, and for typical grasps, at most three distinct object surfaces are contacted by the fingertips. The tactile data from these contacts is usually not rich enough to uniquely estimate the three-dimensional object pose by itself. By fusing the vertex features from the image and the contact points from the fingertips, we expect to correctly estimate the object pose and also the feature correspondence in the two data sets. The fusion of the tactile and image feature data is used to derive an improved estimate of the object pose which guides the manipulation. In general, the object shape must also be identified, from a library of possible object shapes.

In a typical multi-sensor fusion problem in robotics, such as the tactile–visual example above, we can choose from a number of environment model structures, environment model parameters, uncertainty models and correspondence models. The uncertainty and the registration/calibration models are often chosen a priori, whereas the sensor constraints can be obtained from physical modeling. However, in complex robotic environments, the a priori choice of the environment model structure, parameters and correspondences is difficult. The object may belong to some library of parameterized object models with corresponding choices of data scaling and data subsample selection as precursors to the pose estimation itself. In this context, three important model selection issues must be addressed:

- **Environment model class selection:** What is the environment model class? How many parameters are required to specify it? What is the parameter resolution?

- **Environment model parameterization and data scaling:** What are the values of the environment parameters? How should the data from different sensors be scaled to determine these parameter values?
- **Data subsample selection:** Which data features are used to determine the environment model parameters? What subset of the data is consistent in the definition of the pose for a given estimator? What data features should be considered outliers?

Fig. 7.13. The *Anthrobot* five-fingered hand holding an object in the field of view of a fixed camera. The contact data obtained from tactile sensors mounted on the fingertips is fused with the processed image data obtained from the camera, to estimate the position and orientation of the object.

Much of the recent progress in multi-sensor fusion (Luo and Kay 1989; Hager 1990; Abidi and Gonzalez 1992; Kokar and Kim 1993; Dekhil and Henderson 1998; Joshi and Sanderson 1999b) for robotics has been based on the application of existing statistical tools to (a) estimate the position of objects with known geometric models (Smith and Cheeseman 1986; Durrant-Whyte 1987; Nakamura 1992), (b) estimate the parameterized shape of an object from sensor information (Bolle and Cooper 1986; Allen 1988; Porill 1988; Shashank et al. 1988; Eason and Gonzalez 1992) and (c) estimate a probability distribution of objects or object features (e.g., surfaces) based on sensor models (Elfes 1989). These methods require a priori selection of the model class, number of parameters and data subsamples used in estimation. The model selection problem is complementary to the estimation process itself and is intended to choose an effective combination of model structure and estimation method for a given class of problems.

The multi-sensor fusion and model selection framework (Joshi and Sanderson 1994, 1996; Joshi 1996) uses a *minimal representation size* criterion (Segen and Sanderson 1981) to choose among alternative models, number of parameters, parameter resolutions and correspondences. The *representation size* or description length of an entity is defined as the length of the shortest length program that reconstructs the entity. The observed data, thought to be arising from a library of environment models, is encoded with respect to one of these models. The minimal representation size criterion selects the model which minimizes the total multi-sensor data representation size, and leads to a choice among alternative models which trades off between the size of the model (e.g., number of parameters) and the representation size of the (encoded) residuals, or errors. Intuitively, the "smaller", less complex, representation is selected as the preferred model for a given estimator.

Searching for the Best Interpretation

The minimal representation size interpretation is obtained by minimizing the multi-sensor fusion criterion. For the tactile–visual multi-sensor fusion problem (Fig. 7.13), the search space is the cross-product of the continuous six-dimensional pose parameter space (Sect. 7.4.3), the discrete space of possible correspondences between the observed vision data and the object vertices, and the possible correspondences between the observed tactile data and the object faces. Finding the best pose and correspondences is generally recognized to be a difficult search problem (Grimson and Lozano-Pérez 1984; Linnainmaa et al. 1988; Huttenlocher and Ullman 1990; Joshi 1996; Joshi and Sanderson 1999b).

We use a *differential evolution* program to search for the best interpretation from the data collected in the laboratory experiments. As developed in this work, the search problem is difficult because we have posed it as a broad search over many general pose configurations (local minimal), and have chosen not to impose heuristic constraints to simplify the search. In practice, there are many such heuristics which may be imposed for specific problems, and they would often improve the execution time of the search.

The methodology presented here is a general approach to fusion which is not restricted to geometric pose estimation, and in fact may be applied to a variety of problems in model identification from a wide perspective, including model-based identification of parameters from noisy data and prioritization in noisy data sets.

7.4.2 Minimal Representation Multi-Sensor Fusion

Generic Multi-Sensor Fusion Framework

The general model-based multi-sensor fusion problem (Joshi 1996; Joshi and Sanderson 1999b) is summarized in Fig. 7.14 which shows S sensors observing an unknown environment, described by an *environment model*, $q_E = \Xi(\theta)$, where Ξ denotes the model structure, and θ denotes a particular parameter instantiation. In our representative problem, an environment model is a polyhedral-shaped object with six associated pose parameters.

Each sensor observes some set of M *model features*, $Y = \{y\}_{1,M}$, and produces a set of N *data features*, denoted by $Z = \{z\}_{1,N}$. A data feature, z, may be related to a *model feature*, y, by a general constraint equation h:

$$h(y; z) = 0. \tag{7.39}$$

Some observed data features may not be related to the underlying environment model; these are referred to as *outliers* or *unmodeled* data features, with a symbol y_0, and may lie anywhere in the sensor measurement space Υ.

The association between the observed data features and the model features (or y_0) is given by a *correspondence*, ω, that maps data feature indices to model feature indices:

$$\omega : \{1, \ldots, N\} \rightarrow \{0, 1, \ldots, M\} \tag{7.40}$$

where the special index 0 is used to denote outliers. The correspondence is often unknown, and may be *many-to-one*, as is the case for tactile sensors where several contact points may correspond to the same object face; or *one-to-one* as is the case for vision sensors where at most one image vertex point may correspond to an object vertex.

The mapping from an environment model to the model features is denoted by a *model feature extractor*, F, where $\{y\} = F(q_E; \alpha)$ is a function of the environment model and the sensor calibration. Thus, for the tactile sensor, this mapping extracts the object faces from the instantiated shape structure, while for a vision sensor it extracts the visible object vertices.

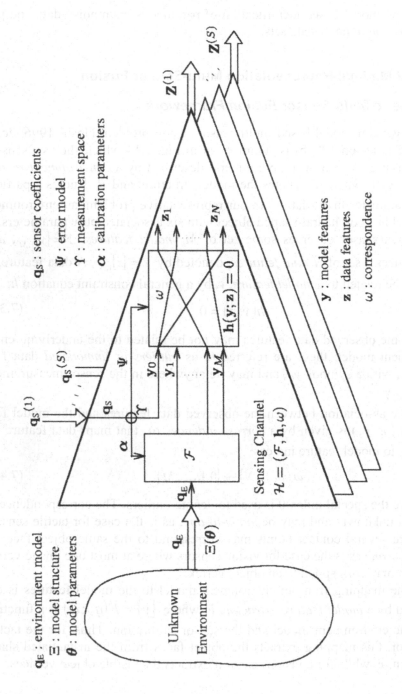

Fig. 7.14. Environment and sensor models

The model feature extractor F, the sensor constraint h and the correspondence ω are collectively referred to as a *sensing channel* $H = (F, h, \omega)$; they define the sensor structure which has the same form for all sensors of a given type. However, the uncertainty, the measurement range and the calibration parameters differ between different sensors of the same type and must be determined for each sensor; they define the *sensor coefficients* $q_S = (\Psi, Y, \alpha)$. The sensing channel and the sensor coefficients constitute a *sensor model*.

Minimal Representation Multi-Sensor Fusion

The minimal representation approach (Joshi 1996; Joshi and Sanderson 1996, 1999b) is based on the principle of building the shortest length program which reconstructs the observed data. The length of this program or *representation size* is well defined for sensors, and may be regarded as the "information", in bits, contained in the data. It depends on the system's "knowledge" of the environment, specified by an environment model library – hence the approach is inherently *model based*. The "best interpretation" minimizes the total representation size of the observed multi-sensor data.

The total representation size is obtained by adding the model representation size and the data representation size for each sensor as shown in Fig. 7.15. Details can be found in Joshi (1996) and Joshi and Sanderson (1996, 1999a, 1999b).

The "best interpretation" is an instantiated environment model and correspondences, found by minimizing the total representation size of observed multi-sensor data. The multi-sensor fusion algorithm searches for the minimal representation size interpretation, in the space of all possible environment models and their parameterizations, and the space of all possible legal correspondences.

The search space is the cross-product of the environment model space and the correspondence spaces, which may be part discrete and part continuous; often the general search problem quickly becomes computationally impractical.

There are three basic components of the search problem:

1. **Environment model instantiation:** An environment model structure, chosen from the library, must be instantiated with specific parameters. Thus, $q_E = \Xi(\theta)$, where the structure, $\Xi(\cdot)$, and the parameters, θ, must be somehow instantiated.

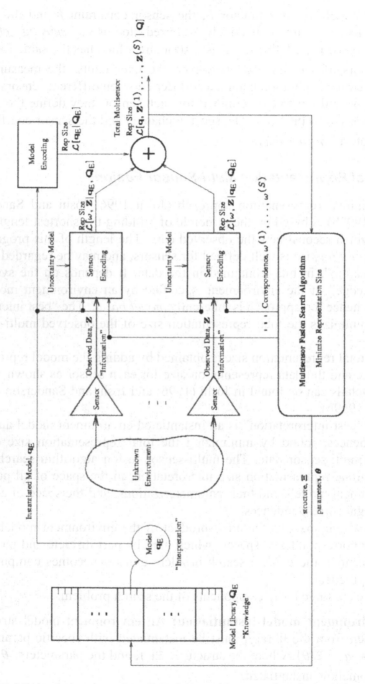

Fig. 7.15. Minimal representation size of the multi-sensor data fusion scheme

2. **Correspondence computation:** The minimal representation size correspondences can be computed, given an instantiated environment model. One approach to computing the correspondence for a given instantiated environment model is based on using a minimum weight assignment algorithm; it operates in cubic time of the number of features (Joshi 1996; Joshi and Sanderson 1999b).
3. **Search engine:** A search engine *systematically* instantiates environment models, computing the minimal representation size correspondences for each one, in order to find the minimal representation size interpretation in the space of possible model parameters and correspondences.

We utilize *evolution programs*, in particular differential evolution (DE), to drive the search, and find the minimal representation interpretation of the observed multi-sensor data.

7.4.3 Differential Evolution for Multi-Sensor Fusion

A search can *sequentially* iterate through the structures in the environment model library, and compute the minimal representation data correspondences for each instantiated environment model (Joshi 1996; Joshi and Sanderson 1999b). A DE program is used to search the space of environment model parameters. The DE as applied to the tactile–visual fusion problem (Fig. 7.13) is summarized below. Details can be found in Joshi (1996) and Joshi and Sanderson (1996, 1999a, 1999b).

Representation. Given a preselected environment model structure, an individual p in the population represents the environment model parameters:

$$S(p) = \theta. \tag{7.41}$$

All points in the parameter space Θ must be representable by the specific form chosen for the environment model parameters.

For the tactile–visual sensor fusion problem (Fig. 7.13), the environment model library consists of rigid polyhedral objects, parameterized by the object pose. The three-dimensional pose is described by a translation, $\vec{t}$, and a rotation, $\exp(\hat{n}\,\phi/2)$, of angle $\phi \in [0, 2\pi)$ about an axis $\hat{n}$ (Fig. 7.16a). This pose representation is expressed as a six-parameter vector in the DE, as shown in Fig. 7.16b, and results in a uniform sampling of the orientation and translation space (Joshi 1996; Joshi and Sanderson 1996, 1999b).

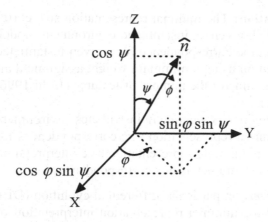

a) Orientation

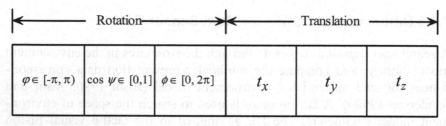

b) Pose representation

Fig. 7.16. The object rotation is given by a rotation of angle ϕ around an axis $\hat{n}$.

Fitness Evaluation. The fitness function is the total multi-sensor representation size (Fig. 7.15):

$$\Phi(S(p)) = L\left[q, Z^{(1)}, \ldots, Z^{(S)} | Q\right]. \tag{7.42}$$

For an individual, $S(p) = \theta$, the fitness is evaluated as follows:

1. If $\theta \notin \Theta$, the total multi-sensor representation size is defined to be ∞.
2. Otherwise, for $\theta \in \Theta$:
    a) For each sensor:
        i. Extract the model features.
        ii. Compute the minimal representation size correspondence, using graph matching algorithms (Gondran and Minoux 1984; Joshi 1996; Joshi and Sanderson 1999b).

b) Compute the total minimal representation size for this parameter vector (for the minimal representation size correspondences just computed).

For the vision sensors (Fig. 7.13) the representation size is calculated by encoding the errors between features predicted by an instantiated model and the observed data as follows. The digital image, obtained from a fixed calibrated camera, is processed to extract vertex features, expressed in pixel coordinates. The intrinsic camera calibration is used to convert these pixel coordinates into image coordinates expressed in the camera frame, and constitute the vision data features. These are related to the object vertices (model features), according to the perspective projection constraint, for a pin-hole camera model. The observation errors are described by an ellipsoidal sensor accuracy (Joshi 1996; Joshi and Sanderson 1999b), equal to the size of a pixel. The measurement space is equal to the image size. At most one vision image vertex can correspond to an object vertex, i.e., the vision correspondence must be *one-to-one* (Joshi 1996; Joshi and Sanderson 1999b).

Similarly, for tactile sensors (Fig. 7.13) the representation size is calculated by encoding the errors between features predicted by an instantiated model and the observed data as follows. Tactile sensors mounted on the hand's fingertips provide contact locations relative to the fingertips. The forward kinematics of the calibrated arm–hand kinematic chain is used to compute the location of these contact points relative to the base of the robot, and constitute the touch data features. These are related to the object faces (model features) according to a containment constraint, i.e., contact points must lie within an object face. The observation errors are described by an ellipsoidal sensor accuracy, whose diameter is equal to the positioning accuracy of the arm–hand manipulator. The measurement space is equal to the workspace volume accessible by the robot hand. Several contacts may be made on an object face, resulting in a many-to-one correspondence (Joshi 1996; Joshi and Sanderson 1999b).

Reproduction. The standard DE crossover operator is used. Typically, $K = 2$ differentials are used, the scale factor $F \in [0,2]$ and the greediness $\gamma \in [0,1]$.

Selection. The standard DE selection operator is used: each offspring competes with its parent and survives only if its fitness is better.

Initialization. The parameter vectors are drawn *randomly* from the space of legal environment model parameters $\theta \in \Theta$. This ensures a diverse initial population.

Termination Condition. A parameter is considered $\eta\%$ converged if at least $\eta\%$ of the population shares the same value (within some pre-specified *parameter tolerance*) as the best individual in the population, for that parameter. The *population* is considered to be $\eta\%$ converged when all the parameters have $\eta\%$ converged.

The evolution is terminated when (a) the population reaches a certain desired level of convergence, or (b) the maximum number of generations is exceeded, or (c) the maximum time limit is exceeded.

7.4.4 Experimental Results

Setup

Experiments were conducted on simulated data and laboratory data. Two different objects, *Lwedge* and *Pedestal* (Fig. 7.17), were used. A 100-member DE/best/2 DE algorithm with $F = 0.8$ and $Cr = 0.8$ was used to search for the minimal representation size pose and the correspondences for each sensor. The search is terminated when the population reaches 50% convergence (parameter tolerance of 0.1), or when the maximum time limit of 900 CPU seconds is exceeded, or when a maximum of 1000 generations has been completed.

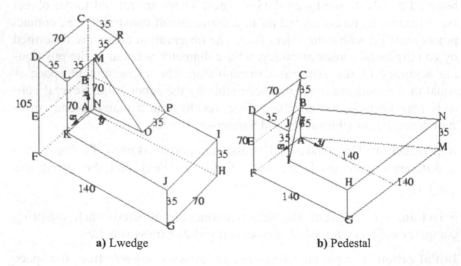

a) Lwedge   b) Pedestal

Fig. 7.17. The object library

For simulated data, a typical problem size was chosen, consisting of 16 vision data features of which 50% were spurious, and 9 tactile data features of which 15% were spurious. The search is considered to have correctly converged (to the global minimum), if it reaches within 14 bits of the simulated multi-sensor data representation size.

For laboratory data, the translation search space was restricted to a $50\times50\times50$ cube, in the interest of computation time. The entire space of orientations and correspondences was searched. The vision preprocessing and the DE search parameters were kept the same for all the experiments with laboratory data.

Quality of Solutions

For simulated data, the results on 100 randomly generated problems for an environment library containing *Lwedge* and *Pedestal* are given in Table 7.2. The mean errors and standard deviations are tabulated among those DE searches which correctly converged, and the correct object shape was chosen.

For laboratory data, the results from Experiment 1, using the *Lwedge* object, are shown in Figs. 7.18a, 7.18b and 7.18c. The interpretation errors for the six experiments, using both vision and touch sensors, touch sensor alone and vision sensor alone, are summarized in Table 7.3.

Table 7.2. Model selection performance of an environment model library containing two object models, on 100 randomly generated problems. Each problem had 16 vision data features with 50% outliers, and 9 touch data features with 15% outliers.

Trials = 100	Correctly converged = 81	Misclassified = 0
	Mean	Standard deviation
Rotation axis error	0.033372°	0.184239°
Rotation angle error	0.028648°	0.165689°
Translation error	0.165613	0.100419
Vision correspondence error	2.222222	1.151086
Touch correspondence error	0.172840	0.380464
Representation size deviation	−6.340123	3.828411

Environment Model Class Selection. As Tables 7.2 and 7.3 illustrate, the object shape was correctly selected in these experiments, thus demonstrating that representation size is effectively used to trade off between the model class and data error residuals.

Environment Model Parameterization and Data Scaling. From Table 7.2 and 7.3, it can be seen that the estimated interpretation produced by the multi-sensor fusion of touch and vision data was close to the reference interpretation for almost all the experiments. The multi-sensor pose estimation errors are comparable in magnitude to the sensor errors themselves.

Table 7.3. Summary of experimental results

Fusing Vision (V) and Touch (T) data

Experiment Id	Sensor	M, Model Feature Count	N, Data Feature Count	Reference Outliers	Estimated Outliers	Modeled-Mis-modeled	Modeled ← Unmodeled	Spurious ← Modeled	Axis Error, degrees	Angle Error, degrees	Translation Error	Rep. Size Deviation	Ref. Pose RMS Error	Generations/CPU Seconds	Object Name
1	V	18	16	8	4	0	0	4	0.06	2.20	1.9	-73.1	1.7	339	Lwedge
	T	13	5	0	0	0	0	0						585	Lwedge
2	V	18	10	2	1	1	0	1	1.71	1.35	26.7	-67.2	1.8	346	Lwedge
	T	13	9	0	0	0	0	0						569	Lwedge
3	V	18	7	1	0	1	0	1	0.13	2.70	8.81	-41.2	1.2	321	Lwedge
	T	13	9	0	0	0	0	0						594	Lwedge
4	V	14	8	4	3	2	1	2	0.33	-17.91	21.2	-117.3	11.8	440	Pedestal
	T	9	9	0	0	6	0	0						596	Pedestal
5	V	14	10	4	2	2	0	2	0.06	-6.67	12.9	-104.0	11.6	407	Pedestal
	T	9	9	0	0	1	0	0						571	Pedestal
6	V	14	11	6	5	0	1	2	0.03	8.91	29.3	-120.2	12.1	448	Pedestal
	T	9	9	0	0	1	0	0						599	Pedestal

Using only Touch (T) data

Experiment Id	Sensor	M, Model Feature Count	N, Data Feature Count	Reference Outliers	Estimated Outliers	Modeled-Mis-modeled	Modeled ← Unmodeled	Spurious ← Modeled	Axis Error, degrees	Angle Error, degrees	Translation Error	Rep. Size Deviation	Ref. Pose RMS Error	Generations/CPU Seconds	Object Name
1	T	13	5	0	0	3	0	0	1.01	70.07	7.7	-47.7	1.7	385	Lwedge
2	T	13	9	0	0	7	0	0	0.90	21.09	13.4	-56.8	1.8	210	Lwedge
3	T	13	9	0	0	9	0	0	0.49	115.51	30.4	-45.9	1.2	239	Lwedge
4	T	9	9	0	0	6	0	0	1.40	10.47	32.6	-110.6	11.8	370	Pedestal
5	T	9	9	0	0	6	0	0	0.05	13.38	20.7	-85.1	11.6	376	Pedestal
6	T	9	9	0	0	7	0	0	1.08	5.58	28.9	-92.1	12.1	267	Pedestal

Using only Vision (V) data

Experiment Id	Sensor	M, Model Feature Count	N, Data Feature Count	Reference Outliers	Estimated Outliers	Modeled-Mis-modeled	Modeled ← Unmodeled	Spurious ← Modeled	Axis Error, degrees	Angle Error, degrees	Translation Error	Rep. Size Deviation	Ref. Pose RMS Error	Generations/CPU Seconds	Object Name
1	V	18	16	8	4	5	2	6	1.35	122.71	19.8	-41.1	1.7	361	Lwedge
2	V	18	10	2	3	3	1	0	0.39	52.08	10.9	-40.8	1.8	511	Lwedge
3	V	18	7	1	0	6	0	1	1.67	59.56	18.6	-21.7	1.2	530	Lwedge
4	V	14	8	4	4	2	1	2	0.09	12.47	29.9	-24.3	11.8	590	Pedestal
5	V	14	10	4	1	5	1	4	0.90	161.20	31.3	-26.0	11.6	501	Pedestal
6	V	14	11	6	5	1	4	2	0.04	11.80	27.1	-33.0	12.1	399	Pedestal

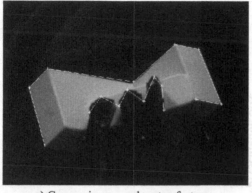

a) Camera image and vertex features

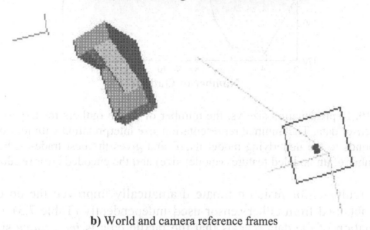

b) World and camera reference frames

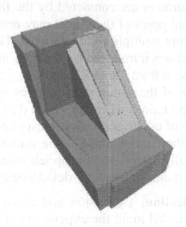

c) Estimation error

Fig. 7.18. Laboratory Experiment 1 results

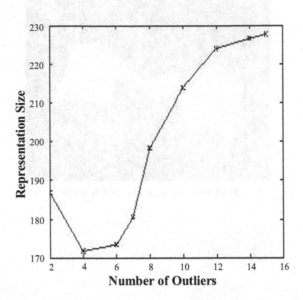

Fig. 7.19. Representation size vs. the number of vision outliers for Experiment 1 multi-sensor data. The minimal representation size interpretation with four outliers corresponds to the underlying model itself, and gives the best trade-off between the number of unmodeled features (model size) and the encoded error residuals.

The multi-sensor fusion estimate dramatically improved the pose estimates obtained from either sensor used independently (Table 7.3). Closer examination of the data reveals that the tactile data is *incomplete* since at most three different surfaces are contacted by the fingers, in these experiments. Several different poses of the object may result in the same contact configuration – therefore multiple solutions exist. The vision data set contains spurious data features introduced by the presence of the hand holding the object, as can be seen from Table 7.3 and Fig. 7.18a. In Experiments 1, 4 and 6, at least 50% of the vision data features were spurious, and can lead to incorrect interpretations when the vision data is used alone.

Multi-sensor fusion of tactile and vision sensors compensates for the deficiencies in using either sensor alone – the minimal representation size framework automatically selects an appropriate combination of the vision and touch sensor constraints to correctly determine the pose parameters.

Data Subsample Selection. The vision and touch data correspondences were meaningfully selected in all the experiments (Tables 7.2 and 7.3). In our implementation, the hidden object vertices were not removed, and some of the spurious vision data features got matched to these "extra" vision model features.

The correspondence selection/data subsampling property of the framework is illustrated by a plot in Fig. 7.19 of minimal representation size vs. the number of outliers. The minimum of this plot corresponds to the underlying interpretation which best explains the observed data, reported in Table 7.3. The minimal representation size solution trades-off between the number of unmodeled data features and the modeled error residuals, and automatically selects a subset of the data features consistent with the object pose.

Convergence

Figure 7.20 shows the progress of the DE algorithm on Experiment 1 multi-sensor data. The algorithm starts with a population of 100 randomly chosen individuals. The population cluster shrinks around the minimal representation size interpretation.

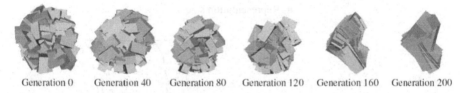

Generation 0   Generation 40   Generation 80   Generation 120   Generation 160   Generation 200

Fig. 7.20. Progress of the 100-member DE algorithm on Experiment 1 multi-sensor data. DE/best/2, $F = 0.8$, $Cr = 0.8$.

Figure 7.21a shows the plots of representation size versus the number of generations, for the DE search on the six laboratory experiments. As can be seen from these plots, the DE fully converged within 400 generations in all cases. The run-times are of the order of a few minutes on a SPARC 20 (Table 7.3). Figure 7.21b shows a plot of the percentage convergence versus the number of generations for the DE. As can be seen from this plot, the orientation parameters converge first followed by the translational pose parameters. This is typical of the DE for this application.

When the observed data is *incomplete* or inconsistent, multiple interpretations result in nearly the same representation size. In such cases, successive runs of the DE may produce varying results, due to the multiple global minima. As the noise in the observed data increases, or as the number of observed data features increases, the DE converges correctly more often, and is more reliable. This is due to the minimal representation size search landscape becoming smoother with deeper valleys, when the observation errors or the number of data features increase.

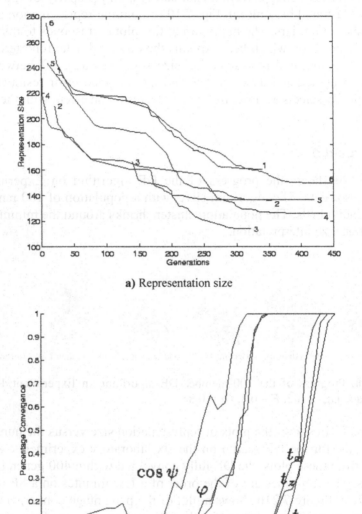

a) Representation size

b) Percentage convergence

Fig. 7.21. DE progress on experimental data. Using a 100-member algorithm DE/best/2, $F = 0.8$, $Cr = 0.8$. The DE ran until the population reached 99% convergence. The parameter tolerance for convergence was set to 0.01, and checked every ten generations.

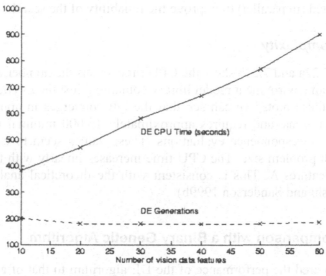

a) Vision data: 20 features, 50% outliers.

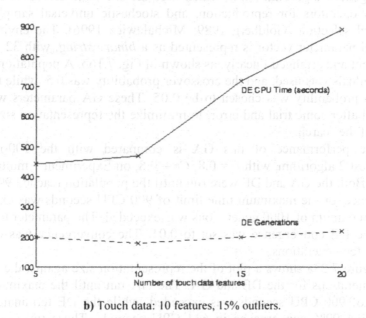

b) Touch data: 10 features, 15% outliers.

Fig. 7.22. CPU time (SPARC 20) and DE generations vs. the number of data features for the DE search algorithm.

As seen from Table 7.2, 81% of the DE searches correctly converged to the global minimum. This number may be further improved by tuning the

DE search parameters Np, F and Cr. In practice, several DE searches may be executed (in parallel) to improve the reliability of the search.

Time Complexity

Figures 7.22a and 7.22b show the CPU time versus the number of data features for an environment model library containing just the *Lwedge* object.

From these plots, we can see that the DE converges in approximately 200 generations, and requires approximately 16,000 minimal representation size correspondence evaluations. These values seem to change very little with problem size. The CPU time increases linearly with the number of data features N. This is consistent with the theoretical analysis (Joshi 1996; Joshi and Sanderson 1999b).

7.4.5 Comparison with a Binary Genetic Algorithm

We compared the performance of the DE algorithm to that of a simple binary genetic algorithm (GA), which uses the one-point crossover and mutation operators for reproduction, and stochastic universal sampling for natural selection (Goldberg 1989; Michalewicz 1996). The environment model parameter vector is represented as a *binary string*, with 32 bits per element and arranged linearly, as shown in Fig. 7.16b. A population of 100 individuals was used, and the crossover probability was 0.5, while the mutation probability was chosen to be 0.05. These GA parameters were selected after some trial and error, to minimize the representation size at the end of the search.

The performance of this GA is compared with the 100-member DE/best/2 algorithm, with $F = 0.8$, $Cr = 0.8$, on Experiment 1 multi-sensor data. Both the GA and DE were run until the population reached 99% convergence, or the maximum time limit of 900 CPU seconds was exceeded, or a maximum of 1000 generations was exceeded. The parameter tolerance for the DE convergence was set to 0.01. The convergence was checked every ten generations.

Figure 7.23a shows a plot of the representation size against the number of generations for the DE and GA. The GA ran until the maximum time limit of 900 CPU seconds was exceeded, while the DE terminated upon reaching 99% convergence in 634 CPU seconds. These times are on a SPARC 20 Sun workstation. The DE found a much smaller value of representation size in less time and a fewer number of minimal representation size correspondence evaluations than the GA. The interpretation errors for the DE and GA are given in Table 7.4. The DE solution had smaller interpretation errors.

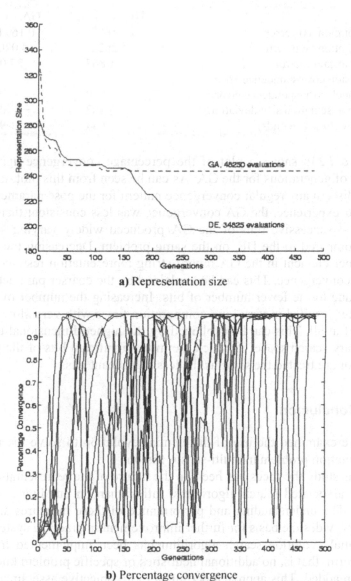

a) Representation size

b) Percentage convergence

Fig. 7.23. DE vs. GA progress on Experiment 1 multi-sensor data, showing a 100-member DE/best/2 algorithm, $F = 0.8$, $Cr = 0.8$, and a 100-member binary GA with crossover probability 0.5 and mutation probability 0.05. The GA ran until the time limit of 900 CPU seconds expired. The convergence was checked every ten generations.

Table 7.4. Comparison of the interpretation error using a DE and GA for Experiment 1 multi-sensor data

	DE	GA
Rotation axis error	0.05605°	0.16949°
Rotation angle error	2.2022°	4.0203°
Translation error	1.867	27.092
Vision correspondence error	4	8
Touch correspondence error	0	0
Representation size deviation	−73.102	−24.064
CPU time (seconds)	634	>900

Figure 7.23b shows a plot of the percentage convergence against the number of generations for the GA. As can be seen from this plot, it is difficult to discern any regular convergence pattern for the pose parameters.

In our experience, the GA convergence was less consistent than that of the DE – successive runs of the GA produced widely varying solutions when compared to the DE, on the same problem. Decreasing the number of bits per element in the GA binary string representation resulted in premature convergence. This can be explained by the coarser parameter resolution, due to the fewer number of bits. Increasing the number of bits per parameter resulted in very long computation times with very slow convergence, if at all. This can be explained by the higher dimensionality of the GA binary search space, in which the more significant bits of the parameter vector are treated the same as the less significant bits.

7.4.6 Conclusion

We have examined the use of differential evolution to solve the minimal representation problem in multi-sensor fusion.

In this study the focus has been on the nature of the representation itself and the associated search algorithms, rather than on building a practical system. The understanding and performance of these measures and algorithms provides the basis for further improvement in practical systems. The differential evolution search algorithm has been implemented in a "generic" form: that is, no additional heuristics or specific problem knowledge has been added. This approach permits a more objective assessment of the minimal representation size criterion for multi-sensor fusion. In practice, problem-specific heuristics might be added to improve the efficiency of implementation for a particular problem domain. For example, in object recognition problems, typical heuristics might include: (a) pose space clustering to reduce the search space (Linnainmaa et al. 1988; Grimson and

Huttenlocher 1990), (b) using transform space geometry to prune pose parameters (Breuel 1992; Cass 1988), (c) pruning matches based on relations which must be satisfied by both data and model features (Fischler and Bolles 1981; Grimson and Lozano-Pérez 1984; Faugeras and Hebert 1986), and (d) using a fixed-size subset of data features for generating pose hypothesis (Chen and Kak 1989; Huttenlocher and Ullman 1990).

The laboratory experiments demonstrate the automatic selection of environment model class (object identity), the environment parameter values (object pose) and the sensor data correspondences (touch and vision correspondences), in the minimal representation size framework for multi-sensor fusion and model selection. Differential evolution search run-times are a few hundred seconds on a SPARC 20 (Table 7.3). In practice, robotic manipulation using multi-sensor data must be done in real time, and it became clear that the general evolutionary algorithms are too slow to estimate pose at sampling speeds for continuous motion. For practical use of the method in "real-time" control applications, this global search procedure should be augmented with an incremental search procedure, which sequentially updates the estimated model as data is acquired during the execution of the real-time control task.

A major advantage of the minimal representation approach is the attainment of consistent results without the introduction of problem-specific heuristics or arbitrary weight factors. The use of an information-based criterion provides a type of *universal yardstick* for sensor data from many disparate sources, and therefore supports efficient implementation to new domains. The model selection properties of this framework are complementary to the estimation process itself, and the framework chooses an effective combination of model structure and estimation method for a given class of problems.

The extension of this approach to many different types of models and estimation problems is an important element of this methodology, and its use for general parametric model estimation with noisy data is open for further research.

References

Abidi MA, Gonzalez RC (eds.) (1992) Data fusion in robotics and machine intelligence. Academic Press, New York

Ali MS, Engler C (1991) System description document for the Anthrobot-2: a dexterous robot hand. Technical memorandum 104535, NASA Goddard Space Flight Center, Maryland

Allen PK (1988) Integrating vision and touch for object recognition tasks. The International Journal of Robotics Research 7(6):15–33

Bolle RM, Cooper DB (1986) On optimally combining pieces of information, with application to estimating 3-D complex-object position from range data. IEEE Transactions on Pattern Analysis and Machine Intelligence 8(5):619–638

Breuel TM (1992) Fast recognition using adaptive subdivisions of transformation space. In: Proceedings of the IEEE Computer Society conference on computer vision and pattern recognition, Champaign, IL, June 1992. IEEE Computer Society Press, Silver Spring, Maryland, pp 445–451

Cass TA (1988) A robust parallel implementation of 2D model-based recognition. In: Proceedings of the Computer Society conference on computer vision and pattern recognition, Ann Arbor, MI, June 1988. IEEE Computer Society Press, Silver Spring, Maryland, pp 879–884

Chen CH, Kak AC (1989) A robot vision system for recognizing 3-D objects in low-order polynomial time. IEEE Transactions on Systems, Man, and Cybernetics 19(6):1535–1563

Dekhil M, Henderson TC (1998) Instrumented sensor system architecture. The International Journal of Robotics Research 17(4):402–417

Durrant-Whyte HF (1987) Consistent integration and propagation of disparate sensor observations. The International Journal of Robotics Research 6(3):3–24

Eason RO, Gonzalez RC (1992) Least-squares fusion of multisensory data. In: (Abidi and Gonzalez 1992), Chap. 9, pp 367–413

Elfes A (1989) Using occupancy grids for mobile robot navigation and perception. IEEE Computer 22(6):46–58

Faugeras OD, Hebert M (1986) The representation, recognition, and locating of 3-D objects. The International Journal of Robotics Research 5(3):27–52

Fischler MA, Bolles RC (1981) Random sample consensus: A paradigm for model fitting with applications to image analysis and automated cartography. Communications of the ACM 24(6):381–395

Goldberg DE (1989) Genetic algorithms in search, optimization, and machine learning. Addison-Wesley, Reading, MA

Gondran M, Minoux M (1984) Graphs and algorithms. Wiley, New York

Grimson WEL, Huttenlocher DP (1990) On the sensitivity of the Hough transform for object recognition. IEEE Transactions on Pattern Analysis and Machine Intelligence 12(3):255–274

Grimson WEL, Lozano-Pérez T (1984) Model based recognition and localization from sparse range or tactile data. The International Journal of Robotics Research 3(3):3–35

Hager GD (1990) Task-directed sensor fusion and planning: a computational approach. Kluwer, Norwell, MA

Huttenlocher DP, Ullman S (1990) Recognizing solid objects by alignment with an image. International Journal of Computer Vision 5(2):195–212

Joshi R (1996) A minimal representation framework for multisensor fusion and model selection. Ph.D. thesis, Rensselaer Polytechnic Institute

Joshi R, Sanderson AC (1994) Model-based multisensor data fusion: A minimal representation approach. In: Proceedings of the 1994 IEEE international conference on robotics and automation, vol 1, San Diego, CA, May, pp 477–484

Joshi R, Sanderson AC (1996) Multisensor fusion and model selection using a minimal representation size framework. In: Proceedings of the 1996 IEEE conference on multisensor fusion, Washington, DC, December, pp 25–32

Joshi R, Sanderson AC (1999a) Minimal representation multisensor fusion using differential evolution. IEEE Transactions on Systems, Man, and Cybernetics, Part A: Systems and Humans 29(1):63–76

Joshi R, Sanderson AC (1999b) Multisensor fusion: a minimal representation framework. Series in intelligent control and intelligent automation, vol 11. World Scientific, Singapore

Kokar M, Kim K (1993) Review of multisensor data fusion architectures and techniques. In: Proceedings of the 1993 IEEE international symposium on intelligent control, Chicago, IL, August

Linnainmaa S, Harwood D, Davis LS (1988) Pose determination of a three-dimensional object using triangle pairs. IEEE Transactions on Pattern Analysis and Machine Intelligence 10(5):634–646

Luo RC, Kay MG (1989) Multisensor integration and fusion in intelligent systems. IEEE Transactions on Systems, Man, and Cybernetics 19(5):901–931

Michalewicz Z (1996) Genetic algorithms + data structures = evolution programs. Springer, Berlin Heidelberg New York

Nakamura Y (1992) Geometric fusion: Minimizing uncertainty ellipsoid volumes. In: (Abidi and Gonzalez 1992), Chap. 11, pp 457–479

Porill J (1988) Optimal combination and constraints for geometrical sensor data. The International Journal of Robotics Research 7(6):66–77

Segen J, Sanderson AC (1981) Model inference and pattern discovery by minimal representation method. Technical report CMU-RI-TR-82-2, The Robotics Institute, Carnegie-Mellon University, Pittsburgh, PA

Shashank S, Oussama K, Makoto S (1988) Object localization with multiple sensors. The International Journal of Robotics Research 7(6):34–44

Smith RC, Cheeseman P (1986) On the representation and estimation of spatial uncertainty. The International Journal of Robotics Research 5(4):56–68

7.5 Determination of the Earthquake Hypocenter: A Challenge for the Differential Evolution Algorithm

Bohuslav Růžek and Michal Kvasnička

Abstract. Determination of the earthquake hypocenter represents a basic problem routinely solved in seismology. The problem belongs to the class of simpler problems in geophysics, but it is still difficult to solve. The dimension of the model space is low (three coordinates of the hypocenter plus origin time, resulting in four parameters to be searched for), but the forward problem exhibits a case-to-case-dependent degree of nonlinearity. The standard solution is based on minimizing the time residuals (differences between observed and computed arrivals of seismic waves) in the common L2 norm. We have compiled a set of 56 synthetic earthquake hypocenter location tasks, which was submitted to three different optimizers for solution: (i) Powell's method, (ii) the downhill simplex algorithm and (iii) the differential evolution (DE) algorithm. Each localization process listed was performed two times using exact and approximate forward modeling. Our analysis has shown that the DE algorithm has worked with 100 % reliability, while other optimizing algorithms have often failed. The accuracy achieved by using the DE algorithm was at least the same or better than that achieved by competing algorithms. The only minor disadvantage of the DE algorithm is a higher computational effort needed to reach the solution.

7.5.1 Introduction

The localization of the focus of an earthquake belongs to the oldest inverse problems ever solved in geophysics. Reliable knowledge of the coordinates of the focus and origin time is extremely important for most of the more advanced studies, e.g., for the determination of earthquake magnitude, focal mechanism, stress conditions around the focus, etc. A proper solution to the location problem is therefore of significant importance. The standard location procedures are based on kinematic principles. The focus of an earthquake is supposed to be a point from which seismic waves radiate in all directions. Basically, two types of seismic waves (P and S waves) can propagate from the source to the receivers at different velocities. The arrival times of these waves can be picked out on seismograms recorded at a number of surface seismic stations distributed in seismogenic regions. Using appropriate knowledge of the geological structure of the region traversed by the rays, synthetic arrival times may be theoretically calculated. The correct position of the hypocenter and origin time are indicated by the close agreement between the observed and computed arrival times of selected seismic phases.

The earthquake location problem is represented as an optimization task in a four-dimensional model space. Let X_0, Y_0, Z_0 be the Cartesian coordinates of the hypocenter, and T_0 the origin time of the earthquake. Then our four unknown parameters form a model vector $\mathbf{m} = (X_0, Y_0, Z_0, T_0)$. Let t_i be the ith observed arrival time (i.e., the time at which some seismic wave has reached a recording station), and τ_i the corresponding calculated (theoretical) counterpart. Both t_i and τ_i can be arranged as components of n-dimensional vectors $\mathbf{t}$ and $\boldsymbol{\tau}$, respectively. Obviously, τ_i depends on all hypocentral coordinates including origin time, $\tau_i = \tau_i(\mathbf{m})$. Observed and calculated arrival times are used to form a data vector $\mathbf{d} = \mathbf{t} - \boldsymbol{\tau}$, whose components (time residuals) should be as small as possible. The dimension of the data space is $n = \dim(\mathbf{d}) \geq m = \dim(\mathbf{m}) = 4$ in order to ensure formally uniqueness of the solution. In real cases, many stations (10–100) contribute to the location process of an earthquake and the problem is overdetermined as a rule. Commonly, the problem is solved in the L2 norm; the following minimization represents the search for the optimum solution:

$$\chi^2 = \frac{1}{v} \sum_{i=1}^{n} [t_i - \tau_i(m)]^2 / \sigma_i^2 = \mathbf{d}^T . \mathbf{C}^{-1} . \mathbf{d} = \min(\mathbf{m}) \tag{7.43}$$

where $\mathbf{C}$ is the data covariance matrix ($C_{ii} = (\sigma_i)^2$, $C_{ij} = 0$ for $i \neq j$) and σ_i are standard (Gaussian) deviations, and $v = n - 4$ is the number of degrees of freedom. The anticipated idea appeared originally in 1910 (Geiger 1910), and has been used widely in many standard location programs (Lee and Lahr 1972; Klein 1978; Herrmann 1979; Lienert et al. 1986). Due to the nonlinearity of the problem ($\boldsymbol{\tau} = \boldsymbol{\tau}(\mathbf{m})$ represents a system of nonlinear equations often available in numerical form only, i.e., the system is not given analytically), Eq. 7.43 should be solved iteratively. The standard approach is based on an iterative least squares search:

$$\mathbf{m} \rightarrow \mathbf{m} + \Delta\mathbf{m}, \Delta\mathbf{m} = (\mathbf{G}^T . \mathbf{G})^{-1} \mathbf{G}^T . \mathbf{C} . \mathbf{d}. \tag{7.44}$$

In Eq. 7.44, $\mathbf{G}$ is the matrix of partial derivatives $G_{ij} = \delta\tau_i / \delta m_j$ and $\Delta\mathbf{m}$ is the correction term. The above outlined approach is fast, but suffers from at least the following drawbacks:

1. Linearization of an essentially nonlinear problem is acceptable only if the available starting model is not very far from the true solution.
2. Real data times t_i are frequently contaminated by reading errors, and the obligatory implicit L2 norm is rather restrictive (Shearer 1997).
3. There are cases (even if not very frequent) when the $\mathbf{G}^T.\mathbf{G}$ matrix is nearly singular – the condition number may be of order 10^{29} (Buland 1976) and numerical difficulties can arise.
4. Evaluation of the derivatives is problematic; in some parts of the model space they may be discontinuous or may not exist at all due to a complicated velocity structure.

There is an alternate approach: Eq. 7.43 may be solved using another search algorithm. Those algorithms that do not rely on the gradient of χ^2 are especially potential candidates for successful implementation. The downhill simplex algorithm (Nelder and Mead 1965; Olsson and Nelson 1975; Press et al. 1992) was used in Rabinowitz (1988). The genetic algorithm (Sambridge and Gallagher 1993; Billings et al. 1994; Bondar 1994), simulated annealing (Billings 1994), interval arithmetic (Tarvainen et al. 1999) and grid search (Sambridge and Kennett 1986; Fischer and Horálek 2000; Janský et al. 2000) are also known to have been utilized in earthquake hypocenter location problems. According to our experience (Růžek and Kvasnička 2001), the DE algorithm represents a suitable compromise between the reliability of getting a true and accurate solution, and the effectiveness of the search.

Nevertheless, the location problem includes a further difficulty. Evaluation of the theoretical arrival times requires perfect knowledge of the velocity model. Unfortunately, the actual geologic medium is very complex. As a consequence, the corresponding velocity model is potentially discontinuous, inhomogeneous and anisotropic. Many rays may then connect selected two points within such a medium and "multi-pathing" takes place. The opposite case is also possible: no ray exists which could connect two points in our model. Despite the problems with the evaluation of arrival times in a complex medium, even more problems are posed by the fact that "perfect knowledge" of the underlying velocity model is very rarely achieved. These problems may be partially overcome by introducing a formal, artificial model for evaluation of the arrival times as described by Xie et al. (1996). The latter approach gives a relatively good solution for the position of the epicenter, but the depth of the focus is rather imperfect. In many situations, such a restricted solution is valuable at least as a first step in the data processing sequence. Simplified and approximate formulas have another unwanted property: the optimized χ^2 functional (Eq. 7.43) is probably multimodal, has flat valleys and is surely difficult to solve.

Originally, the use of a genetic algorithm was recommended. In the following, we shall discuss the possibility of introducing a much more powerful DE algorithm.

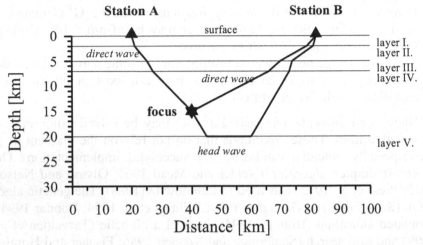

Fig. 7.24. An example demonstrating the propagation of direct and head waves from the focus to the surface stations within a layered medium. Station A may be illuminated by a direct wave due only to the relatively small epicentral distance. Station B may record both direct and head waves. Scales for information only.

Table 7.5. The one-dimensional P-wave model used for location

Depth (km)	Velocity (km/s)
0–5	5.7
5–18	6.0
18–39	6.4
>39	7.9

7.5.2 Brief Outline of Direct Problem Solution

Probably the best way to study and demonstrate the properties of a selected inverse problem is to generate synthetic examples with a priori known solutions and submit them to the inversion. We shall follow this way as well. In order to keep our analysis as simple as possible, we shall use only one-dimensional layered velocity models (i.e., models consisting of a sequence of homogeneous isotropic layers separated by horizontal interfaces). The focus of the earthquake may be anywhere inside the model, and recording seismic stations will be distributed irregularly along the surface. Each station will report only one fastest P-wave arrival. Under such simplifications, the ray corresponding to the direct (refracted) wave and the appropriate time of propagation may be calculated for any possible source–station pair. Under specific conditions (sufficiently great epicentral distance, high-velocity layer below the focus) another "head" wave may exist and may reach the same station along a different ray and at a different arrival time. Any one of the direct or head waves may be the faster one (see Fig. 7.24). While the head wave may be determined analytically in a finite number of steps, the direct wave must be calculated iteratively using the shooting or bending method. A rigorous explanation of how to construct the rays and how to calculate the propagation times is beyond the scope of this section, and may be found in, for example, Červený et al. (1977).

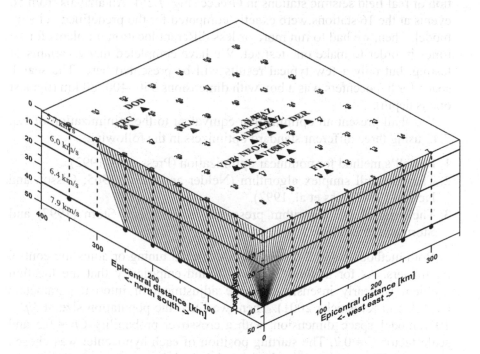

Fig. 7.25. Configuration of our synthetic location test. Earthquake foci are numbered 1–56 (shown in projection on the surface) and are at a depth of 40 km. Seismic stations are marked by full triangles and labeled by three-character codes. Some rays from focus 1 are drawn in two perpendicular vertical planes at the borders of the model.

7.5.3 Synthetic Location Test

In order to keep our numerical experiments on a realistic basis we tried to simulate typical location problems for seismic events originating in the Gulf of Corinth, Greece. The routinely used velocity model recommended in Tselentis et al. (1996) consists of three homogeneous isotropic layers and a homogeneous isotropic half-space with the parameters given in Table 7.5. The lowermost interface of our model corresponds to the Moho discontinuity separating the Earth's crust and the upper mantle, therefore the velocity model is of regional validity. As long as the velocity in all layers increases with depth, all interfaces potentially generate head waves. We decided to initiate 56 fictitious earthquakes with foci in a regular 50×50 km grid at a depth of 40 km. Hypothetical recordings were enabled by 16 seismic stations, distributed on the surface according to the actual distribution of real field seismic stations in Greece (Fig. 7.25). All arrivals from 56 events at the 16 stations were exactly computed for the predefined velocity model. Then, we had to run more or less different location problems for 56 times in order to make one test set. We have completed many variants of testing, but only a few typical results will be presented here. The search space for hypocenters was a box with dimensions 350×400×80 km (the last one is depth).

We shall present an optimization equivalent to the minimization of Eq. 7.43 using three different standard optimizers in the following:

1. Powell's method for nonlinear optimization (Press et al. 1992).
2. The downhill simplex algorithm (Nelder and Mead 1965; Olsson and Nelson 1975; Press et al. 1992).
3. The original DE algorithm presented in Price and Storn (1997) and Storn and Price (1997).

Both methods 1 and 2 do not require special tuning or adjusting control parameters. As for method 3, it was found empirically that the location problem is nearly insensitive to the adjustment of internal parameters (Růžek and Kvasnička 2001). Then, we fixed the population size at $Np = 10 \times$ model_space_dimension, with a crossover probability $Cr = 0.5$ and scale factor $F = 0.7$. The starting position of each hypocenter was chosen randomly. We calculated the theoretical arrival times τ_i using two methods:

1. Arrivals were calculated exactly using the same velocity model and the same algorithm employed in the preparation of the synthetic problem
2. Arrivals were calculated approximately using the algorithm given in Xie et al. (1996), which has the great practical advantage that the actual velocity model need not be known. On the other hand, two extra

artificial parameters must be added to the parameter space and the inversion becomes more difficult due to the unfavorable topography of the optimized functional.

In conclusion, we shall present results obtained by the three optimizers applied to two clones of the optimized functional; that is, six different modes will be discussed.

7.5.4 Convergence Properties

We monitored the current value of χ^2 in the course of minimization, depending on the number of misfit evaluations. The corresponding graphs are arranged in triplets in Fig. 7.26. Each graph contains 56 curves connected with the location of 56 synthetic earthquakes.

The left part of Fig. 7.26 represents the location of error-free data using consistent forward modeling. Ideally, each curve should then reach exactly zero at the end (perfect fit achieved). Due to rounding errors, the iterative method of refracted wave computations and the scaling, final values of misfit around 10^{-4}–10^{-3} may be considered as sufficiently accurate solutions.

- If Powell's method is used (Fig. 7.26, upper right), only 46 events out of 56 are satisfactorily located. In 10 cases, the search terminates far from the correct position. In successful cases, this method needs a broad range of 5.10^2–10^4 function evaluations in order to achieve convergence.
- Downhill simplex exhibits fast convergence (up to 3.10^2 function evaluations), but the probability of getting the correct solution is even lower (45 cases out of 56). Often the algorithm is unable to make any improvement on the current position in the model space and the procedure stagnates.
- DE convergence curves show a relatively uniform pattern (Fig. 7.26, lower left). The solution is found in all cases, but the computational effort is higher: we need to evaluate the forward problem around 5.10^3 times in order to decrease the misfit value to a satisfactory level $\sim 10^{-4}$.

The same triplet of optimizers was then applied to the same physical problem but formulated differently (right part of Fig. 7.26). Now the arrivals of seismic waves are calculated approximately. This is useful in practice, because we do not speculate on the validity of the velocity model used. On the other hand, the dimension of the model space is now six (four for the hypocenter and two additional formal parameters for the velocity model approximation). It can be shown that the optimization is much harder. Because forward and inverse modeling are now not exactly consistent, good solutions are indicated by final misfit values around 1.

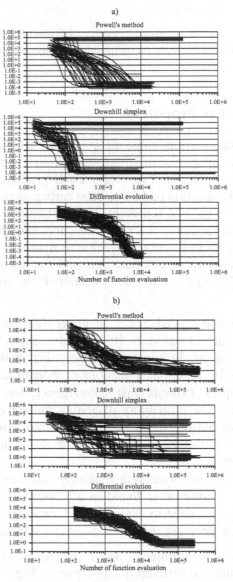

Fig. 7.26. Convergence curves (dependence of the χ^2 value on the number of function evaluations) for two variants of forward modeling: a) consistent (exact) forward modeling, b) approximate forward modeling. A label indicates the optimizer used. Approximate forward modeling is much harder, but does not require perfect knowledge of the velocity model. The stagnation of the optimization process is apparent if the χ^2 value is not reduced appropriately.

- As we can see from the upper right part of Fig. 7.26, Powell's algorithm needs around 10^4 misfit function evaluations in order to achieve convergence. Nevertheless, there are two cases when the procedure fails completely and the solution is not found. The convergence is also rather irregular: from time to time the misfit value falls quickly and then stays at nearly the same level for a long time and for a great number of calculations.
- The downhill simplex procedure (Fig. 7.26, middle right) behaves worse than Powell's method. The solution is found in 20 cases only; for the others the inversion stagnates. Convergence curves are even more irregular than in the case of Powell's optimization. In a successful case, downhill simplex requires $\sim 10^4$ misfit function evaluations until convergence.
- DE yields quite another picture (Fig. 7.26, lower right). The optimization is regular, all curves are relatively smooth and they decrease very similarly to each other. The convergence is 100% reliable. The only disadvantage in comparison with the previous two optimizing algorithms (if any comparison is at all possible) is a higher number of function evaluations until convergence: the minimized function has to be called $\sim 5.10^4$ times (i.e., the computations may be up to five times slower).

Table 7.6. Probability of discovering the solution

Optimizing algorithm	Method of forward modeling	
	Exact calculations	Approximate calculations
Powell's algorithm	82%	96%
Downhill simplex	79%	64%
DE	100%	100%

A summary of the reliability of obtaining good solutions for different kinds of optimization is given in Table 7.6. Further, we may discuss the accuracy in those lucky cases when a "satisfactory solution" is found. This accuracy may be quantified by the Euclidean distance between the determined hypocenter and the true hypocenter (which is known by solving the synthetic problem). This approach would be consistent if our model space contained parameters of the same physical units only. The actual model space has parameters of either two physical units (distance and time in the case of exact forward modeling) or three physical units (distance, time and velocity in the case of approximate forward modeling). The concept of Euclidean distance is therefore rather problematic. However, an alternate and consistent measure of the precision is given by discussing the magnitude of the final χ^2 value.

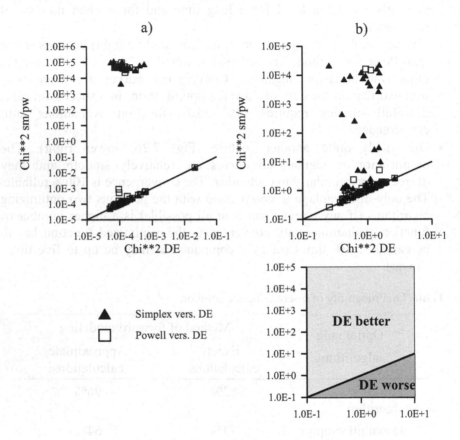

Fig. 7.27. Final measures of the accuracy obtained for exact forward modeling (left) and approximate forward modeling (right). The display compares the accuracy reached by DE and the downhill simplex algorithm or by DE and Powell's method. It can be clearly seen that DE is never worse than either of the two competing optimizing methods.

We have plotted χ^2 values for corresponding solutions obtained in a pair-wise manner by Powell's method and DE or by downhill simplex and DE in Fig. 7.27. It can be seen in the figure that DE never yields worse results than either of the two competing methods. Especially when approximate forward modeling is used, DE behaves excellently in finding very accurate solutions.

7.5.5 Conclusions

Inverse problems in geophysics are mostly hard problems to solve. They are nearly always nonlinear, non-unique, medium or high dimensional, and applied to noisy data. Calculation of the forward problem is time consuming and sometimes only approximate. Therefore good minimizing/optimizing algorithms are of great importance.

The problem of earthquake hypocenter determination belongs to the relatively simple geophysical problems. Nevertheless, we can see that "classic" or "standard" optimizing algorithms sometimes give inaccurate or incorrect results. We have demonstrated that the DE algorithm introduces an attractive way to perform geophysical inversions. A set of 56 synthetic location problems was submitted for solution by three different optimizing algorithms. The DE algorithm output correct results in all cases, while the competing algorithms often failed. Of course, there is a price for DE reliability: a greater computational effort is needed to achieve convergence. Nevertheless, continuous development of computing technologies makes this aspect of minor importance. Even if the problem solutions were found by all the tested optimizers, the DE solution would nearly always be characterized by the lowest χ^2 value and therefore by the best fit.

DE has many pleasant and elegant features: it is simple, easy to use, does not insist on evaluations of derivatives, works with continuous parameters which need not be bound and their starting positions are not necessary. The only disadvantage of DE is the already-mentioned slightly higher number of misfit evaluations. We believe that DE will be broadly used in geophysics in the near future, especially when considering that DE itself will probably be further developed as well.

References

Billings SD (1994) Simulated annealing for earthquake location. Geophysical Journal International 118:680–692

Billings SD, Kennett BLN, Sambridge MS (1994) Hypocenter location: genetic algorithms incorporating problem-specific information. Geophysical Journal International 118:693–706

Bondar I (1994) Hypocenter determination of local earthquake using genetic algorithm. Acta Geodaetica et Geophysica Hungarica 29:39–56

Buland R (1976) The mechanics of locating earthquakes. Bulletin of the Seismology Society of America 66:173–187

Červený V, Molotkov IA, Pšenčík I (1977) Ray method in seismology. Charles University Press, Prague

Fischer T, Horálek J (2000) Refined locations of the swarm earthquakes in the Nový Kostel focal zone and spatial distribution of the January 1997 swarm in Western Bohemia, Czech Republic. Studia Geophysica et Geodaetica 44:210–226

Geiger L (1910) Herdbestimmung der Erdbeben aus den Ankunftzeiten. Der Königlichen Gesellschaft der Wissenschaften zu Göttingen 4:331–349

Herrmann RB (1979) FASTHYPO – A hypocentre location program. Earthquake Notes 50:25–37

Janský J, Horálek J, Málek J, Boušková A (2000) Homogeneous velocity models of the West Bohemian swarm region obtained by grid search. Studia Geophysica et Geodaetica 44:158–174

Klein RW (1978) Hypocenter location program HYPOINVERSE part I: Users guide to versions 1, 2, 3 and 4. US Geological Survey Open-File Report 78-694

Lee WHK, Lahr JC (1972) HYPO71: A computer program for determining hypocenter, magnitude, and first motion pattern of local earthquakes. US Geological Survey Open-File Report 75-311

Lienert BR, Berg E, Frazer LN (1986) HYPOCENTER: An earthquake location method using centered, scaled, and adaptively damped least squares. Bulletin of the Seismology Society of America 76:771–783

Nelder JA, Mead R (1965) A simplex method for function minimization. Computer Journal 7:308-313

Olsson DM, Nelson LS (1975) The Nelder-Mead simplex procedure for function minimization. Technometrics 17:45–51

Press WH, Teukolsky SA, Vetterling WT, Flannery BP (1992) Numerical recipes in C: The art of scientific computing, 2nd ed. Cambridge University Press

Price K, Storn R (1997) Differential evolution. Dr. Dobb's Journal April:18–24

Rabinowitz N (1988) Microearthquake location by means of nonlinear simplex procedure. Bulletin of the Seismology Society of America 78:380–384

Růžek B, Kvasnička M (2001) Differential evolution algorithm in the earthquake hypocenter location. Pure and Applied Geophysics 158:667–693

Sambridge M, Gallagher K (1993) Earthquake hypocenter location using genetic algorithm. Bulletin of the Seismology Society of America 83:1467–1491

Sambridge MS, Kennett BLN (1986) A novel method of hypocenter location. Geophysical Journal of the Royal Astronomical Society 87:679–697

Shearer PM (1997) Improving local earthquake locations using the L1 norm and waveform cross correlation: Application to the Whittier Narrows, California, aftershock sequence. Journal of Geophysics Research 102:8269–8283

Storn R, Price K (1997) Differential evolution – a simple and efficient heuristic for global optimization over continuous spaces. Journal of Global Optimization 11:241–354

Tarvainen M, Tiira T, Husebye ES (1999) Locating regional seismic events with global optimization based on interval arithmetic. Geophysics Joural International 138:879–885

Tselentis GA, Melis NS, Sokos E, Papatsimpa K (1996) The Egion June 15, 1995 (6.2 M_L) earthquake, Western Greece. PAGEOPH 147(1):83–97

Xie Z, Spencer TW, Rabinowitz PD, Fahlquist DA (1996) A new regional hypocenter location method. Bulletin of the Seismology Society of America 86:946–958

7.6 Parallel Differential Evolution: Application to 3-D Medical Image Registration

Michel Salomon, Guy-René Perrin, Fabrice Heitz and J.-P. Armspach

Abstract. A common framework for 3-D image registration consists in minimizing a cost (or energy) function that expresses the pixel or voxel similarity of the images to be aligned. Standard cost functions, based on voxel similarity measures, are highly nonlinear, non-convex, exhibit many local minima and thus yield hard optimization problems. Local, deterministic optimization algorithms are known to be sensitive to local minima. Global optimization methods (like simulated annealing or evolutionary algorithms) yield better solutions often close to the optimal ones, but are time consuming. In this section we consider the parallelization of a general-purpose global optimization algorithm based on random sampling and evolutionary principles: the differential evolution algorithm. The inherent parallelism of evolutionary algorithms is used to devise a data-parallel implementation of differential evolution. The performances of the parallel version are assessed on a 3-D medical image registration problem. Besides yielding accurate registrations, parallel differential evolution exhibits fast convergence and a speedup almost growing linearly with respect to the number of processors.

7.6.1 Introduction

Global optimization problems are encountered in many areas of science and engineering. In particular, many problems in 2-D and 3-D image processing and computer vision have been expressed as global optimization problems. The general issue is to find the global minimum of an objective function (also called cost or energy function) describing the interactions between the different variables modeling the image features in a given application (Heitz et al. 1994). Image restoration (Geman and Geman 1984), image segmentation, image registration, image motion measurement, texture analysis as well as scene interpretation have for instance been recast into this framework. Due to the large volume of data, the large number of variables involved and the generalized use of nonlinear interaction models, the optimization problems under consideration are generally hard ones, involving non-convex objective functions and many local minima.

A typical example is *image registration*. The purpose of image registration (also called image matching) is to geometrically align one image (the floating or source image) with another (the reference or target image) so that voxels (or pixels) representing the same underlying structure may be superimposed. A standard framework for image registration consists in minimizing a cost function or maximizing a "similarity measure" that expresses the pixel or voxel similarity of the images to be aligned. During the last decade, image registration has become an important preliminary step in a wide range of image analysis and computer vision tasks. Due to its large variety of sensors, 3-D medical imaging is certainly one of the first application fields, as are remote sensing, military imaging or multi-sensor robot vision. Applications range from computer-assisted surgery to the analysis of sequences of functional images used for instance to follow the evolution of diseases.

Standard cost functions, based on voxel similarity measures (Woods et al. 1993; Nikou et al. 1998), are highly nonlinear, non-convex and exhibit many local minima. Like *NP*-complete problems, they yield challenging optimization problems, with large and irregular search spaces, and thus require computationally demanding global optimization algorithms to compute close to optimal solutions. Less CPU-intensive deterministic algorithms may be used instead, but they are sensitive to local minima and require a good initial guess (Nikou et al. 1998).

In this section we consider differential evolution as an appealing candidate for general-purpose global optimization (Storn and Price 1996; Storn 1996). We show that differential evolution is particularly suited for the registration of medical images using similarity measures. To register medical images with computational times suited to clinical applications, we develop comprehensive parallelization schemes for this class of algorithm. Standard parallelizations of evolutionary algorithms consist in distributing the population of candidate solutions maintained by the algorithms, making it evolve in parallel. In our approach we have one potential solution per processor, thereby defining a novel, fully synchronous data-parallel algorithm. The parallel algorithm has been applied successfully to register 3-D magnetic resonance images of the brain, exhibiting a nearly linear speedup.

The remainder of this section is organized as follows. After introducing the similarity-based registration model (Sect. 7.6.2), we briefly describe the particular differential evolution algorithm used here and show the results obtained with a sequential version of this algorithm (Sect. 7.6.3). In Sect. 7.6.4, comprehensive parallel solutions are proposed for evolutionary algorithms and the adopted parallelization scheme, relying on a data-parallel programming model, is described. Section 7.6.5 is devoted to the experimental assessment of the performances of the proposed parallel algorithm, compared to the reference, sequential version.

7.6.2 Medical Image Registration Using Similarity Measures

A general review of image registration may be found in Brown (1992). Recent updates of the state of the art in this field, specifically devoted to medical image analysis, have been proposed in van den Elsen et al. (1993), Hill and Hawkes (2000) and Woods (2000b).

To define a medical image registration algorithm, several points must be considered:

1. The imaging modalities involved.
2. The feature space used to describe the image content.
3. The similarity measure used.
4. The nature and domain of transformation.
5. The algorithm used to find the optimal transformation.

The registration may concern images from the same modality (single modal image matching) or images stemming from different modalities (multi-modal image matching). To compare these images, many features (voxels, edges, surfaces, etc.) and similarity measures are now available. Besides, transformations range from rigid transformations, with a small number of parameters (see Fig. 7.32 below), to deformable image warps, depending on several thousands or millions of parameters (Christensen et al. 1996; Musse et al. 1999; Woods 2000a).

Imaging Modalities and Feature Space

Medical imaging modalities can be divided into two categories: anatomical and functional. The first category primarily depicts morphology, whereas the second one highlights metabolic information of the underlying anatomy. The classical medical imaging modalities include:

- Anatomical modalities: X-ray, CT (Computerized Tomography), MRI (Magnetic Resonance Imaging).
- Functional modalities: PET (Positron Emission Tomography), SPECT (Single-Photon Emission Computed Tomography), fMRI (functional Magnetic Resonance Imaging).

A typical application of single modality image registration is the following one on the evolution of lesions on temporal sequences (see Fig. 7.33 below). On the other hand, multi-modal registration is generally used to integrate anatomical and functional information. Obviously the registration problem is far more difficult when considering images from different modalities (which generally may not be compared on a voxel-by-voxel basis).

The features used in image registration may be raw pixel/voxel intensities, characteristic points, edges or lines, surfaces or volumes, or even high-level image representations such as statistical models or graph-based representations. In this work, we consider the standard case of single modal image registration based on voxel similarity measures, which is now used routinely on a daily basis in many hospitals.

Similarity Measures and Image Transformations

Similarity measure-based approaches rely on the minimization of cost functions that express the pixel (2-D) or voxel (3-D) similarity of the images to be aligned. Similarity measures have been introduced for both single and multi-modal image registration (for recent reviews see Hill and Hawkes 2000; Woods 2000b). Standard similarity metrics used in single modal image registration are related to least squares estimation or to the maximization of the correlation function. For multi-modal images, the definition of similarity generally relies on first- or second-order image statistics such as mean, variance, entropy or mutual information.

Many transformations, from rigid transforms to elastic warps, may be considered in image registration (Woods 2000a). In this work the transformation is assumed to be rigid, therefore the output of the registration process is a set of six independent parameters: three rotation parameters and three translation parameters (see Fig. 7.32 below). Rigid registration is routinely used in order to compensate for the difference of patient position between successive scans. Within this framework, voxel similarity metric-based registration consists in estimating the parameter vector Θ of the rigid transformation T_Θ minimizing a cost function C that expresses the similarity between the single or multi-modal image pair:

$$\Theta_{\min} = \arg \min_{\Theta}[C(I(.),J(T_\Theta(.)))] , \qquad (7.45)$$

where (for 3-D images)

$$\Theta = \left(t_x, t_y, t_z, \theta_x, \theta_y, \theta_z\right)^{\mathrm{T}} \qquad (7.46)$$

is a vector containing the 3-D translation parameters (t_x, t_y, t_z) with respect to the X, Y and Z axes and the Euler rotation angles $(\theta_x, \theta_y, \theta_z)$. $I(.)$ represents the reference image and $J(.)$ the floating image, to be registered on the reference image. A widely used similarity measure for the registration of single modal images is the quadratic similarity measure (Brown 1992).

This similarity measure assumes that the images to be matched differ only by an additive Gaussian noise, leading to the following cost function:

$$C(I(.),J(T_\Theta(.))) = \sum_s [I(s) - J(T_\Theta(s))]^2 \qquad (7.47)$$

where s designates the voxel (or pixel) coordinates. More sophisticated similarity measures may be found in Nikou et al. (1998), Hill and Hawkes (2000) and Woods (2000b).

Global Optimization Algorithm

The similarity measure presented above is highly nonlinear, non-convex and has multiple local minima. In most image registration methods, local deterministic optimization algorithms, such as gradient descent or Newton optimization algorithms, are applied. They are known to be sensitive to local minima, unless they are initialized close to the optimal solution. In Nikou et al. (1998), for example, a first step consists in a fast global search using simulated annealing, followed by a deterministic descent using a Gauss–Seidel-like algorithm.

In this work we consider differential evolution as an appealing candidate for the global optimization of the objective functions involved in image processing. To reduce the computational burden, the optimization is conducted on a multi-grid sequence of images of increasing resolution. In practice, the cost function is calculated by successively considering one voxel out of 81, 27, 9, 3 and finally every voxel in the 3-D MR images (for 128^3 images). Multi-grid algorithms are known to be less sensitive to local minima than single-resolution implementations, yielding fast convergence toward good solutions (Heitz et al. 1994).

The choice of differential evolution results from a comparison of several global search algorithms: simulated annealing (Kirkpatrick et al. 1983), tabu search (Glover and Laguna 1993) and evolutionary algorithms (Bäck 1996). A careful experimental assessment (Salomon 2001) has shown that evolutionary algorithms, particularly evolution strategies and differential evolution, are best suited here. This point is also emphasized by other work on image registration using evolutionary algorithms (Fischer et al. 1999). In our experiments, differential evolution has significantly outperformed the other global optimization approaches: it is faster, yields better solutions, has fewer control parameters, and is easier to implement than evolution strategies.

7.6.3 Optimization by Differential Evolution

It is assumed that the reader is already familiar with the differential evolution algorithm presented in Chap. 2 of this book. Therefore we will only describe the reproduction operator (the so-called crossover operator by Storn et al (Storn 1996), as it differs slightly from the original one.

Reproduction Operator

Among the different variants of differential evolution (DE), throughout this investigation we used the DE/rand-to-best/1 scheme. This scheme is briefly presented below, considering the medical image registration optimization problem. We chose this scheme after preliminary experiments; in fact it outperformed the three other schemes that we considered: DE/rand/1, DE/best/1 and DE/best/2 (Storn 1996).

For the problem of rigid 3-D registration, a new individual

$$\left(x'_{i,j}\right)^{(G+1)}_{j\in\{1,...,6\}} = \left(t'_{x_i},t'_{y_i},t'_{z_i},\theta'_{x_i},\theta'_{y_i},\theta'_{z_i}\right)^{(G+1)} = \Theta'^{(G+1)}_{i},$$

where $i \in \{1, ..., n_{pop}\}$ and G with range $\{0, ..., G_{max}\}$ is the population index, is generated according to

$$x'^{(G+1)}_{i,j} = \begin{cases} x^{(G)}_{i,j} + \lambda \cdot \left(x^{(G)}_{min,j} - x^{(G)}_{i,j}\right) + F \cdot \left(x^{(G)}_{S,j} - x^{(G)}_{T,j}\right) \\ \quad \text{if } (\chi_1 \leq \chi_2) \wedge (\chi_1 \leq j \leq \chi_2) \\ \quad \text{or } (\chi_1 > \chi_2) \wedge (1 \leq j \leq \chi_2 \ \vee \ \chi_1 \leq j \leq n_{par}) \\ x^{(G)}_{i,j} \quad \text{otherwise} \end{cases} \tag{7.48}$$

where $x^{(G)}_{min}$ represents the current minimum; n_{pop} is the number of individuals and n_{par} the number of parameters; $S, T \in \{1, ..., n_{pop}\}$ satisfying $(S \neq T) \neq i$ denote two randomly chosen individuals; and χ_1, χ_2 are the two crossover points. The last variables are obtained in the following way:

$$\chi_1 = \left(\chi \bmod n_{par}\right) + 1, \chi_2 = \left((\chi + L - 1) \bmod n_{par}\right) + 1 \tag{7.49}$$

where χ and L are two integers uniformly drawn from set $\{1, ..., n_{par}\}$, for each individual. χ defines the first position of the crossover whereas L is the number of parameters to be exchanged. Integer L is chosen using the algorithm given in Storn and Price (1996), according to the crossover probability C_r given by the user. The control parameters λ and F have the range $[0.1, 1.0]$.

Sequential Differential Evolution Algorithm

To evaluate the DE algorithm on a representative set of 3-D MRI/MRI registration problems of 128^3 voxels, we generate a set of 20 randomly defined registration problems. Each registration problem is obtained by applying a random rigid transformation on a reference MRI provided by the Institut de Physique Biologique (Strasbourg University Hospital, UMR CNRS 7004). The translation parameters are within the range [−20.0,+20.0] voxels, whereas the rotation angles are within [−20.0,+20.0] degrees. Let us emphasize that large rotations are usually difficult to handle, leading to objective function landscapes with many local minima. Further, many voxel interpolations are involved in the registration process: trilinear interpolation is adopted here, as a good compromise between reconstruction quality and computational cost.

As stated in Sect. 7.6.2, the DE algorithm is applied on a sequence of multi-resolution grids, using a standard top-down approach, starting from the coarsest resolution level. We generate the same number of populations at all resolutions, except at the final one, in which case the fitness (cost) of each individual will be simply computed again, as is the case when forwarding the population from a given resolution to a finer one. This step is necessary because the cost of an individual changes with resolution. A suitable value for λ and F has been found in some preliminary experiments, whereas the termination criterion, i.e., the maximum number of generated populations, was chosen in order to evaluate about 900 individuals. The total number of populations to be evaluated is thus $G_{max} + 5$.

Figure 7.28 presents the average evolution of the minimum cost for the 20 registration problems, during the optimization process. In Fig. 7.28 the influence of the population size according to the crossover probability (Fig. 7.28a: $Cr = 0.3$, Fig. 7.28b: $Cr = 0.8$) is shown. It appears that a small population is sufficient to get a good convergence: 8 individuals for a small crossover probability, 16 individuals for a large crossover probability. We also notice that these population sizes induce convergence curves that are almost similar.

To evaluate the ability of the DE algorithm to achieve accurate registrations (errors less than 1 voxel in translation and 1 in rotation), we compute various error statistics on the estimated parameters, as well as the root mean square (RMS) error. RMS corresponds to the average misregistration between voxels in the proposed solution and the optimal one. It can be seen from the statistics collected from the set of 20 images (Table 7.7) that the algorithm achieves good accuracy. Considering a sampling of voxels, we obtain an RMS error of 0.35 voxels, showing subvoxel accuracy.

The CPU time obtained on an MIPS R12000 processor (300 MHz) is approximately 11 minutes. Compared to other approaches, the registration of medical images using DE is fairly competitive.

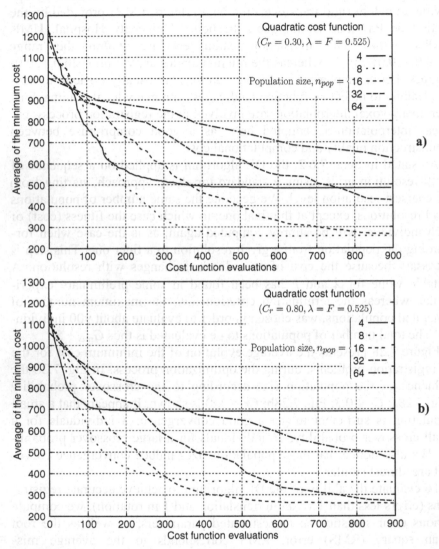

Fig. 7.28. Average evolution of the minimum cost considering different population sizes (**a** $C_r = 0.3$ and **b** $C_r = 0.8$; $\lambda = F = 0.525$).

Table 7.7. Single modal (MRI/MRI) registration (3-D) $\{\lambda = F = 0.525\}$: error statistics (mean ± standard deviation)

$n_{pop} = 8$; $C_r = 0.3$ and $G_{max} = 108$		$n_{pop} = 16$; $C_r = 0.8$ and $G_{max} = 52$	
Δt_x	0.19 ± 0.15	Δt_x	0.20 ± 0.15
Δt_y	0.14 ± 0.10	Δt_y	0.15 ± 0.10
Δt_z	0.16 ± 0.09	Δt_z	0.16 ± 0.09
$\Delta \theta_x$	0.005 ± 0.004	$\Delta \theta_x$	0.030 ± 0.040
$\Delta \theta_y$	0.004 ± 0.003	$\Delta \theta_y$	0.020 ± 0.030
$\Delta \theta_z$	0.009 ± 0.006	$\Delta \theta_z$	0.040 ± 0.080

Note: The translation errors are given in voxels and the rotation errors in degrees.

7.6.4 Parallelization of Differential Evolution

Parallelizing an algorithm can be motivated by several reasons. The main ones are: reducing the computation time by equally dividing the computation cost over all the processors; the ability to solve larger problems than possible in sequence; last but not least, designing new optimization strategies based on sequential methods. Let us notice that the computation time needed to solve an optimization problem is mainly related to the number of parameters to be optimized and the smoothness of the cost function landscape.

How To Parallelize an Evolutionary Algorithm

Evolutionary algorithms are intrinsically parallel, since, as in nature, individuals evolve simultaneously. There are three levels of parallelization that can be considered for an evolutionary algorithm (Tomassini 1999): the fitness evaluation level, the population level and the individual level. The first level will not be discussed, as it is a problem-dependent parallelization, whereas the two other levels are general parallelization.

Population-based parallelization consists in dividing the population into as many subpopulations as processors, each processor running the evolutionary algorithm on its own subpopulation. This kind of parallelization is said to be coarse-grained. The subpopulations are usually called "islands" (or demes). They can be connected following a defined neighborhood enabling the migration of individuals from a subpopulation to another one, replacing the worst individuals in the target subpopulation: usually it is the individual having the best fitness that migrates. The interacting scheme is called the "connected island model". The background idea for allowing migration of individuals is to ensure diversity over the subpopulations. In the scheme without interactions, the only communications take place at the end in order to select the best individual among all subpopulations. Unfortunately, the lack of communications can result in a subpopulation exploring a bad region of the search space, ignoring more promising regions explored by other processors.

Introducing the parallelization at the individual level consists in distributing the population so that each individual is assigned to one virtual processor, resulting in a data-parallel implementation with a finer granularity. In this scheme, the processor's network topology is very important, since it induces the isolation degree across the population and in this way the diversity of the whole population. The diffusion of individuals is done by selection and reproduction operators locally on each processor, considering a neighborhood structure. Consequently the neighborhood is the major component in this parallel algorithm: the four or eight nearest neighbors of an individual form the most typical neighborhoods used. Local interactions between neighboring processors prevent premature convergence, since the diffusion of the individuals across the population is very slow, reducing the probability of seeing a superindividual emerge.

Parallelization of evolutionary algorithms has been mainly studied on genetic algorithms, since they are the most popular evolutionary algorithms and the ones having the widest range of applications. Our contribution is to study the parallelization of DE.

Parallelizing Differential Evolution: A Data-Parallel Implementation

In this work, we propose an individual-based parallelization for the DE algorithm. In fact, as shown by the study of the sequential algorithm (Sect. 7.6.3), a population with a small size (less than 20 individuals) was enough to give good results. Therefore a coarse-grained parallelization is irrelevant: the subpopulations would have been too small.

Consider a population of 16 individuals and a grid of virtual processors (a real processor usually simulates several virtual ones), having a size 4×4. We distribute the population on this grid in order to have one individual per virtual processor. Thus parallelizing DE consists in parallelizing the different steps of the algorithm, namely: initialization, evaluation, reproduction and selection.

Clearly initialization and evaluation can be simply done in parallel. For example, the first individual on each processor results from a random sampling in the search space. Furthermore, each individual $x_i'^{(G)}$, $i \in \{1, ..., n_{pop}\}$, where $n_{pop} = 4 \times 4$, resulting from reproduction will be located on the same virtual processor as $x_i^{(G)}$. Consequently, as the selection operator consists in a comparison of these two individuals for each value of i, the parallelization is communication free. Finally we have to parallelize the reproduction operator. Equation 7.48 shows that generating a new individual $x_i'^{(G)}$ requires three other individuals, apart from $x_i^{(G)}$:

- the individual of minimum cost in the current population, denoted $x_{\min}^{(G)}$;
- two individuals $x_S^{(G)}$ and $x_T^{(G)}$ randomly chosen in the population, mutually different and also different from $x_i^{(G)}$ ($(S \neq T) \neq i$).

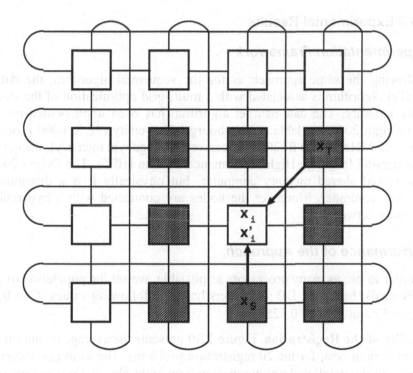

Fig. 7.29. Transmission of x_S and x_T (the neighborhood is in gray).

The broadcasting of the current minimum across the grid results in irregular communications, but its real cost depends on the architecture of the parallel machine. In fact the compiler should optimize broadcast, since it is a basic operation in data-parallelism. To draw the two individuals $x_S^{(G)}$ and $x_T^{(G)}$ in parallel, we define around each virtual processor a neighborhood corresponding to the eight-nearest neighbors of the considered processor. $x_S^{(G)}$ and $x_T^{(G)}$ are then randomly chosen in this neighborhood and communicated to the processor where the corresponding $x_i^{(G)}$ is located, as presented in Fig. 7.29. The choice of the two individuals is done in the same way by all the processors, yielding a regular communication pattern. Obviously, introducing a neighborhood to choose $x_S^{(G)}$ and $x_T^{(G)}$ gives a parallel algorithm semantically different from the sequential one.

The main question can be stated in the following words: "Has the semantic modification introduced in the parallel algorithm any effect on its results?" The next section will answer this, showing clearly the relevance of our approach.

7.6.5 Experimental Results

Experimentation Framework

Following the same approach as for the sequential algorithm, the data-parallel algorithm is associated with a multi-grid optimization of the similarity measure. The data-parallel algorithm has been implemented on an SGI Origin 2000 available at Strasbourg I University (32 R10000 processors, 195 MHz and 20 R12000 processors, 20 gigabyte memory), using the data-parallel language High Performance Fortran (HPF). The Origin 2000 is a virtual shared memory computer, but physically it is a distributed memory computer. Moreover the nodes are connected with a hypercube topology network.

Performance of the Approach

In order to use as many processors as possible, we set the population to 16 individuals; hence the DE parameters have the following values: $C_r = 0.8$, $G_{\max} = 52$ and $\lambda = F = 0.525$.

Quality of the Registration. Figure 7.30 presents the average evolution of the minimum cost, for the 20 registration problems. The solid curve corresponds to the parallel algorithm executed on eight physical processors; the dashed line depicts the sequential version. Clearly, the sequential and the parallel algorithms have the same average behavior, resulting in a final parameter vector with similar cost. In practice, the semantic modification introduced by the data-parallel approach is without any effect. Consequently, as highlighted by Table 7.8, the parallel algorithm is also able to achieve registrations with subvoxel accuracy.

The good results of the data-parallel implementation may be explained by the choice made for the neighborhood during the reproduction step, since it is this operator that differentiates the parallel version from the sequential one. For the sequential version, a population of nine individuals is sufficient to enable convergence toward a final parameter vector corresponding to an accurate registration (in fact, as we have seen in Sect. 7.6.3, eight individuals are enough). As we choose the eight nearest neighbors to form the neighborhood, we see that in the parallel version during the reproduction on a virtual processor a subpopulation of nine individuals is considered. Thus the "parallel" reproduction is locally (on each virtual processor) almost equivalent to the sequential version with a population of nine individuals, and, as noted above, this one yields very good results.

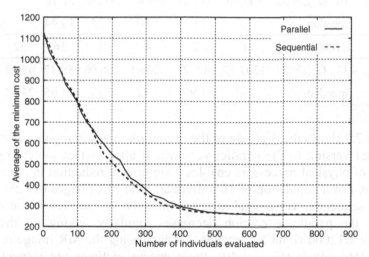

Fig. 7.30. Average evolution of the minimum cost

Table 7.8. Single modal (MRI/MRI) registration (3-D): Average and standard deviation of the registration errors

$n_{pop} = 16;\ C_r = 0.8$ and $G_{max} = 52$	
Δt_x	0.20 ± 0.14
Δt_y	0.14 ± 0.10
Δt_z	0.16 ± 0.09
$\Delta \theta_x$	0.020 ± 0.020
$\Delta \theta_y$	0.030 ± 0.040
$\Delta \theta_z$	0.040 ± 0.030

Performance. The parallel algorithm has been executed on 2, 4, 8, 16 and 32 physical processors with different population sizes. Indeed, as can be seen from Fig. 7.28b, a population size of 32 individuals is competitive when the crossover probability is close to one. Note that for the population of 16 individuals, the experiments were made on the R12000, whereas for the 32 size population, they were done on the R10000. As a reference, we present the sequential execution time for one processor.

Table 7.9. Single modal 3-D (MRI/MRI) registration: average run-times (seconds) and corresponding speedup values

R12000		1	2	4	8	16	
$n_{pop} = 16$ $G_{max} = 52$		700.80	349.01	184.47	95.54	52.16	
Speedup			2.01	3.80	7.34	13.44	
R10000		1	2	4	8	16	32
$n_{pop} = 32$ $G_{max} = 24$		1207.57	609.27	324.91	166.06	88.47	49.02
Speedup			1.98	3.72	7.27	13.65	24.63

Table 7.9 shows the decrease of the execution time when going from the sequential version to the parallel versions. It appears that increasing the number of physical processors enables a significant reduction in computation time. For a population of 16 individuals and MR images of 128^3 voxels, the sequential version takes approximately 11 minutes 40 seconds, whereas the parallel execution decreases regularly, taking on the 16-processor R12000 about 52 seconds. On increasing the MR image resolution to 256^3 voxels ($G_{max} = 40$), these execution times are respectively equal to 1 hour 51 minutes and 8 minutes 32 seconds, which are compatible with clinical routines.

The good performances are also highlighted by the speedup curves (Figs. 7.31a and 7.31b), showing a slow decrease in performance when increasing the number of processors: for two processors the speedup is almost linear; for more processors the curves only moderately move away from the ideal speedup (see the dashed line). For instance, on 16 processors the speedup is about 13.44 for 16 individuals and 24.63 for 32 32 individuals (32 processors).

The small loss in performance observed when increasing the number of processors originates in the increasing number of irregular communications between physical processors needed to broadcast the current minimum, even if the compiler optimizes this operation. This problem could become cumbersome for applications requiring larger populations than the ones considered here and that could be executed on more processors. We think that there is certainly an upper bound on the population size, beyond which an island-based parallelization should be preferred, and vice versa.

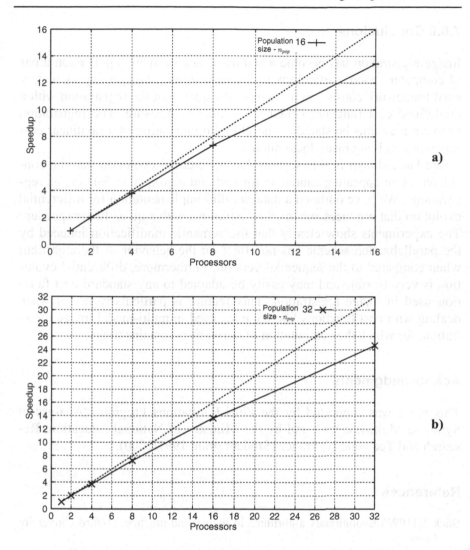

Fig. 7.31. Speedup for parallel DE. Population size of **a** 16 individuals (R12000 processors) and **b** 32 individuals (R10000 processors).

Another potential problem is the fact that the crossover length L is chosen randomly by each virtual processor. Therefore the crossover does not have the same cost across the processor grid, so when the individuals contain many parameters, a load balancing problem may appear. However, for the single modal registration application considered in this section, we were not faced with this problem.

7.6.6 Conclusions

Image registration has become a standard issue in many application areas of computer vision, particularly in the field of medical imaging. A widely used framework consists in measuring the quality of the registration with a predefined cost function, called the similarity measure. The registration problem may thus be stated as the global optimization of a nonlinear cost function involving many local minima.

We have shown in this section that differential evolution may be considered as an appealing candidate for fast and high-quality 3-D image registrations. We have outlined a data-parallel implementation for differential evolution that produces subvoxel registrations with a quasi-linear speedup. The experiments show clearly that the semantic modification induced by the parallelization scheme has no effect on the behavior of the algorithm when compared to the sequential version. Furthermore, differential evolution is very flexible and may easily be adapted to any standard cost function used in image registration. This feature is particularly useful when dealing with cost functions like the ones used in multi-modal image registration, for which the computation of derivatives remains tricky.

Acknowledgments

This work was supported by the Pluri-Formations Program "Analyse et Synthèse Multi-Images", and by the Ministry of National Education, Research and Technology, France (Student grant 1997/2000).

References

Bäck T (1996) Evolutionary algorithms in theory and practice. Oxford University Press
Brown LG (1992) A survey of image registration techniques. ACM Computing Survey 24(4):325–376
Christensen G, Miller M, Vannier M, Grenander U (1996) Individualizing neuro-anatomical atlases using a massively parallel computer. IEEE Computer, January:32–38
van den Elsen PA, Paul EJD, Viergever MA (1993) Medical image matching – a review with classification. IEEE Engineering in Medicine and Biology 12(4):26–39
Fischer D, Kohlhepp P, Bulling F (1999) An evolutionary algorithm for the registration of 3D surface representations. Pattern Recognition 32:53–69

Geman D, Geman S (1984) Stochastic relaxation, Gibbs distribution, Bayesian restoration of images. IEEE Transactions on Pattern Analysis and Machine Intelligence 6:721–741

Glover F, Laguna M (1993) Tabu search. In: Reeves C (ed.), Modern heuristic techniques for combinatorial problems. Blackwell Scientific, Oxford

Heitz F, Perez P, Bouthemy P (1994) Multiscale minimization of global energy functions in some visual recovery problems. Computer Vision, Graphics and Image Processing 59(1):125–134

Hill DLG, Hawkes DJ (2000) Across-modality registration using intensity-based cost functions. In: Bankman IN (ed.), Handbook of medical imaging. Academic Press, New York, Chap. 34

Kirkpatrick S, Gelatt CD, Vecchi MP (1983) Optimization by simulated annealing. Science 220:671–680

Musse O, Heitz F, Armspach J-P (1999) 3D deformable image matching using multiscale minimization of global energy functions. In: IEEE international conference on computer vision pattern recognition, CVPR'99, Fort Collins, USA, June

Nikou C, Heitz F, Armspach J-P, Namer I-J, Grucker D (1998) Registration of MR/MR and MR/SPECT brain images by fast stochastic optimization of robust voxel similarity measures. Neuroimage 8:30–43

Salomon M (2001) Étude de la parallélisation de méthodes heuristiques d'optimisation combinatoire. Application au recalage d'images médicales. Ph.D. thesis, Université Louis Pasteur (Strasbourg I University)

Storn R (1996) On the usage of differential evolution for function optimization. In: Conference of the North American Fuzzy Information Processing Society (NAFIPS 1996), Berkeley, CA, pp 519–423

Storn R, Price K (1996) Minimizing the real functions of the ICEC'96 contest by differential evolution. In: IEEE international conference on evolutionary computation, Nagoya , Japan, May, pp 842–844

Tomassini M (1999) Parallel and distributed evolutionary algorithms. In: Miettinen K (ed.), Evolutionary algorithms in engineering and computer science. Wiley, New York, Chap. 7

Woods RP (2000a) Spatial transformation models. In: Bankman IN (ed.), Handbook of medical imaging. Academic Press, New York, Chap. 29

Woods RP (2000b) *Within*-modality registration using intensity-based cost functions. In: Bankman IN (ed.), Handbook of medical imaging. Academic Press, New York, Chap. 33

Woods RP, Mazziotta JC, Cherry SR (1993) MRI-PET registration with automated algorithm. Journal of Computerized Assisted Tomography 17(4):536–546

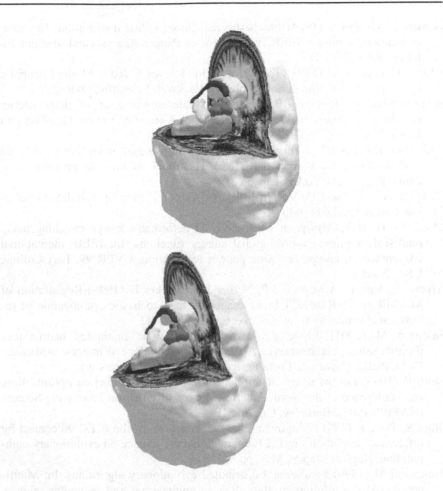

Fig. 7.32. Rigid transformation between two 3-D MR images, with segmented brain structures. Rigid registration is used to compensate for the difference of patient position between successive scans. The goal of the registration procedure is to estimate the six independent parameters of the rigid spatial transformation model, in order to obtain an accurate alignment of anatomical and functional information.

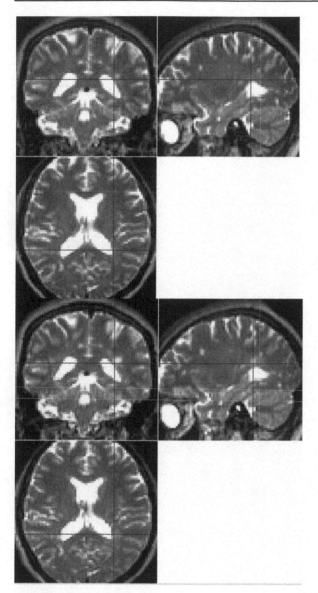

Fig. 7.33. The automatic analysis of changes between MRI scans is an important tool for assessing the evolution of lesions. Most systems use registration as a preliminary step to obtain accurate geometrical alignment of the images before image comparison (sequence of two 3-D MRI scans of a patient suffering from relapsing remitting multiple sclerosis, after registration; the evolution of a lesion is visible at the cross-hairs).

7.7 Design of Efficient Erasure Codes with Differential Evolution

Amin Shokrollahi and Rainer Storn

Abstract. The design of practical and highly powerful codes for protection against erasures in digital communication can be reduced to optimizing solutions of a highly nonlinear constraint satisfaction problem (Luby et al. 1997). In this section we will attack this problem using the differential evolution approach (Storn and Price 1997) and significantly improve results previously obtained using classical optimization procedures.

7.7.1 Introduction

The communication revolution initiated by the phenomenal explosion of the Internet and wireless communication has led to an increased use of error correcting codes as a means of protecting information against communication errors. Nowadays, codes are being used in many diverse communication media such as wireless phones, satellites, hard disks, CDs, modems and the Internet, to name a few.

The general idea of coding theory is to partition the data into blocks and augment the blocks with redundant information so that an error recovery is possible if some part of the information is lost. The aim is to add as little redundancy as possible, and, at the same time, to protect against as many errors as possible. These two requirements are obviously conflicting, and this makes the design of good codes a challenging task.

Most of the common communication channels lead to corruption of data: a random magnetization of a hard disk, atmospheric electric discharges during a satellite digital broadcast, or interference of signals of different cellular phones can flip bits in random positions. The task of decoding becomes a hard problem, since the receiver does not know the positions of the errors. There are, however, other communication channels in which data is "erased" and the receiver knows the position of the erased data. One of the prime examples of such a channel is the Internet. Data sent over the Internet is partitioned into so-called packets. Each such packet has a header which identifies the entity which the packet belongs to, as well as the position of the packet inside that entity. Packets are then routed through the network to a designated receiver. Typically some packets are lost during the transmission, while others may be corrupted once they arrive at their destination. There are usually so-called checksums associated with each packet which are used to detect corrupted packets. Once a packet has been identified as corrupted, it is declared missing. So, one can concentrate on erased packets only.

A standard solution to packet loss is retransmission of data that is not received. When some of this retransmission is lost, another request is made, and so on. In some applications, this introduces technical difficulties. For real-time transmission this solution can lead to unacceptable delays caused by several rounds of communication between sender and receiver. For a multicast protocol with one sender and many receivers, different sets of receivers can lose different sets of packets, and this solution can add significant overhead to the protocol.

An alternative solution based on using codes is sometimes desirable. It is a challenge to design fast enough encoding and decoding algorithms to make coding solutions feasible in real time for high-bandwidth applications. Practical codes with these properties were first introduced in Luby et al. (1997). Those codes can be encoded and decoded in linear time while providing near-optimal loss protection.

Based on the theoretical results proved in Luby et al. (1997), we will in this section attack a nonlinear constrained satisfaction problem the solutions of which correspond to highly efficient codes. The optimization problem involved will be attacked by differential evolution, a robust optimizer which has proved quite effective for similar types of problems (Storn 1996a).

The section is organized as follows. In the next subsection we will recall the basic construction of the codes in Luby et al. (1997). In Sect. 7.7.3 we will introduce the optimization problem related to these codes and describe different choices for the cost function of that problem. Afterwards, we recall basic properties of differential evolution. Finally, in Sect. 7.7.5 we will present our optimization results.

7.7.2 Codes from Bipartite Graphs

Construction of the Codes

The codes from Luby et al. (1997) are built from sparse bipartite graphs and generalize a classic construction of Gallager (1963). A bipartite graph is a graph whose set of nodes is a disjoint union of a set of left nodes and a set of right nodes.

Suppose that the graph B has n nodes on the left and r nodes on the right. We enumerate the left nodes of G by the numbers 1, ..., n. The code associated to G is defined as follows: it consists of all binary n-tuples $c = (c_1, \ldots, c_n)$ such that for any right node of G, the sum modulo 2 of the co-ordinates of c that are connected to that node equals zero. An example is given in Fig. 7.34. In that example, a binary vector $(x_1, \ldots, x_7)$ belongs to the code *iff* the sums $x_1 + x_4 + x_5 + x_7$, $x_2 + x_4 + x_6 + x_7$ and $x_3 + x_5 + x_6 + x_7$ are zero modulo 2. Hence, for instance, the binary vector $(1, 1, 0, 0, 1, 1, 0)$ belongs to the code, while $(1, 1, 0, 0, 1, 0, 1)$ does not.

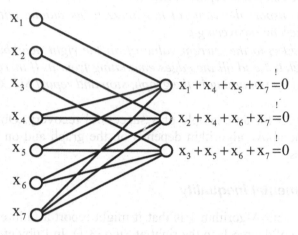

Fig. 7.34. Example of a code from a graph. The equality sign with exclamation mark denotes that equality is a requirement.

It is easy to see that any codeword has at least $k = n - r$ "free" bits of in-formation: since we are imposing r linear conditions on the coordinates, the code is the intersection of r hyperplanes and hence has dimension at least $n - r$. We call the quantity $\mathfrak{R} = k/n$ the *rate* of the code. In practice, the rate of the code is dictated by the specific application. For instance, if a particular computer network has a peak loss rate of 60%, the rate of the code has to be at most 0.4. This is because one cannot recover from a frac-tion of bits which is less than the rate, i.e., one cannot recover from a loss fraction larger than one minus the rate.

The main contribution of Luby et al. (1997) is the design and analysis of the bipartite graph G so that the following simplistic decoding operation recovers all the missing bits.

Algorithm 1 [Loss recovery]. *We assume that a word $(c_1, \ldots, c_n)$ is re-ceived where each c_i is either 0/1, or is erased. The algorithm works with two sets of registers, one for the left nodes and one for the right nodes in the graph. We will identify these registers with the corresponding nodes:*

(1)   Initialize the contents of the right-hand nodes of the graph with zero.

(2)   Collect all the non-erased coordinate positions, add their value to the current value of their right neighbors, and delete the corresponding left node and all edges emanating from it.

(3)   Repeat the following step:

   (3.1)   Look for a right node in the graph of degree 1, i.e., a node that has only one edge coming out of it. If no such node is found, then stop and report FAILURE.

   (3.2)   Transport the value of that node to its unique left neighbor l thereby recovering c_l.

   (3.3)   Add c_l to the current value of all the right neighbors of l and delete l and all the edges emanating from it. If there are no left nodes remaining in the graph, stop and report SUCCESS.

See Fig. 7.35 for an example of a successful recovery. Obviously, the success of the above algorithm depends on the graph and on the specific set of erasures.

The Fundamental Inequality

The problem with Algorithm 1 is that it might report a failure if it cannot find any node of degree 1 on the right at Step (3.1). In Luby et al. (1997) it is proved that, under the assumption that losses occur at random locations, it is only the structure of the graph that decides on the success or the failure of the decoding. To express the result, we need one last piece of notation. An edge in the graph is called *an edge of left degree i* if it is connected to a left node of degree i, and it is called of *right degree i* if it is connected to a right node of degree i. Let λ_i denote the fraction of edges of left degree i, and let ρ_i denote the fraction of edges of right degree i. Consider the generating functions $\lambda(x) := \sum_i \lambda_i x^{i-1}$ and $\rho(x) := \sum_k \rho_k x^{k-1}$. Note that $\lambda(x)$ and $\rho(x)$ are polynomials depending only on the graph. We call them the *edge degree distributions* of the graph.

The following theorem is essentially from Luby et al. (1997). The formulation given here has been taken from Luby et al. (1998).

Theorem 1. *The above loss recovery algorithm recovers a δ-fraction of erased nodes with high probability if and only if the graph has a degree distribution given by $\lambda(x)$ and $\rho(x)$ such that*

$$\delta\lambda(1-\rho(1-x)) < x \qquad (7.50)$$

holds true on the interval $(0, \delta]$.

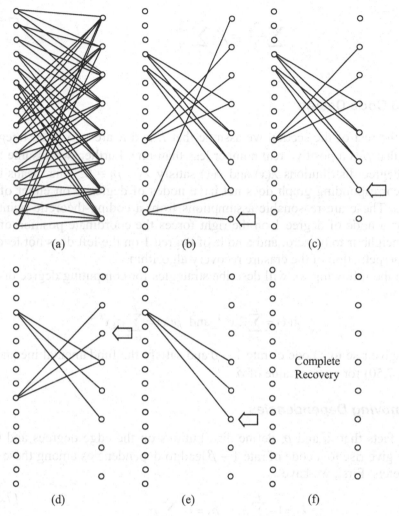

Fig. 7.35. Recovery from erasures; the arrow at a right node indicates that that node is used in Step (3.1).

The task at hand is now to find appropriate polynomials λ and ρ with non-negative coefficients that give rise to a code of a given rate such that the fundamental inequality (Eq. 7.50) is satisfied for a large value of δ.

For future reference, we record the following formula which relates the rate of the code with the degree distributions (Luby et al. 1997). In this formula, and in the sequel, we denote by β the quantity $1 - \mathfrak{R}$.

$$\sum_k \frac{\rho_k}{k} = \beta \cdot \sum_i \frac{\lambda_i}{i}. \qquad (7.51)$$

7.7.3 Code Design

For the rest of the section we assume that L and R are two fixed integers and that β is a positive real number less than one. Further, we assume that the degree distributions $\lambda(x)$ and $\rho(x)$ satisfy $\lambda_1 = \rho_1 = 0$. This means that the corresponding graph does not have nodes of degree 1 on either of its sides. These are reasonable assumptions from a coding-theoretic point of view: a node of degree 1 on the right forces the coordinate position of its left neighbor to be zero; and a node of degree 1 on the left does not lead to the perpetuation of the erasure recovery algorithm.

In the following we will describe strategies for computing degree distributions

$$\lambda(x) = \sum_{i=2}^{L} \lambda_i x^{i-1} \quad \text{and} \quad \rho(x) = \sum_{k=2}^{R} \rho_k x^{k-1}$$

that give rise to a code of rate $1 - \beta$ and satisfy the fundamental inequality (Eq. 7.50) for a large value of δ.

Removing Dependencies

The facts that λ_i and ρ_k define distributions on the edge degrees and that they give rise to a code of rate $1 - \beta$ lead to dependencies among these parameters. First, we have

$$\lambda_2 = 1 - \sum_{i=3}^{L} \lambda_i, \quad \rho_2 = 1 - \sum_{k=3}^{R} \rho_k. \qquad (7.52)$$

Next, the condition in Eq. 7.51 defines a further relation among these parameters. Solving for λ_L and using Eqs. 7.52 gives

$$\lambda_L = \frac{\dfrac{\beta-1}{2} + \sum_{k=3}^{R} \rho_k \left(\dfrac{1}{k} - \dfrac{1}{2} \right) - \sum_{i=3}^{L-1} \lambda_i \left(\dfrac{1}{i} - \dfrac{1}{2} \right)}{\dfrac{1}{L} - \dfrac{1}{2}}. \qquad (7.53)$$

The next simplification of the problem consists of discretizing the interval $(0, \delta]$ into N equidistant points $\delta j / N, j = 1, \ldots, N$. Condition 7.50 is then transformed into

$$\delta\lambda\left(1 - \rho\left(1 - x_j\right)\right) < x_j, \quad x_j = \frac{\delta \cdot j}{N}, \quad j = 1,...,N. \tag{7.54}$$

By choosing N moderately large and slightly decreasing the value of δ obtained from solving the above inequality (if necessary), we can find a δ that satisfies condition 7.50. For this reason, we will concentrate on the above set of inequalities. The results we will present later will all satisfy inequality 7.50, however.

Condition (Eq. 7.54) can be compactly written as

$$f\left(\lambda_i, \rho_k, \delta, x_j\right) := \delta\lambda\left(1 - \rho\left(1 - x_j\right)\right) - x_j < 0 \tag{7.55}$$

for $j = 1, ..., N$, where we have suppressed the implicit dependency of f on β to save a notational explosion. After going through some algebra and using Eq. 7.52, the function f can be rewritten as

$$f\left(\lambda_i, \rho_k, \delta, x\right) = \delta\left(\sum_{i=3}^{L-1} \lambda_i\left(\alpha^{i-1} - \alpha\right) + \alpha + \lambda_L\left(\alpha^{L-1} - \alpha\right)\right) - x \tag{7.56}$$

where

$$\alpha = \sum_{k=3}^{R} \rho_k\left(\left(1 - x\right) - \left(1 - x\right)^{k-1}\right) + x \tag{7.57}$$

and λ_L has the value defined in Eq. 7.53.

The Linear Programming Approach

Suppose that α, the value of δ and the point x are fixed. Then the function $f(\lambda_i, \rho_k, \delta, x)$ is a linear form in the variables $\lambda_3, ..., \lambda_{L-1}$. This leads to the following approach for computing a good value of δ. Fix δ and the ρ_k, $k = 3, ..., R$, and set up a linear program given by the inequalities 7.55 and $\lambda_3 ,..., \lambda_{L-1} \geq 0$. If there exists a feasible solution, then increase the value of δ, otherwise decrease that value. The optimal δ is that found by binary search. This value corresponds to the optimal δ for the specifically chosen $\rho(x)$. This approach was suggested in Luby et al. (1997). The disadvantage of this idea is that one has to know a good choice for $\rho(x)$ to start with. This problem was partly solved in Luby et al. (1998) which derived a condition equivalent and similar to Eq. 7.50, but with the nesting of $\lambda(x)$ and $\rho(x)$ reserved. The idea is to start with a fixed $\rho(x)$ and optimize $\lambda(x)$ by linear programming as described above. Holding this $\lambda(x)$ constant allows linear programming to be applied to a similar equation, but this time with the ρ_k as parameters. Using linear programming back and forth this way will ultimately lead to a good value of δ.

Table 7.10. Degree distributions for rate 1/2 codes with the linear programming approach

δ	$\lambda(x)$ and $\rho(x)$	$\hat{\delta}$	$\delta/\hat{\delta}$
0.4845	$\lambda(x) = 0.25594x + 0.37910x^2 + 0.127423x^{10} + 0.237538x^{11}$ $\rho(x) = x^6$	0.4959	0.9770
0.4834	$\lambda(x) = 0.25800x + 0.32950x^2 + 0.01056x^4 + 0.07637x^7 + 0.32557x^8$ $\rho(x) = 0.47013x^5 + 0.41733x^6 + 0.11253x^{20}$	0.4958	0.9749
0.4886	$\lambda(x) = 0.11286x + 0.06153x^2 + 0.12936x^3 + 0.25559x^7 + 0.24226x^8 + 0.02550x^{24} + 0.17290x^{25}$ $\rho(x) = 0.17263x^5 + 0.28658x^6 + 0.14805x^{28} + 0.39274x^{29}$	0.4998	0.9775
0.4946	$\lambda(x) = 0.038029x + 0.195063x^3 + 0.140799x^4 + 0.160451x^{11} + 0.037743x^{12} + 0.172622x^{20} + 0.206781x^{57} + 0.048513x^{58}$ $\rho(x) = 0.180714x^6 + 0.122209x^7 + 0.329968x^{18} + 0.104520x^{49} + 0.043951x^{70} + 0.218638x^{175}$	0.4999	0.9892
0.4943	$\lambda(x) = 0.196050x + 0.257821x^2 + 0.191453x^8 + 0.046831x^9 + 0.063126x^{23} + 0.059209x^{24} + 0.060652x^{62} + 0.124858x^{63}$ $\rho(x) = 0.820342x^8 + 0.177571x^9 + 0.002087x^{199}$	0.4992	0.9903

However, even this scheme has a major disadvantage: one has to start with a fixed initial distribution and it is absolutely not clear which distribution leads to an optimal value of δ. Furthermore, fixing the initial distribution fixes the average degrees of the distributions involved for the course of the entire optimization. Hence, many potentially good degree distributions are not at all visited by the procedure. For this reason, this approach does not have good chances of producing extraordinarily good degree distributions. Nevertheless, we have included some results of this approach for purposes of comparison with the method described in the next section, see Table 7.10. In that table, the last two columns correspond to a theoretical upper bound $\hat{\delta}$ on δ given in Eq. 7.63.

The δ values obtained from the linear programming approach above are close to their optimal values. However, as one can see from the last two examples in that table, very good values for δ would require an increase in the degrees of the nodes on the left- and right-hand sides of the graph. This prevents their realization for constructing short codes. To obtain excellent values of δ while preserving small node degrees, a different optimization procedure has to be applied. This will be the topic of the remainder of this section.

Code Design by Cost Function Minimization

Cost functions like the function f where the goal to be reached is explicitly known belong to the class of so-called constraint satisfaction problems. A beneficial property of this problem class is that the stopping criterion for any optimization algorithm is unambiguous. Let $D := L + R - 5$ denote the number of free parameters among the λ_i and ρ_k. We can define a D-dimensional parameter vector $p = (p_0, p_1, \ldots, p_{D-1})^{\mathrm{T}}$ with the mapping $p_i = \lambda_{3+i}$ for $i = 0, \ldots, L - 4$ and $p_k = \rho_{k-L+6}$ for $k = L - 3, \ldots, L + R - 6$. The goal is to optimally choose the elements of this parameter vector in order to satisfy condition 7.55 with a large value of δ.

This problem can be phrased in terms of minimizing a cost function. We chose two approaches to this problem. These methods differed in the choice of the cost function.

Approach A. In this approach we consider for a fixed δ the largest value among $f(\lambda_i, \rho_k, \delta, x_j)$. The aim of the optimization is to find a setting of the parameters that make this value non-positive.

Approach B. In this approach the cost function is given by the largest value of δ such that $f(\lambda_i, \rho_k, \delta, x_j) < 0$ for $j = 1, \ldots, N$. The aim of the optimization is to find a setting of the parameters that made the value of δ as large as possible. We solved these problems using differential evolution, a robust optimizer which is described in the next section.

7.7.4 Differential Evolution

The code design problem as described above is a nonlinear constraint satisfaction problem with continuous space parameters, a problem class where differential evolution (DE) (Storn and Price 1997) has proven to be very effective (Storn 1996a, 1996b). The main properties of DE are repeated here for convenience.

a) Initialization. DE is an evolution strategy that uses NP D-dimensional parameter vectors

$$P_{i,G}; \quad i = 0, 1, 2, \ldots, NP - 1 \tag{7.58}$$

in a generation G, with NP being constant over the entire design process. Hence DE is similar to a (μ, λ) evolution strategy (ES) (Bäck et al. 1997) where $\mu = \lambda = NP$. We will see later, however, that there are several important differences from a standard ES approach. At the start of the procedure, i.e., generation $G = 0$, the population of vectors is usually chosen randomly. As a rule, we will assume a uniform probability distribution for all random decisions unless otherwise stated.

b) **Mutation.** The classical variant of DE (Storn and Price 1997) uses the following vector generation scheme: for the following generation $G + 1$, new vectors $\mathbf{v}_{i,G+1}$ are generated according to the following mutation scheme:

$$\mathbf{v}_{i,G+1} = \mathbf{p}_{i,G} + F \cdot (\mathbf{p}_{r_1,G} - \mathbf{p}_{r_2,G}) \tag{7.59}$$

for

$$i = 0,1,2,\ldots,NP-1.$$

The integers r_1 and r_2 are chosen randomly over $[0,NP-1]$ and should be mutually different as well as different from the running index i. F is a real constant factor which controls the amplification of the differential variation $(\mathbf{p}_{r_1,G} - \mathbf{p}_{r_2,G})$ and is usually taken from the range $[0.1,1]$.

For the problem described here, however, it has been found that another variant of DE yielded better results with less computational expense. This variant follows the mutation scheme

$$\mathbf{v}_{i,G+1} = \mathbf{p}_{best,G} + 0.5 \cdot (\mathbf{p}_{r_1,G} - \mathbf{p}_{r_2,G} + \mathbf{p}_{r_3,G} - \mathbf{p}_{r_4,G}). \tag{7.60}$$

The vector $\mathbf{p}_{best,G}$ corresponds to the vector which has the lowest cost function in generation G. The usage of two vector differentials instead of one shifts the probability distribution of the perturbation used to mutate $\mathbf{p}_{best,G}$ more into the Gaussian. Therefore the variance of the perturbation is increased, which helps to prevent the algorithm from getting stuck in a local minimum. A notable difference of DE to known ESs, however, is the fact that mutation is not done via some separately defined probability density function (PDF). Instead the mutation is solely derived from positional information of the current population. This scheme provides for automatic self-adaptation and eliminates the need to adapt standard deviations of a PDF.

c) **Recombination.** In order to increase the diversity of the new parameter vectors, discrete recombination is introduced, a common ingredient in ESs.

There exist many variants of recombination mechanisms (Bäck et al. 1997). The one used here is to form the vector

$$\mathbf{u}_{i,G+1} = (u_{0i,G+1}, u_{1i,G+1}, ..., u_{(D-1)i,G+1})^{\mathrm{T}} \tag{7.61}$$

with

$$u_{ji,G+1} = \begin{cases} v_{ji,G+1} & \text{if } \left(randb(j) \le CR\right) \quad \text{or} \quad j = rnbr(i) \\ p_{ji,G} & \text{otherwise} \end{cases} \tag{7.62}$$

$j = 0,1,..., D-1.$

Here, $randb(j) \in [0,1]$ is the jth evaluation of a uniform random number generator. CR is the crossover constant $\in [0,1]$, which was always chosen to be equal to 1 in our examples. The value $rnbr(i)$ is a randomly chosen index in $\{0, 1, ..., D - 1\}$ which ensures that $\mathbf{u}_{i,G+1}$ gets at least one parameter from $\mathbf{v}_{i,G+1}$.

d) Selection. The selection scheme in DE is deterministic but still different from the methods that are generally employed in standard ESs. DE's selection scheme is based on local competition only, i.e., a child $\mathbf{u}_{i,G+1}$ will compete against one population member $\mathbf{p}_{i,G}$ and the survivor will enter the new population. Explicitly, if $\mathbf{u}_{i,G+1}$ yields a smaller cost function value than $\mathbf{p}_{i,G}$, then $\mathbf{p}_{i,G+1}$ will be set to $\mathbf{u}_{i,G+1}$. Otherwise the old value $\mathbf{p}_{i,G}$ is retained. In other words, DE can be regarded as a (μ, λ) ES with $\mu = NP$ and $\lambda = NP$, local competition, and a differential-based mutation scheme.

e) Stopping criterion. The stopping criterion for DE depends on the type of problem. If, as in our case, the goal is to find just one parameter set which meets the constraint of the cost function, the design procedure can be stopped as soon as one member of the vector population meets the requirements.

7.7.5 Results

Results for Approach A

Our approach for finding good solutions to condition 7.54 consisted of first starting with values δ, β, L and R where results found with linear programming methods are available. This way we could test whether DE was capable of reproducing these results. After having achieved this we gradually increased δ and tested whether new solutions were still possible. Once this seemed highly improbable we slightly increased L and R. Since an increase in L and R means an increase in the number of parameters D, we also increased NP. As described above, the DE variant defined by Eqs. 7.60–7.62 turned out to be the most effective one. In all our examples, CR was always set to 1. That way, the only control variable used in DE was the population size NP which was varied in the range $10D, ..., 20D$.

In all our calculations we worked with the value $\beta = 0.5$, i.e., with codes of rate 0.5. This resulted in a fair comparison of our results and those available in the literature. As stated before, we wanted to find solutions to Eq. 7.54 with δ as large as possible. At the same time it was desired that L and R be as small as possible, so as to facilitate construction of short codes. This is due to the results of Luby et al. (1998) which suggest that the performance of a short code is better, the smaller the degrees involved in the graph. Finding good solutions required some experimentation but the crucial point here is that it was possible to find good solutions at all. Evaluating the quality of the solutions can be done either by comparing the solution to ones existing in the literature, or by comparing the optimal value of δ with a theoretical upper bound $\hat{\delta}$ derived in Shokrollahi (1999).

For the convenience of the reader we briefly recall this upper bound: $\hat{\delta}$ is the unique real number in the interval $(0,1)$ such that

$$\delta \leq \hat{\delta}, \quad \hat{\delta} = \beta\left(1 - \left(\hat{\delta}\right)^{\gamma}\right), \quad \gamma = \frac{1}{\int_0^1 \rho(x)dx}. \tag{7.63}$$

Our best results were obtained after 1,897,145 cost function evaluations with a population size of 700, and the constants $R = L = (D + 5)/2 = 20$, $\delta = 0.494$. The initial values for the parameters were randomly drawn from the interval $[-0.1, 0.1]$. The cost function used in approach A can best be described by the C-style pseudo-code shown in Fig. 7.36.

It describes a minimax formulation of the cost function with simple penalty terms. The penalty terms help to ensure that the parameters stay positive. Although this cost function is very straightforward, very good results, i.e., large values of δ, could be achieved. For instance, we found degree distributions $\lambda(x)$ and $\rho(x)$ having degrees less than 20 on both sides with a maximal δ value of 0.494. All the good degree distributions we found had the property of having negligible λ_i and ρ_k for most values of i, k. Therefore, we experimented in approach B with degree distributions for which certain node degrees were forced to zero; this considerably reduced the dimension of the problem and the running time of the algorithm. These results and other strategies are described in the next subsection.

```
cost = 0;
for ( i=2; i <= L; i++ )
{
    if (λ  < 0) cost:= cost + 100 - 10* λ
         i                                i
// penalty for negativity
}
for (i=2; i<=R; i++)
{
    if (ρ  < 0) cost:= cost + 100 - 10* ρ
         i                                i
// penalty for negativity
}
if (cost == 0) cost =max(f(λ  , ρ  , δ , x  ),j=1,…,N);
                            i     k        i
```

Fig. 7.36. Cost function used in approach A

Results for Approach B

This approach differs from the previous one in the choice of the objective function as described in Sect. 7.7.3. Also, we incorporated some modifications to the phase of the DE which computes the initial population: note that the conditions relating the coefficients of $\lambda(x)$ and $\rho(x)$ force the free coefficients of these polynomials to lie in a finite polytope. Choosing the free parameters randomly does not necessarily result in choosing a random point from the polytope. In order to achieve the latter task, we implemented a different strategy, known as the "Queen's move". We started with some point inside the polytope constructed deterministically, and repeated the following procedure between 50 and 100 times: we randomly selected a line through the point, and randomly selected a point on that line inside the polytope. This gave us one population member. For the next members, we repeated the whole procedure again, until all the population members were generated.

Another modification with respect to approach A was that we did not let the node degrees on the left and the right take on *all* possible node degrees below L and R. As stated before, a closer inspection of the results of approach A reveals that many of the node distributions are close to zero or at least very small. In fact only the larger values for the node distributions are of practical interest for the construction of codes. As a further refinement of approach A, we experimented in approach B with the idea of forcing to zero those λ_i and ρ_k which have small values and not to treat them as free parameters subject to optimization. This idea yielded further improvements in the codes being constructible as well as in running times of the DE optimization, since it resulted in a major reduction in the dimension of the problem, and hence a decrease in the population size. Typically, we chose the node degrees in the following way: on the left-hand side, we chose the degrees 2, 3, a highest degree (between 20 and 30) and one degree in between. On the right-hand side, we chose two consecutive degrees, either 7 and 8, or 8 and 9. This sort of choice was suggested by the results we obtained from approach A. We find it very remarkable that the δ values obtained here are extremely close to their above-mentioned upper bound (Eq. 7.63). Some results and comparisons are given in Table 7.11. The columns *NP* and *NFE* in that table correspond to the population size and the number of function evaluations, respectively.

Table 7.11. Some good degree distributions for codes of rate 1/2 obtained with DE

δ	$\lambda(x)$ and $\rho(x)$	*NP*	*NFE*	*D*	$\hat{\delta}$	$\delta/\hat{\delta}$
0.4939	$\lambda(x) = 0.29730x + 0.17495x^2 + 0.24419x^5 + 0.28353x^{19}$ $\rho(x) = 0.33181x^6 + 0.66818x^7$	50	640	3	0.4974	0.9929
0.4948	$\lambda(x) = 0.27692x + 0.20256x^2 + 0.26207x^6 + 0.25843x^{24}$ $\rho(x) = 0.89468x^7 + 0.10531x^6$	100	1400	3	0.4979	0.9939
0.4955	$\lambda(x) = 0.26328x + 0.18020x^2 + 0.27000x^6 + 0.28649x^{29}$ $\rho(x) = 0.63407x^7 + 0.36593x^8$	100	1400	3	0.4985	0.9941

Acknowledgments

We would like to thank Peter Winkler and Graham Brightwell for suggesting to us the Queen's move and other strategies to uniformly sample points from a polytope.

References

Bäck T, Hammel U, Schwefel H-P (1997) Evolutionary computation: comments on the history and current state. IEEE Transactions on Evolutionary Computation, April:3–17

Gallager RG (1963) Low density parity-check codes. MIT Press, Cambridge, MA

Luby M, Mitzenmacher M, Shokrollahi MA, Spielman D, Stemann V (1997) Practical loss-resilient codes. In: Proceedings of the 29th annual ACM symposium on theory of computing, pp 150–159

Luby M, Mitzenmacher M, Shokrollahi MA, Spielman D (1998) Analysis of low density codes and improved designs using irregular graphs. In: Proceedings of the 30th annual ACM symposium on theory of computing, pp 249–258

Shokrollahi MA (1999) New sequences of linear time erasure codes approaching the channel capacity. In: Proceedings of AAECC'13, Lecture Notes in Computer Science 1719, pp 65-76. Springer-Verlag, New York

Storn R (1996a) Differential evolution design of an IIR-filter with requirements of magnitude and group delay. In: Proceedings of the IEEE international conference on evolutionary computation, ICEC'96, pp 268–273

Storn R (1996b) On the usage of differential evolution for function optimization. In: NAFIPS'96, pp 519–523

Storn R, Price K (1997) Differential evolution - a simple and efficient heuristic for global optimization over continuous spaces. Journal of Global Optimization 11:341–359

7.8 FIWIZ – A Versatile Program for the Design of Digital Filters Using Differential Evolution

Rainer Storn

Abstract. FIWIZ is a constraint-based design program for recursive (IIR) as well as transversal (FIR) digital filters which is geared toward features which are difficult, if at all, to find in other filter design programs. The main design tasks are elaborated, and the approach via differential evolution (DE) with emphasis on objective function design and achieving computational efficiency is explained. Some design examples are presented to show results that have been achieved with the current implementation of FIWIZ.

7.8.1 Introduction

Digital filters are ubiquitous in signal processing circuitry and in general have the task to shape signals in a certain desired way or remove unwanted components. These filters come in two flavors: there are recursive filters that have an infinite impulse response when excited with a single impulse (IIR filters) and purely transversal filters that have a finite impulse response (FIR filters). A more detailed introduction into digital filtering can be found, for example, in Rabiner and Gold (1975), Antoniou (1993), Mitra and Kaiser (1993) and Corne et al. (1999). In this section the background and application of digital filters will be of less concern; the focus will be more on the mathematical intricacies of filter design and the ensuing optimization task. To lay the foundation for this endeavor the mathematical structure of digital filters will be described in the following.

The transfer function, $H(z)$, of an IIR filter in its most general form is

$$H(z) = \frac{U(z)}{D(z)} = \frac{\sum_{n=0}^{N} a_n \cdot z^{-n}}{1 + \sum_{m=1}^{M} b_m \cdot z^{-m}} = A_0 \frac{\prod_{n=0}^{N-1} (z - z_{0,n})}{\prod_{m=0}^{M-1} (z - z_{p,m})} \tag{7.64}$$

with

$$z = e^{\iota 2\pi\Omega} = \cos(2\pi\Omega) + \iota \cdot \sin(2\pi\Omega), \quad \iota = \sqrt{-1} \tag{7.65}$$

and

$$\Omega = \frac{f}{f_s} \tag{7.66}$$

with Ω stemming from the interval [0,0.5]. The variables f and f_s denote the natural frequency and the sampling frequency respectively (Rabiner and Gold 1975). The degree of $H(z)$ is defined as max(N,M).

The parameters a_n and b_n are called the coefficients of the filter while $z_{0,n}$ and $z_{p,m}$ denote the zeros and poles of the filter respectively. In the majority of cases the coefficients a_n and b_n are real valued, for which the poles and zeros show up in complex conjugate pairs. This property is always assumed in FIWIZ. For stability reasons

$$\left| z_{p,m} \right| < 1; \quad m = 0,1,\ldots, M-1 \tag{7.67}$$

must always hold. Note that FIR filters are a special case of Eq. 7.64 where $M = 0$, i.e., the denominator is 1. Since there are only poles at $z = 0$ in FIR filters, these filters are always stable.

In digital filter design the magnitude $A(\Omega)$ which is computed via

$$A(\Omega) = \left| H\left(e^{\iota 2\pi \cdot \Omega}\right) \right| = \sqrt{\mathrm{Re}\left(H\left(e^{\iota 2\pi \cdot \Omega}\right)\right)^2 + \mathrm{Im}\left(H\left(e^{\iota 2\pi \cdot \Omega}\right)\right)^2} \tag{7.68}$$

is usually subject to constraints. A simple example for a low-pass filter is provided in Fig. 7.37 which shows the set of constraints, the tolerance scheme, which has to be met by $A(\Omega)$ in the logarithmic domain. It can be seen that for some interval in Ω the values of $A(\Omega)$ are fairly high. This region is commonly called the pass-band. In some other region the values of $A(\Omega)$ are low with respect to the pass-band which is why this region is called the stop-band. In between the pass-band and the stop-band is the transition-band. Since the pass-band in Fig. 7.37 is located in the lower region of Ω, $A(\Omega)$ is said to have a low-pass characteristic.

Other quantities that are derived from $H(z)$ and which sometimes are subject to constraints are the phase

$$\begin{aligned} \varphi(\Omega) &= arc\left(H\left(e^{\iota 2\pi \cdot \Omega}\right)\right) \\ &= arc\left(U\left(e^{\iota 2\pi \cdot \Omega}\right)\right) - arc\left(D\left(e^{\iota 2\pi \cdot \Omega}\right)\right) \\ &= \varphi_u(\Omega) - \varphi_d(\Omega) \end{aligned} \tag{7.69}$$

where

$$arc(c) = \begin{cases} \arctan \dfrac{Im(c)}{Re(c)} & \text{if} \quad Re(c) \geq 0 \\[3mm] \pi + \arctan \dfrac{Im(c)}{Re(c)} & \text{if} \quad Re(c) < 0 \end{cases} \qquad (7.70)$$

or the group delay

$$G(\Omega) = -\frac{d\varphi(\Omega)}{d\Omega} = \frac{d\varphi_d(\Omega)}{d\Omega} - \frac{d\varphi_u(\Omega)}{d\Omega}. \qquad (7.71)$$

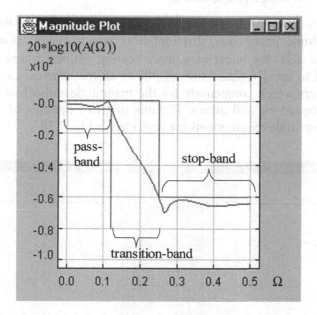

Fig. 7.37. Example of a tolerance scheme for a low-pass digital filter

Most digital filter design tasks can be described as constraint satisfaction problems where one or more function graphs have to fit into a tolerance scheme. Also FIWIZ regards all design tasks exclusively as constraint satisfaction problems.

There is a vast body of literature (e.g., Rabiner and Gold 1975; Cappellini et al. 1978; Antoniou 1993; Mitra and Kaiser 1993; Rorabaugh 1993) and many freely or commercially available programs (QED-2000™, ScopeFIR™, FilterExpress™, SPTool for MATLAB™, etc.) dealing with standard filter design problems like low-pass, high-pass, bandstop and bandpass filters. While FIWIZ can also design these filter types, it was set up to tackle unconventional problems that are difficult, if at all, to find in available software.

7.8.2 Unconventional Design Tasks

Conventional filter design mostly deals with tolerance schemes that are piecewise constant and exhibit only one pass-band (Mitra and Kaiser 1993). There are, however, designs which do not fit into this simple scheme as will be shown by example. Among the more common but still unconventional design problems are:

- **Arbitrary magnitude constraints.** These include multi-band filters with multiple pass-bands, differentiators which exhibit sawtooth-like shapes of $A(\Omega)$ for linear magnitude scaling, Hilbert filters which are supposed to apply a phase shift of 90° to all frequencies, sinc compensated filters which compensate for the magnitude roll-off of digital-to-analog converters, and others. Pictures of two examples of unconventional magnitude requirements are provided in Fig. 7.38.

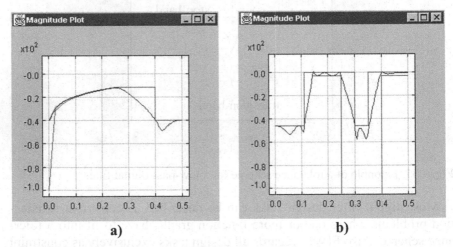

Fig. 7.38. Differentiator **a** and multi-band filter **b**

- **Additional group delay constraints.** Applications are mainly classical low-pass, high-pass, bandpass and bandstop filters which should exhibit approximately linear phase in the pass-band but not necessarily in the stop-band. The filter degree can often be reduced considerably compared to exactly linear phase FIR filters. An example for an IIR filter with constrained group delay is shown in Fig. 7.39.

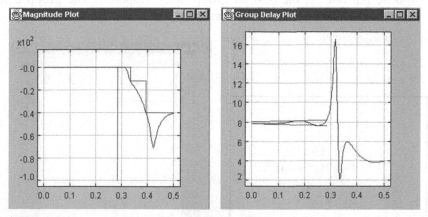

Fig. 7.39. Example for constraints in the magnitude as well as the group delay

- **Minimum phase filters.** Some applications, e.g., in speech processing, require minimum delay, which can be achieved by using minimum phase filters. Minimum phase filters (Hess 1988) have the property $z_{0,n} \leq 1$ and can generally be constructed by reflecting all zeros of a given filter which are outside the unit circle $|z| = 1$ to its inside. In order to do this, however, the zeros must be known explicitly.

- **Filter design with quantized coefficients.** In real-world applications the coefficients a_n and b_n are represented with finite precision. In FIWIZ the quantization can be incorporated into the design as opposed to quantizing the coefficients after the filter has been designed with high-precision coefficients (Table 7.12). The filter structures currently supported are the direct forms 1 and 2, as well as second-order sections (Mitra and Kaiser 1993).

- **Definition of a pre-filter with constant coefficients.** Defining a pre-filter has many applications, like presetting specific zeros to suppress a constant voltage bias or the 50/60Hz power line frequency, accommodating filters which are already in a given design and cannot be removed, or setting a frequency response for equalization. A well-known example of the last is sinc compensation needed for digital-to-analog conversion. An equalizer example is shown in Fig. 7.40 and Fig. 7.41.

- **All-pass filters.** For existing IIR filters sometimes the phase must be linearized using all-pass filters (Hess 1988; Mitra and Kaiser 1993). All-pass filters have a magnitude response $A(\Omega) = $ const., which is guaranteed by having $N = M$ and $z_{0,n} = 1/z_{p,n}$. Figures 7.42 and Fig. 7.43 show an example for a pre-filter already exhibiting the desired magnitude response but not yet the desired group delay response. The group delay tolerance scheme is allowed to float here.

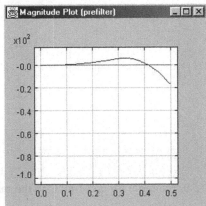

$$H_{\text{prefi}}(z) = \frac{1+z^{-1}}{1+0.75z^{-1}+0.4z^{-2}}$$

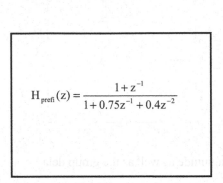

Fig. 7.40. Magnitude response which has to be equalized

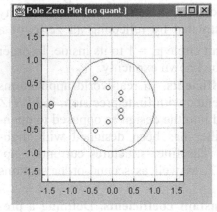

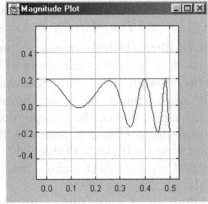

Fig. 7.41. Characteristics for the filter which equalizes the filter specified in Fig. 7.40 with an error of ±0.2 dB.

Table 7.12. Coefficients of an IIR filter which meets prescribed magnitude constraints but fails on the group delay constraints

i	a_i	b_i
0	0.19881558418273926	1.0
1	0.6713742017745972	0.5888696908950806
2	0.956078052520752	1.0049885511398315
3	0.7260297536849976	0.14867675304412842
4	0.29759085178375244	0.18157732486724854
5	0.0613178014755249	−0.014554023742675781

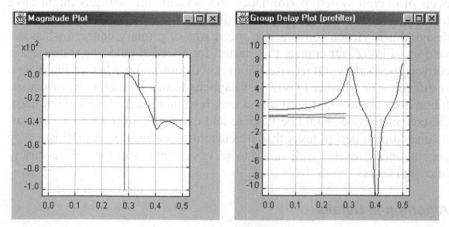

Fig. 7.42. Specification of the IIR filter the group delay of which has to be linearized

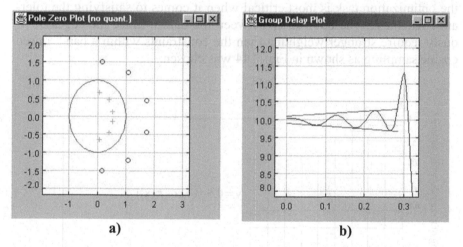

a)   b)

Fig. 7.43. Pole–zero plot of the all-pass filter **a** and resulting group delay for the entire filter **b**. The poles are indicated by plus signs while the zeros are indicated by small circles.

7.8.3 Approach

Requirements

Filter design generally requires quite a bit of experimentation since the specifications themselves, i.e., the tolerance schemes, are often not carved in stone but instead need to be determined in order to optimize various aspects of overall system performance. As several filter designs might have to be tried, a filter design program that is useful should exhibit reasonable design times. Hence convergence speed was of primary concern when developing FIWIZ. Of course, this design for speed should not compromise convergence safety. Otherwise the user will not know whether the design is infeasible or whether FIWIZ just happens to mis-converge. In case a solution does not exist, at least a best match should result. Last but not least, the computer code should be flexible and extensible in order to accommodate additional design tasks that are currently not built in. As DE is a very general optimizer it is well suited to serve as a platform for a wide variety of design tasks.

Algorithmic Details

Sampling the Frequency Axis. An important choice within FIWIZ concerns the way the frequency axis Ω is sampled. It has been observed that the optimization task is most critical when it comes to satisfying the tolerance schemes at the boundaries between the bands. Since the edges obviously require stronger weighting than the constraints within a band, raised cosine sampling as shown in Fig. 7.44 was chosen.

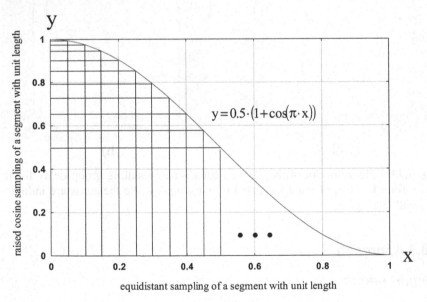

Fig. 7.44. Illustration of raised cosine sampling

Raised cosine sampling strikes a good balance between emphasis of the edges and prevention of "sampling holes" within the band. FIWIZ currently provides the choice of coarse (10 samples), medium (28 samples) and fine (64 samples) sampling granularity of a segment on the frequency axis.

If the filter design of FIWIZ converges, i.e., the cost function goes down to zero, but the resulting filter still violates the constraints set forth in the magnitude and/or group delay screens, the sampling density in the design control screen has been chosen too low. The user can pursue two basic strategies. In the first strategy the sampling density is increased, e.g., from coarse to medium or fine. In the second strategy the constraints are divided into more segments. Figure 7.45 provides an example where a constant magnitude constraint of x dB between the normalized frequencies Ω_1 and Ω_2 is changed into a sequence of two constraints of x dB between the frequencies Ω_1 and Ω_3 as well as Ω_3 and Ω_2 respectively. As FIWIZ applies its sampling density to each segment individually, this procedure also increases the overall sampling density.

This second strategy has also proven to be very valuable.

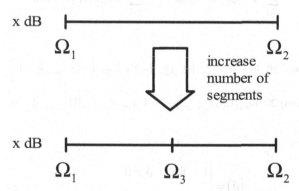

Fig. 7.45. Example for increasing the number of constraint segments

Choice of Parameters. When looking at Eq. 7.64 it is evident that the potential parameters that can be varied in the optimization are either the coefficients a_n and b_n, or the zeros and poles $z_{0,n}$ and $z_{p,m}$. Using the coefficients appears to be attractive at first because the coefficients are the values finally wanted, and if the coefficients are quantized the quantization can be readily applied. A great disadvantage, however, is that the stability criterion of Eq. 7.67 is computationally intensive to check (Antoniou 1993). Using the zeros and poles as parameters makes the stability check almost trivial. Zeros and poles have another advantage: their positioning with respect to the unit circle $|z| = 1$ is directly related to pass-bands and stop-bands. Zeros gather around stop-bands and poles gather around pass-bands, which is why these parameters exhibit a direct relationship to the magnitude constraints being set forth. Coefficient quantization can still be applied easily when it comes to the computation of the objective function which is done via $H(z)$. The coefficients a_n and b_n are easily calculated from $z_{0,n}$ and $z_{p,m}$ while the reverse is not true. For FIWIZ it was chosen to use zero and pole radii $|z_{0,n}|$ and $|z_{p,m}|$, the zero and pole angles $arc(z_{0,n})$ and $arc(z_{p,m})$, as well as the gain A_0 (see Eq. 7.64) as parameters.

The Cost Function. The cost function is made up of two components, the constraint-based part and the penalty part. The constraint-based part is simply made up of a magnitude and a group delay component according to

$$f(\mathbf{x}) = f_m(\mathbf{x}) + f_{gd}(\mathbf{x}) \tag{7.72}$$

where the magnitude component is

$$f_m(\mathbf{x}) = \sum_{i=1}^{i_{max,upper}} s(\Delta_{i,upper}) \cdot \Delta_{i,upper}^2 + \sum_{j=1}^{j_{max,lower}} s(\Delta_{j,lower}) \cdot \Delta_{j,lower}^2 \tag{7.73}$$

with

$$\Delta_{i,upper} = \max\left(20 \cdot \log_{10}\left(C_{m,upper}(\Omega_{i,upper})\right) - 20 \cdot \log_{10}\left(A(\Omega_{i,upper})\right),\ 0 \right) \tag{7.74}$$

$$\Delta_{j,lower} = \max\left(20 \cdot \log_{10}\left(A(\Omega_{j,lower})\right) - 20 \cdot \log_{10}\left(C_{m,lower}(\Omega_{j,lower})\right),\ 0 \right)$$

and

$$s(\delta) = \begin{cases} 1 & \text{for} \quad \delta \geq 0 \\ 0 & \text{otherwise.} \end{cases} \tag{7.75}$$

The equations above simply state that the magnitude $A(\Omega)$ is sampled in the frequency domain, and wherever a sample violates the tolerance scheme the difference from the tolerance scheme is squared and added to an error term. The functions $C_{m,upper}()$ and $C_{m,lower}()$ denote the upper and lower magnitude constraints of the tolerance scheme respectively. $\Delta_{i,upper}$ and $\Delta_{j,lower}$ are computed in the logarithmic domain so that pass-band and stop-band have comparable weight. The group delay component $f_{gd}(\mathbf{x})$ is computed using the same principles as $f_m(\mathbf{x})$. The deviations to the group delay tolerance scheme, however, are not computed in the logarithmic domain but are provided in normalized time

$$T = \frac{t}{t_s} = t \cdot f_s . \tag{7.76}$$

Also in contrast to magnitude constraints the group delay constraints will be floating, i.e., an arbitrary but constant group delay T_0 might be added to the given constraints by the filter design procedure as indicated in Fig. 7.46. Floating is simply done by first computing the maximum group delay which may occur at frequency Ω_{gd_max}. Then the tolerance scheme is shifted such that the upper constraint curve touches the group delay graph at Ω_{gd_max}.

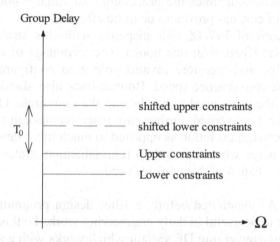

Fig. 7.46. The filter design procedure has the freedom to shift the group delay tolerance scheme along the ordinate.

Coefficient quantization is incorporated into the cost function simply by quantizing $A(\Omega)$ and $G(\Omega)$ appropriately. One choice is to quantize the parameters a_n and b_n in Eq. 7.64, which will lead to a "direct form" representation of the filter (Mitra and Kaiser 1993). Another choice is to build the filter from a concatenation of first- and second-order filter sections. Each second-order section consists of a complex conjugate pole–zero pair, and each first-order section must have its pole and zero on the real axis. The coefficients of these sections can also be quantized for the computation of $A(\Omega)$ and $G(\Omega)$.

Constraints. FIWIZ tries to prevent unstable filter solutions, i.e., the poles must always lie within the unit circle of the z domain. In earlier versions this was obtained by applying a penalty for each pole that had a radius greater than one. The penalty was applied such that the value of the cost function $f(x)$ was amplified for each pole radius that exceeded the unit circle. This amplification is the stronger the more a pole radius exceeds the unit circle and the more the pole radii are violating the stability constraint. The final cost function $f_{M-1}(x)$ was recursively computed via

$$f_{m+1}(\mathbf{x}) = f_m(\mathbf{x}) + s\big(\big|z_{pm}\big| - 1\big) \cdot \big(4 \cdot f_m(\mathbf{x}) + 400\big) \cdot \big|z_{pm}\big| \tag{7.77}$$

where $f_0(x) = f(x)$. It can be seen that each time a pole radius is greater than one the step function $s(|z_{pm}| - 1)$ becomes unity and hence adds some penalty that is at least four times the preceding cost value $+ 400$. In practice this escalation of cost has proven to be quite effective.

Future versions of FIWIZ will dispense with this strategy and use bounce-back (see Chap. 4 in this book). The advantage of not having to evaluate unstable and therefore invalid pole–zero configurations should help to increase convergence speed. Bounce-back also should prove very effective when the poles need to be close to the unit circle. Once the optimization is almost converged, further parameter vectors can stay in the vicinity of the converged solution as opposed to much more randomized vectors that can emerge when the random reinitialization strategy for bounds constraints (see Chap. 4 in this book) is used.

DE Strategy. As mentioned before, a filter design program needs to be reasonably fast to be useful in daily engineering work. To this end the goal was to use a fast-converging DE variant which works with a small number of population members. Eventually the following mutation method proved to be the best performer for the particular class of problems occurring in digital filter design:

```
for (i=0; i<dim; i++)
{
    xtrial[i] = pop_old_best[i] +
    ((0.001*rnd()+0.85)-Np*0.0005) *
    (pop_old[r1][i] - pop_old[r2][i]);
}
```

One can see that it is basically the strategy DE/best/1 with jitter to prevent stagnation. Also the weighting factor of the vector difference becomes smaller the larger *NP* gets. There is no special crossover treatment, i.e., *CR* = 1 all the time, and the selection method is DE's standard selection where the trial vector has to compete against the target vector. Two populations are used and each member of the current population has to serve once as the target vector. The winner of target vector and trial vector goes into the new population. In order to speed up convergence the parameters are not quantized until the cost function belonging to the best-so-far vector is below or equal to 1.E-6. If this cost value is reached then the quantization is applied (if quantization has been chosen) which can temporarily increase the cost function value again. For this refinement phase a number of population members may be selected. The increase in population size is simply done by starting from the current population and generating the additional vectors by applying the above DE strategy. Once the additional vectors have been produced, DE proceeds as usual.

Computational Issues. FIWIZ contains several numerical evaluations that need to be computed not only with sufficient precision but also with the best possible efficiency in order to facilitate a speedy convergence.

a) *Coefficients a_n and b_m.* Since FIWIZ employs zeros and poles as parameters for DE but coefficient quantization must be possible to be included, the coefficients a_n and b_m have to be computed first before magnitude $A(\Omega)$ and group delay $G(\Omega)$ can be evaluated. Since only filters with real-valued coefficients (as opposed to complex-valued) are computed, zeros and poles must appear in complex conjugate pairs or otherwise be on the real axis. For a complex conjugate zero pair the relationship

$$(z - z_{0,n}) \cdot (z - z_{0,n}^{*}) = z^2 - 2|z_{0,n}| \cdot z \cdot \cos(arc(z_{0,n})) + |z_{0,n}|^2 \qquad (7.78)$$

is used, where the superscripted asterisk denotes complex conjugation. This procedure already yields the real-valued subcoefficients

$$1, \quad 2|z_{0,n}| \cdot \cos(arc(z_{0,n})), \quad |z_{0,n}|^2$$

which either can be used directly for second-order section realization of the filter, or are further used in polynomial multiplications to obtain the direct form coefficients a_n. The treatment of the poles happens in just the same way.

b) *Magnitude.* Magnitude computation at the discrete points Ω_k as defined by the sampling described in Sect. 7.8.1 happens via Eq. 7.64 according to

$$(7.79)$$

$$A(\Omega_k) = \left| \frac{\sum_{n=0}^{N} a_n \cdot z^{-n}}{1 + \sum_{m=1}^{M} b_m \cdot z^{-m}} \right|_{z=e^{i2\pi\Omega_k}}$$

$$= \left| \frac{\sum_{n=0}^{N} a_n \cdot e^{-i2\pi n\Omega_k}}{1 + \sum_{m=1}^{M} b_m \cdot e^{-i2\pi m\Omega_k}} \right|$$

$$= \sqrt{ \frac{\left(\sum_{n=0}^{N-1} a_n \cdot \cos(2\pi n\Omega_k) \right)^2 + \left(\sum_{n=0}^{N-1} a_n \cdot \sin(2\pi n\Omega_k) \right)^2}{\left(\sum_{m=0}^{M-1} b_m \cdot \cos(2\pi m\Omega_k) \right)^2 + \left(\sum_{m=0}^{M-1} b_m \cdot \sin(2\pi m\Omega_k) \right)^2} }.$$

The evaluation of the trigonometric sums is obviously computationally expensive so it must be made as efficient as possible. An attractive method seemed to be Horner's method (Hildebrand 1987) which computes a polynomial

$$p(z) = \sum_{j=0}^{J-1} p_j \cdot z^j$$

via the recurrence formula

$$u_k = z \cdot u_{k+1} + p_k; \quad k = J-1, J-2, \ldots, 0; \quad u_J = 0 \quad \text{and} \quad p(z) = u_0. \quad (7.80)$$

If

$$z = e^{i2\pi\Omega_k} = \cos(2\pi\Omega_k) + i \cdot \sin(2\pi\Omega_k)$$

and the magnitude operation is done at the end of the complex-valued polynomial computation then the costly trigonometric evaluations are kept at a minimum. Unfortunately the numerical properties are not satisfactory due to error propagation, especially when N and M become large, which is why the trigonometric sums of Eq. 7.79 have to be computed directly. The direct way of computing $A(\Omega_k)$ corresponds to an evaluation method called the discrete Fourier transform (DFT) (Mitra and Kaiser 1993).

For plotting $A(\Omega)$ the sampling points are not tied to the sampling grid as defined in Sect. 7.8.1 but can be chosen to be at equidistant points. This fact is utilized to be able to employ a fast version of the DFT computation called the radix-2 fast Fourier transform (FFT) (Mitra and Kaiser 1993). To this end the series of polynomial coefficients must be zero padded according to

$$
H(\Omega_k) = \frac{\displaystyle\sum_{n=0}^{K-1} a_n \cdot e^{-\imath 2\pi n \Omega_k}}{\displaystyle\sum_{m=0}^{K-1} b_m \cdot e^{-\imath 2\pi n \Omega_k}} = \frac{\displaystyle\sum_{n=0}^{K-1} a_n \cdot e^{-\imath 2\pi n \cdot k / K}}{\displaystyle\sum_{m=0}^{K-1} b_m \cdot e^{-\imath 2\pi n \cdot k / K}} ;
\tag{7.81}
$$

$$
b_0 = 1; \underset{n>N}{a_n} = 0; \underset{m>M}{b_m} = 0; \quad k = 0,1,2,\dots,K/2.
$$

The FFT can be applied separately to the numerator and denominator of $H(\Omega_k)$ yielding $K-1$ complex points for both. K is chosen to be a power of 2. Complex division of the numerator and denominator values corresponding to the same frequency yields the complex result $H(\Omega_k)$, the magnitude of which can be computed by

$$
A(\Omega_k) = \sqrt{\mathrm{Re}(H(\Omega_k))^2 + \mathrm{Im}(H(\Omega_k))^2} .
$$

c) *Group Delay.* The group delay computation starts from Eq. 7.69 through Eq. 7.71 and uses the shorthand notation

$$
I_u = \mathrm{Im}\big(U\big(e^{\imath 2\pi \cdot \Omega}\big)\big), \quad R_u = \mathrm{Re}\big(U\big(e^{\imath 2\pi \cdot \Omega}\big)\big)
\tag{7.82}
$$

to yield

$$
\frac{d\varphi_u(\Omega)}{d\Omega} = \frac{1}{1 + \left(\dfrac{I_u}{R_u}\right)^2} \cdot \frac{d\left(\dfrac{I_u}{R_u}\right)}{d\Omega}
\tag{7.83}
$$

$$
= \frac{1}{1 + \left(\dfrac{I_u}{R_u}\right)^2} \cdot \frac{d\left(\dfrac{I_u}{d\Omega}\right) \cdot R_u - d\left(\dfrac{R_u}{d\Omega}\right) \cdot I_u}{(R_u)^2}
$$

$$
= \frac{d\left(\dfrac{I_u}{d\Omega}\right) \cdot R_u - d\left(\dfrac{R_u}{d\Omega}\right) \cdot I_u}{(R_u)^2 + (I_u)^2} .
$$

By evaluating Eq. 7.83 explicitly it follows that

$$\frac{d\varphi_u(\Omega)}{d\Omega} = \frac{\frac{d}{d\Omega}\left(\sum_{n=0}^{N-1} a_n \cdot \sin(2\pi n\Omega)\right) \cdot \left(\sum_{n=0}^{N-1} a_n \cdot \cos(2\pi n\Omega)\right) - \frac{d}{d\Omega}\left(\sum_{n=0}^{N-1} a_n \cdot \cos(2\pi n\Omega)\right) \cdot \left(\sum_{n=0}^{N-1} a_n \cdot \sin(2\pi n\Omega)\right)}{\left(\sum_{n=0}^{N-1} a_n \cdot \cos(2\pi n\Omega)\right)^2 + \left(\sum_{n=0}^{N-1} a_n \cdot \sin(2\pi n\Omega)\right)^2} \qquad (7.84)$$

$$= \frac{\left(2\pi \sum_{n=0}^{N-1} a_n \cdot n \cdot \cos(2\pi n\Omega)\right) \cdot \left(\sum_{n=0}^{N-1} a_n \cdot \cos(2\pi n\Omega)\right) + \left(2\pi \sum_{n=0}^{N-1} a_n \cdot n \cdot \sin(2\pi n\Omega)\right) \cdot \left(\sum_{n=0}^{N-1} a_n \cdot \sin(2\pi n\Omega)\right)}{\left(\sum_{n=0}^{N-1} a_n \cdot \cos(2\pi n\Omega)\right)^2 + \left(\sum_{n=0}^{N-1} a_n \cdot \sin(2\pi n\Omega)\right)^2}$$

and naturally an equivalent formula for the derivative of $\varphi_d(\Omega)$ can be written. A close look at Eq. 7.84 reveals that for the evaluation of the denominator, results from the magnitude computation in Eq. 7.79 might be reused. Yet this can only be done if the sampling points of the frequency grid are the same in the magnitude and group delay domain.

Table 7.13. Filter design examples run on a Pentium III 650 MHz PC with the Java interpreter from JDK 1.1.8. Bounds constraints were treated with the penalty-based approach.

Filter	No. of zeros	No. of poles	No. of para-meters	NP1	NP2	No. of function evalua-tions	Running time
Low-pass (Fig. 7.37)	4	4	9	30 (medium)	100 (12 bit)	5760	10 s
Differentiator (Fig. 7.38)	3	4	8	30 (coarse)	60 (16 bit)	4800	28 s
Multi-band filter (Fig. 7.38)	12	12	25	100 (coarse)	100 (12 bit)	64,100	7 min 30 s
Graphics codec (Fig. 7.39)	18	8	27	30 (fine)	60 (24 bit)	197,100[1]	27 min
Graphics codec (Fig. 7.39 but linear-phase FIR)	60	0	61	30 (fine)	60 (32 bit)	333,900	112 min 20 s
Equalizer (Figs. 7.40 and 7.41)	14	6	21	30 (fine)	–	1830	5 s
All-pass filter (Figs. 7.42 and 7.43)	12	12	25	50 (medium)	–	18,050[1]	2 min 11 s

1) Took several runs to converge.

7.8.4 Examples

In Table 7.13 several examples are provided which give an impression of the running times of typical filter design tasks undertaken with FIWIZ which has been implemented in the Java® (Chan and Lee 1997; Coad and Mayfield 1997) language.

7.8.5 Conclusion

FIWIZ has proven to be a versatile tool for the design of digital filters, especially if unconventional designs are encountered. The usage of DE allows FIWIZ to be very flexible so that a wide variety of design tasks can be tackled. If a filter design is not feasible at least a best approximation in the least squares sense will be provided. The toughest problems so far pose filters that have restrictions in both the magnitude and the group delay. These types of problems will be the topic of future research in order to increase convergence safety.

References

Antoniou A (1993) Digital filters - analysis, design, and applications. McGraw-Hill, New York

Cappellini V, Constantinides AG, Emiliani P (1978) Digital filters and their applications. Academic Press, New York

Chan P, Lee R (1997) The Java class libraries. Addison-Wesley, Reading, MA

Coad P, Mayfield M (1997) Java design - building better apps & applets. Yourdon Press, Englewood Cliffs, NJ

Corne D, Dorigo M, Glover F (1999) New ideas in optimization. McGraw-Hill, New York

Hess W (1988) Digitale Filter. Teubner, Stuttgart

Hildebrand FB (1987) Introduction to numerical analysis. Dover, New York

Mitra SJ, Kaiser JF (1993) Handbook for digital signal processing. Wiley, New York

Rabiner LR, Gold B (1975) Theory and application of digital signal processing. Prentice Hall, Englewood Cliffs, NJ

Rorabaugh CB (1993) Digital filter designer's handbook. McGraw-Hill, New York

7.9 Optimization of Radial Active Magnetic Bearings by Using Differential Evolution and the Finite Element Method

Gorazd Štumberger, Drago Dolinar and Kay Hameyer

Abstract. Magnetic bearings are a system of electromagnets, which makes possible contact-less suspension of a rigid body. The work presented here deals with the optimization of radial active magnetic bearings for a spindle drive. The bearings are optimized by differential evolution. The optimization aim is to achieve a maximum force at a minimum mass of the entire construction. Predefined design parameters are the bearing outer diameter, the shaft diameter, the air gap and the minimal generated force. The dependency of the objective function on the design parameters is not known in analytical form due to the magnetically nonlinear properties of the iron core. Therefore, the objective function of each individual parameter set in the population of the optimization algorithm is evaluated by the finite element method.

7.9.1 Introduction

An actively controlled magnetic bearing system is an indispensable element when we have to satisfy the machine-tool industry's demand for high-speed and high-precision machining. A typical system of active magnetic bearings (AMBs) (Schweitzer et al. 1994) consists of controlled electromagnets which control five degrees of freedom (DOFs). A driving motor controls the sixth DOF. Two pairs of radial bearings, which control four DOFs, are placed at the end of each rotor. The fifth DOF is controlled by a pair of axial bearings.

Two electromagnets on opposite sides of the ferromagnetic rotor pull the rotor in opposite directions. The total force acting on the rotor is equal to the vector sum of forces of all electromagnets. Such a system of electromagnets together with a ferromagnetic rotor is unstable in open-loop operation. It can be stabilized by an appropriate current and position control assuring the contact-less suspension of the rotor.

The total force of two electromagnets is a nonlinear function of the current, the rotor position and the point of saturation of the ferromagnetic iron (Antila et al. 1998). The nonlinear current–force dependency is efficiently linearized by the bias and control current, while the position–force dependency and iron magnetization remain nonlinear. The design of the control is usually based on a linearized dynamic model.

The design of AMBs is expected to satisfy the static and dynamic requirements in the best possible way. This can be done either by experience and trials or, as done here, by applying numerical optimization methods. AMBs are nonlinear systems. The dependency of the objective function and its gradients on the design parameters is unknown. For the optimization of such constrained, nonlinear electromagnetic problems, the use of stochastic search methods in combination with an analysis based on the finite element method (FEM) is recommended (Hameyer 1994).

In this work the numerical optimization of radial AMBs using differential evolution (DE) (Storn and Price 1996; Štumberger et al. 2000) is presented. The objective of the optimization is to achieve a maximum force at a minimum mass of the entire construction. The objective function is evaluated by FEM-based two-dimensional computations. This includes the determination of the nonlinear solution of the magnetic vector potential and the determination of forces by Maxwell's stress tensor method.

7.9.2 Radial Active Magnetic Bearings

AMBs are a system of controlled electromagnets acting on a single rigid body, in this case a rotor. An eight-pole radial AMB (RAMB) is schematically shown in Fig. 7.47a. It is constructed from the shaft, the ferromagnetic rotor, the ferromagnetic stator and the stator winding. The coils are wound around the stator poles. The magnetic field, excited by the currents in the stator coils, crosses the air gap between stator and rotor, and generates forces which act on the ferromagnetic rotor. To achieve a stable contact-less suspension of the rotor, the force acting on the rotor must be controlled. This can be realized by closed-loop control of the stator currents.

The stator coils wound around adjacent poles are commonly arranged in pairs in order to generate four (almost) independent magnetic loops as shown in Fig. 7.47b. A pair of coils connected in series produce the magneto motive force (mmf) $2Ni$, where N is the number of turns of an individual coil and i is the coil current. The mmf "forces" the flux ϕ through the magnetic circuit. The coil connections, the directions of currents in the coil pairs and the directions of the corresponding fluxes can be taken from Fig. 7.47b.

a)

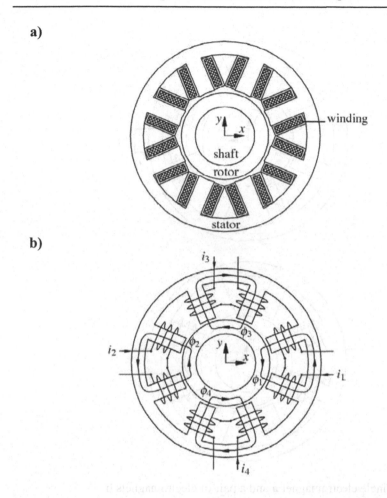

b)

Fig. 7.47. An eight-pole radial active magnetic bearing **a** and its coils connections, currents and flux patterns **b**.

In the first approximation, each of the four magnetic loops shown in Fig. 7.47b can be treated as an independent electromagnet shown in Fig. 7.48a. The voltage balance in the coil of an electromagnet is described by Eq. 7.85:

$$u_1 = R i_1 + L \frac{d i}{d t} + k_u \frac{d x}{d t} \tag{7.85}$$

where u_1 is the voltage, i_1 is the current, R is the resistance, L is the inductance, k_u is the coefficient of induced voltage and $d x / d t$ is the derivative of the rotor displacement along the x axis.

a)

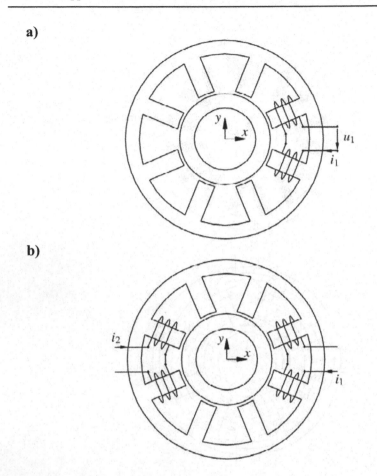

b)

Fig. 7.48. A single electromagnet **a** and a pair of electromagnets **b**

An electromagnet can only produce an attractive force F, which is generated at the boundaries of materials with different permeability μ. To derive the force equation, let us first determine the energy W stored in the air gap shown in Fig. 7.49. The air gap is a part of the magnetic loop and is enclosed by the high permeability of the ferromagnetic iron core. The volume V of the air gap is given by the air gap length g and the air gap cross-sectional area A. If the magnetic field in the air gap is homogeneous then the energy stored in the air gap is given by Eq. 7.86:

$$W = \tfrac{1}{2} B H_g V = \tfrac{1}{2} B H_g A g \qquad (7.86)$$

where H is the magnetic field strength and B is the flux density in the air gap. It can be assumed that for small displacements dg the flux $\phi = B A$ remains constant. To increase the air gap g by dg, the attractive force F acting on the ferromagnetic body has to be applied. Simultaneously, in the system without any loss of energy, the energy stored in the air gap increases by dW due to the increasing volume of the air gap, which yields:

$$F = \frac{dW}{dg} = \tfrac{1}{2} B H_g A \qquad (7.87)$$

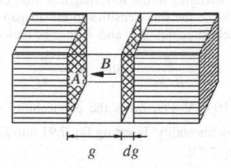

Fig. 7.49. Air gap enclosed by ferromagnetic material

In the case of the electromagnet in Figs. 7.48a and 7.50a, the force F_{x1} (Eq. 7.88) acting in the direction of x can be calculated by Eq. 7.87, considering that both poles are displaced by the angle φ referring to the x axis. The angle φ for an eight-pole radial bearing is $\varphi = \tfrac{\pi}{8}$, while $k = \cos(\tfrac{\pi}{8})$.

$$F_{x1} = 2 F \cos(\varphi) = B H_g A \cos(\varphi) = k B H_g A \qquad (7.88)$$

Let the cross-section A be uniform everywhere in the magnetic circuit shown in Fig. 7.50b and let the flux ϕ run entirely within the magnetic loop with uniform cross-section A. In this case the flux ϕ and the flux density $B = \tfrac{\phi}{A}$ are constant everywhere in the magnetic loop. If the magnetic field within the magnetic loop shown in Fig. 7.50b is supposed to be homogeneous in the ferromagnetic iron core and in the air gap then the general equation, Eq. 7.89, describing magnetic conditions along the integration path l can be simplified to Eq. 7.90:

$$\oint_l H \, dl = \sum iN \tag{7.89}$$

$$2 l_{Fe} H_{Fe} + 2g H_g = 2i_1 N \tag{7.90}$$

where $\sum iN$ are the ampere turns enclosed by the closed integration path l, H is the magnetic field strength, l_{Fe} is the mean length of the magnetic path in the ferromagnetic iron core, g is the air gap length, $2i_1 N$ is the mmf generated by the both coils shown in Fig. 7.50b, and H_{Fe} and H_g are the magnetic field strengths in the ferromagnetic iron core and in the air gap, respectively. Since the flux densities in the air gap and the iron core are identical, the field strengths H_{Fe} and H_g can be replaced by Eq. 7.91:

$$H_{Fe} = \frac{B}{\mu_r \mu_0} \quad \text{and} \quad H_g = \frac{B}{\mu_0} \tag{7.91}$$

where $\mu_0 = 4\pi \times 10^{-7}$ V s/(A m) is the permeability of vacuum, while μ_r is the relative permeability. Inserting Eq. 7.91 into Eq. 7.90 yields Eq. 7.92 and further Eq. 7.93:

$$l_{Fe} \frac{B}{\mu_r \mu_0} + g \frac{B}{\mu_0} = i_1 N \tag{7.92}$$

$$B = \mu_0 \frac{i_1 N}{l_{Fe}/\mu_r + g}. \tag{7.93}$$

Since in the iron core the relative permeability $\mu_r \gg 1$, the magnetization of the iron core is often neglected and Eq. 7.93 is simplified to Eq. 7.94:

$$B = \mu_0 \frac{i_1 N}{g}. \tag{7.94}$$

Inserting Eqs. 7.91 and 7.94 into Eq. 7.88 yields the force equation

$$F_{x1} = k \frac{B^2}{\mu_0} A = k \mu_0 \frac{(i_1 N)^2}{g^2} A. \tag{7.95}$$

a)

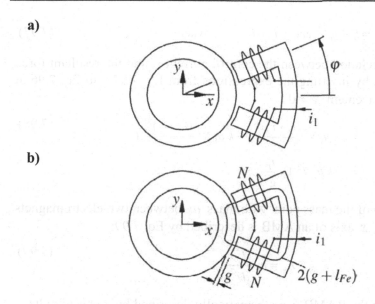

b)

Fig. 7.50. Electromagnet pole displacement by the angle φ referring to the x axis **a** and the magnetic circuit of an electromagnetic pole pair **b**

In the RAMB pairs of electromagnets are normally used. A pair of electromagnets is shown in Fig. 7.48b. Each electromagnet can generate only an attractive force acting on the ferromagnetic rotor. Therefore, two electromagnets on the same bearing axis pull the rotor in opposite directions. Neglecting the nonlinearity of the iron core, the force F_x generated by a pair of electromagnets on the bearing axis x is given by Eq. 7.96 as the difference of two expressions, Eq. 7.95:

$$F_x = F_{x1} - F_{x2} \qquad (7.96)$$
$$= k\frac{B_1^2}{\mu_0}A - k\frac{B_2^2}{\mu_0}A$$
$$= k\mu_0\frac{(i_1 N)^2}{(g-x)^2}A - k\mu_0\frac{(i_2 N)^2}{(g+x)^2}A$$

where x is the rotor displacement, and B_1 and B_2 are the flux densities in the air gaps of both electromagnets excited by the currents i_1 and i_2, respectively. Let us introduce the bias current i_b and the control current i_p. The same bias current i_b is supplied to the coils of both electromagnets. Force control is done by superposing a control current i_p on the bias current in the coils of one electromagnet and subtracting it from the bias current in the coils of the other one (Eq. 7.97):

$$i_1 = i_b + i_p \quad \text{and} \quad i_2 = i_b - i_p \quad \text{where} \quad i_p \le i_b. \tag{7.97}$$

The linear relation between the control current i_p and the resultant force F_x is obtained by inserting the expressions from Eq. 7.97 into Eq. 7.96 at the rotor displacement $x = 0$:

$$
\begin{aligned}
F_x &= k\mu_0 N^2 \frac{(i_b + i_p)^2}{g^2} A - k\mu_0 N^2 \frac{(i_b - i_p)^2}{g^2} A \\
&= 4k\mu_0 N^2 A i_b \frac{i_p}{g^2}.
\end{aligned}
\tag{7.98}
$$

The motion of the mass point with mass m between two electromagnets located on the x axis of an AMB is described by Eq. 7.99:

$$F_x = m \frac{d^2 x}{d t^2}. \tag{7.99}$$

One axis of the RAMB is mathematically described by a pair of voltage equations (Eq. 7.85), the force equation (Eq. 7.96) and the equation of motion (Eq. 7.99). The complete RAMB model contains two pairs of electromagnets which are placed on the x and y axes. However, the described model given by Eqs. 7.85, 7.96 and 7.99 can be simplified: the magnetic field distribution is considered to be homogeneous, the cross-sectional areas of the magnetic loops are uniform, the coupled magnetic circuits shown in Fig. 7.47b are considered to be independent, the leakage flux and the magnetization of the iron core are neglected, and the magnetically nonlinear properties of the iron core are considered to be linear. This simplified model can be used to study the operational principles of the RAMB. It is appropriate for control synthesis and the initial bearing design, but it is not accurate enough to be used for optimization of the RAMB design.

7.9.3 Magnetic Field Distribution and Force Computed by the Two-Dimensional FEM

The simplified model presented in the previous subsection can be used to determine the initial RAMB design. The magnetic field distribution in the real RAMB is quite different from the one supposed in the simplified model. To optimize the RAMB geometry a realistic magnetic field distribution is required. The two-dimensional magnetic field distribution is close to the realistic one and can be computed using the FEM by solving Poisson's equation (Eq. 7.100) numerically:

$$\nabla \bullet (v \nabla A) = -\mathbf{J} \tag{7.100}$$

where ∇ is the Laplace operator, $\bullet$ denotes the dot product, v is the magnetic reluctivity, A is the magnetic vector potential and $\mathbf{J}$ is the applied current density. The nonlinear solution of the magnetic vector potential is required for computation of the force. The force $\mathbf{F}$ acting on the ferromagnetic rotor can be computed by Maxwell's stress tensor method:

$$\mathbf{F} = \oint_S \boldsymbol{\sigma}\, dS = \oint_S \left[\frac{1}{\mu_0} (\mathbf{B} \bullet \mathbf{n})\mathbf{B} - \frac{1}{2\mu_0} \mathbf{B}^2\, \mathbf{n} \right] dS \tag{7.101}$$

where $\boldsymbol{\sigma}$ is Maxwell's stress tensor, $\mathbf{n}$ is the unit vector normal to the integration surface S and $\mathbf{B}$ is the magnetic flux density. In the case of two-dimensional computations, the closed integration path is a contour placed in the air gap, which encloses the ferromagnetic rotor.

Experienced designers can improve the RAMB design by employing their own intuition and experience in combination with FEM computations.

7.9.4 RAMB Design Optimized by DE and the FEM

The task is to design an RAMB for a spindle drive. The simplified RAMB model is used to find the initial bearing design. The design is improved by FEM computations using "trial and error" in order to achieve maximum force at minimum mass of the entire construction. For any further improvement in the bearing design a numerical magnetic field computation in combination with an optimization method is applied.

In the authors' opinion, DE in combination with the FEM is at present one of the most suitable and powerful tools for the optimization of electromagnetic devices, which are usually nonlinear and entail constrained optimization problems. Commonly, for such a problem class, the dependency of the objective function on the design parameters is unknown. According to Pahner (1998), for optimization of electromagnetic devices in combination with the FEM, DE is superior to other stochastic direct search algorithms such as simulated annealing and self-adaptive evolution strategies. DE converges faster and is more stable when compared to these other methods. The DE strategies "DE/best/1/exp" and "DE/best/2/exp" (Storn and Price 1997) are favored for most technical problems. For details and to study the theory of such optimization methods, the authors refer to the literature (Hameyer 1994; Storn and Price 1996, 1997; Pahner 1998).

If the number of design parameters is larger than 2, the number of population members NP should be at least 15 (Pahner 1998). In our experience, however, technical problems with up to 25 parameters can be successfully solved with a population size of less than 40. The recommended setting for the DE step size F is 0.5 and that for the crossover probability constant CR is between 0.5 and 0.9. DE in combination with the FEM is well suited for parallel implementation in a computer network, which can substantially reduce the time required for optimization (Pahner et al. 1998).

To optimize an electromagnetic device with DE and the FEM, both of them must be adapted to function and operate together. A good convergence can be achieved if the optimization function parameters are independent. To optimize an electromagnetic device, it is important to improve the device design with respect to the given objective function. This function represents the technical device. Therefore, particular attention has to be paid to an appropriate formulation of this function, otherwise it is possible for the algorithm get stuck in a local minimum. To allow the algorithm to reach the global optimum, contradictory partial aims formulated in the objective function have to be avoided. This can be realized by choosing appropriate constraints for the optimization problem.

As mentioned earlier, the aim here is to achieve a maximum force at a minimum mass of the entire construction. The bearing must generate a maximum force of at least 500 N at the prescribed shaft diameter of 35 mm, a stator outer diameter of 105 mm and an air gap of 0.4 mm. For the optimization, the rotor is placed in the center of the air gap. According to Eq. 7.98, the maximum force F_x is generated if one of the electromagnets on the x axis is supplied with the current $i_b + i_p$, while the other one is supplied with the current $i_b - i_p$. Zero force is generated on the y axis. To achieve realistic operating conditions, both electromagnets on the y axis are supplied with the current i_b as shown in Fig. 7.51a. The RAMB design is optimized for the currents $i_b = 5$ A and $i_p = 5$ A.

The dependency of the objective function $f(\mathbf{X}):R^4 \to R$ on its parameters $\mathbf{X}$ is unknown, while the value of the objective function can be computed by the FEM for each generated set of parameters $\mathbf{X}$. The objective of the optimization is to find the set of objective function parameters $\mathbf{X}$ for which the value of the objective function is a minimum. The design parameters, whose values are to be optimized, are the stator yoke height s_y, the rotor yoke height r_y, the leg width l_w and the bearing's axial length l. The first three of these design parameters are shown in Fig. 7.51b.

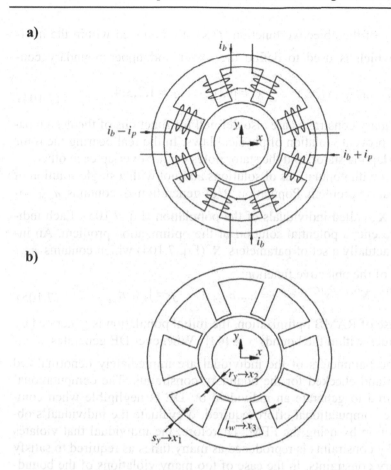

Fig. 7.51. Optimization of the radial bearing: **a** currents supplied to the coils, **b** the objective function parameters.

The design parameters (s_y, r_y, l_w, l) are transformed by normalization (Eq. 7.102) into the set of objective function parameters $\mathbf{X} = (x_1, x_2, x_3, x_4)$, which are used in DE; the inverse process is denormalization (Eq. 7.103):

$$x_j = \frac{d_j - d_j^{(L)}}{d_j^{(U)} - d_j^{(L)}} \qquad j = 1,2,3,4 \qquad (7.102)$$

$$d_j = d_j^{(L)} + x_j (d_j^{(U)} - d_j^{(L)}) \qquad j = 1,2,3,4. \qquad (7.103)$$

Here d_j denotes the j th design parameter, while $d_j^{(L)}$ and $d_j^{(U)}$ are its lower and upper boundary constraints, respectively. In this way, all parameters x_j of the objective function $f(\mathbf{X})$ are bounded within the interval [0,1], which is used to define the lower and upper boundary constraints:

$$\mathbf{X} = (x_1, x_2, x_3, x_4) \qquad x_j \in [0,1] \qquad j = 1,2,3,4. \tag{7.104}$$

The boundary constraints are also defined for functions of the design parameters to prevent violation of physical laws. In the real bearing the rotor cannot overlap the stator, and the stator poles cannot overlap each other.

DE works with populations of solutions and not with a single solution of the optimization problem. Population $\mathbf{P}$ of generation G contains n_{pop} solution sets $\mathbf{X}$, called individuals of the population (Eq. 7.105). Each individual represents a potential solution of the optimization problem. An individual is actually a set of parameters $\mathbf{X}$ (Eq. 7.104) which contains n_{par} parameters of the objective function:

$$\mathbf{P}^{(G)} = \mathbf{X}_i^{(G)} = x_{i,j}^{(G)} \qquad i = 1,...,n_{pop} \qquad j = 1,...,n_{par}. \tag{7.105}$$

In the case of RAMB optimization, the initial population is generated by random values within the bounds $x_j \in [0,1]$. Whenever DE generates an individual, the parameters of the individual are immediately denormalized (Eq. 7.103) and checked for the boundary constraints. The computational effort required to generate an individual by DE is negligible when compared to the computational effort required to evaluate the individual's objective function by using the FEM. Therefore, the individual that violates the boundary constraints is reproduced as many times as required to satisfy the boundary constraints. In the case of too many violations of the boundary constraints, the DE settings such as the DE step size F and the crossover probability constant CR can be adjusted.

To evaluate the objective function of a design parameter set (individual), the FEM package must be able to accept parameters generated by DE, to perform the FEM computation automatically, and to return the value of the objective function to the DE algorithm. The parameters of the individual define the temporary bearing design. They are passed to the FEM package where the parametrically defined FEM model of the RAMB is updated according to the temporary design. For the bearing geometry obtained, the material, the current densities and the FEM boundary conditions are defined. The stator and the rotor iron core are constructed from laminated electric steel M36, whose magnetization characteristic is plotted in Fig. 7.52a.

a)

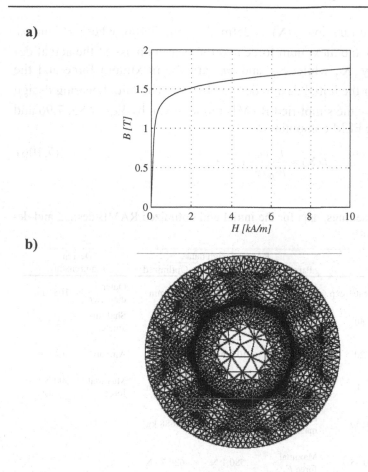

b)

Fig. 7.52 *B–H* characteristic for laminated electric steel M36 **a** and the discretization of the model **b**

The mesh shown in Fig. 7.52b is automatically generated by dividing the geometry into discrete elements. Standard triangular elements are applied here. For the two-dimensional problem described by Eq. 7.100, the nonlinear solution of the magnetic vector potential is computed by using a conjugated gradient and the Newton–Raphson method. The errors in the solution obtained are analyzed, the mesh is refined, and the problem is solved again. The procedure is repeated until the solution error is smaller than a predefined value. The force F_x is computed from the obtained nonlinear solution of the magnetic vector potential by Maxwell's stress tensor (Eq. 7.101). The closed contour in the middle of the air gap is used as the path of integration. The mass of the entire construction is calculated from the bearing geometry and the specific mass of the materials used.

The objective function $f(\mathbf{X})$ is defined by Eq. 7.106, where F_x and m are respectively the maximum force and the bearing mass of the actual design defined by $\mathbf{X}$, while F_{x0} and m_0 are the maximum force and the bearing mass of the initial design defined by $\mathbf{X}_0$. The initial bearing design is determined by the simplified RAMB model given by Eqs. 7.85, 7.96 and 7.99 and by the FEM computations:

$$f(\mathbf{X}) = \frac{F_x\, m_0}{F_{x0}\, m} + p_1 + p_2. \tag{7.106}$$

Table 7.14. DE settings, data for the initial and optimized RAMB design, and design requirements

DE settings		RAMB design data			Design requirements	
		Parameter	Initial	Optimized		
DE strategy	DE/best/1/exp	Stator yoke s_y	8.5 mm	7.2 mm	Outer diameter	105 mm
Number of generations	60	Rotor yoke r_y	9.0 mm	7.8 mm	Shaft diameter	35 mm
Population size	20	Leg width l_w	10.0 mm	9.0 mm	Air gap	0.4 mm
Number of parameters	4	Axial length l	53.0 mm	56.3 mm	Maximal force	500 N at least
DE step size	0.50	Bearing mass m	2.691 kg	2.688 kg		
Crossover probability constant	0.75	Maximal force F_x	580.1 N	629.74 N		
		Objective function $f(\mathbf{X})$	1.00	0.92		

The penalties p_1 and p_2 are used to ensure that the bearing mass of the optimized design is not larger than the initial mass and that the maximum force is not smaller than the initial force. The penalties are calculated by the constraint functions (Eq. 7.107):

$$\text{if } m > m_0 \;\Rightarrow\; p_1 = \frac{m}{m_0} \quad \text{and} \quad \text{if } F_x < F_{x0} \;\Rightarrow\; p_2 = \frac{F_{x0}}{F_x}. \tag{7.107}$$

The obtained value of the objective function is returned to the DE algorithm. DE proceeds to generate new generations of individuals until the desired value of the objective function or the maximum number of generations is reached.

The RAMB data and the DE settings used for optimizing the RAMB design are summarized in Table 7.14.

7.9.5 Conclusion

This work deals with the optimization of an RAMB for a spindle drive. A simplified model of the RAMB is presented first. It is used to determine the initial bearing design, which is improved by using experience in combination with FEM computations. In this way the RAMB design obtained is optimized by DE. Optimizing an RAMB is a nonlinear and constrained optimization problem, where the dependency of the objective function on the design parameters (objective function parameters) is unknown. For the optimization of the bearing geometry, exact knowledge of the magnetic field distribution inside the nonlinear ferromagnetic iron core and in the bearing's air gap is required. The magnetic field distribution can be computed by the FEM. The RAMB is optimized by DE, while the objective function of each individual is evaluated by the FEM. To work together, the DE algorithm and the FEM package must be adapted. The FEM package must be able to accept parameters generated by DE and to return the value of the objective function. The FEM package must also be suitable for use with parameterized models and to work with user-defined procedures. Automatic mesh generation, solvers for different problems, error estimation, mesh refinement and automatic acquisition of results must be a part of the FEM package.

The results presented here show that the performance of the electromagnetic devices can be substantially improved if DE in combination with the FEM is applied for the optimization. In the case of the RAMB, the value of the objective function was improved by about 8%, which means that the force increased by 49 N at the same bearing mass.

Acknowledgments

The optimization of the RAMB was performed in a special environment called Olympos, an FEM program package, which is tuned for numerical optimizations (Pahner et al. 1998). Olympos was developed at Katholieke Universiteit Leuven, Department E.E. (ESAT), Division ELEN, Leuven, Belgium. The laboratory prototype of the optimized bearing has been realized and constructed within the scope of the project "Active magnetic bearings" supported by the Ministry of Science and Technology of the Republic of Slovenia and by the company Domel, now Indramat elektromotorji, Proizvodnja električnih motorjev d.o.o. Železniki, Slovenia.

References

Antila M, Lantto E, Arkkio A (1998) Determination of forces and linearized parameters of radial active magnetic bearings by finite element technique. IEEE Transactions on Magnetics 34(3):684–694

Hameyer K (1994) Numerical optimization of finite element models in electromagnetics with global evolution strategy. In: Adaptive computing in engineering control 1994, Plymouth, USA, pp 61–66

Pahner U (1998) A general design tool for the numerical optimization of electromagnetic energy transductors. PhD thesis, Catholic University Leuven

Pahner U, Mertens R, De Gersem H, Belmans R, Hameyer K (1998) A parametric finite element environment tuned for numerical optimization. IEEE Transactions on Magnetics 34(5):2936–2939

Schweitzer G, Bleuler H, Traxler A (1994) Active magnetic bearings. Vdf Hochschulverlag AG an der ETH Zürich, ETH Zürich

Storn R, Price KV (1996) Minimizing the real functions of the ICEC'96 contest by differential evolution. In: IEEE international conference on evolutionary computation, Nagoya, May 1996, IEEE, New York, pp 842–844

Storn R, Price KV (1997) Differential evolution – a simple evolution strategy for fast optimization. Dr. Dobb's Journal 22(4):18–24 and 78

Štumberger G, Dolinar D, Pahner U, Hameyer K (2000) Optimization of radial active magnetic bearings using the finite element technique and the differential evolution algorithm. IEEE Transactions on Magnetics 36(4):1009–1013

7.10 Application of Differential Evolution to the Analysis of X-Ray Reflectivity Data

Matthew Wormington, Kevin M. Matney and D. Keith Bowen

Abstract. X-ray reflectivity (XRR) is a technique for characterizing the structure of thin, multi-layer devices. This section describes the application of the differential evolution (DE) algorithm to automatically and reliably analyze XRR data. A data-fitting method is presented that is conceptually simple, easy to implement and is capable of converging to a global minimum in the parameter space even when there are many additional local minima. The method is quite general and could be applied to many problems in science and engineering.

7.10.1 Introduction

X-ray reflectivity (XRR) is a technique used to characterize the structure of thin (1–1000 nm), multi-layer devices. Such devices are used in numerous high-technology products, most notably products made by the microelectronics, optoelectronics and magnetic storage industries. XRR is nondestructive and can be used to characterize crystalline, polycrystalline and amorphous materials. The technique accurately and reproducibly measures layer thickness, density and roughness/grading.

Until recently, the XRR technique was mainly used for research and development because an experienced scientist was required to manually analyze the data. The development of an automated and robust data analysis method, which is the subject of this section, has allowed the technique to be used in manufacturing environments. For example, XRR is now being used by the microelectronics industry to measure layer thickness in the latest generation of microprocessors – from the semiconductor layers that form the transistors on the silicon substrate to the conductors and insulators that are used to connect these transistors together. The technique is also being used by the magnetic storage industry to characterize the structure of the read-heads used in modern hard disk drives.

Figure 7.53 schematically illustrates the XRR technique. A collimated, monochromatic beam of X-rays illuminates the surface of a sample at a very small angle of incidence (0–5°). The specularly reflected beam (i.e., the beam with reflection angle equal to the incidence angle) is detected and its intensity is measured. The reflected intensity as a function of the incidence angle is related to the structure of the sample. There is, however, no direct method for determining a sample's structure from such a measurement and so indirect methods must be used.

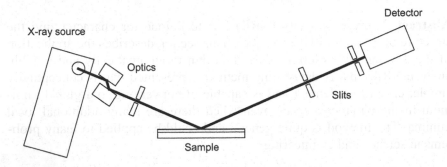

Fig. 7.53. Schematic of an XRR measurement. The X-ray beam is generated by a laboratory source and passes from left to right. The optics produce a collimated, monochromatic beam (i.e., small spread in both angle and wavelength). The beam illuminates the sample at a small angle of incidence and is specularly reflected from the sample. The intensity of the reflected beam is measured as the sample and detector are scanned in a ratio of 1:2.

The most common way to determine the structure of a sample from its measured XRR data is to create a model that we hope reasonably describes its structure and from which we can simulate the reflected X-ray intensity. Using the model, we then simulate the XRR intensity and calculate the difference between the experimental and simulated curves using some objective function, f. The model parameters are then adjusted by some optimization method in order to minimize the difference between the two curves. This procedure is repeated until the difference between the two curves is judged to be sufficiently small, at which point we accept the model to be an accurate representation of the structure.

The field of data-fitting and parameter optimization has a long and fruitful history. The earliest successes were for linear problems that possessed a single minimum in the objective function. The mean-squared difference between the experimental and simulated data was commonly used as the objective function because of its computational simplicity in the days before fast digital computers. More recent research has focused on nonlinear problems, and on those with local minima in the objective function in addition to the global minimum. A variety of data-fitting and parameter optimization strategies have been developed for such systems (Bevington 1969; Press et al. 1989) and those most commonly encountered are:

1. *Direct search.* The parameter space is divided up into small, but finite, regions. The objective function is calculated for each region and the region that gives the smallest value for f is said to give the best-fit (optimum) parameter values.
2. *Downhill simplex.* An initial guess at the parameter values is made. The simplex (a geometrical construction) then moves in directions that decrease the value of f. The parameters that yield the smallest value of f in the neighborhood of the initial guess are said to be the best-fit parameters.
3. *Levenberg–Marquardt method.* An initial guess at the parameter values is made by the user. The algorithm then combines linearization and gradient searching of the objective function to minimize f in the neighborhood of the initial guess. The parameter values giving the smallest value for f are then selected as the best-fit parameters.
4. *Monte Carlo method* The parameter space is again divided into small regions. Regions are selected at random and the objective function is evaluated. After a certain number of regions have been chosen, or when f is smaller than some specified value, the algorithm is stopped. The region with the smallest value for f is said to yield the best-fit parameter values.
5. *Simulated annealing.* This uses the physical principles governing annealing (i.e., the heating of a material and subsequent slow cooling such that the material forms a crystal) to search the objective function and obtain the best-fit parameters. There is a finite probability in any step that the parameters can move in a direction so as to increase f, so the method does escape from local minima, but slowly.

All of the above methods run into severe difficulties when fitting XRR data. The parameter space is simply too vast for direct searches and becomes uncomputable for all but the simplest cases. The downhill simplex and Levenberg–Marquardt methods work well for nonlinear problems because they are guided by the geometry of the objective function in parameter space. However, the initial estimate of the parameter values needs to be very close to the optimized values if local minima are present, as they will become trapped in the first local minimum that they encounter. These two methods are therefore only effective when the parameters are initially contained within the multi-dimensional "well" of the global minimum, and in most practical cases in X-ray scattering, we have found them to be of little use. The Monte Carlo and simulated annealing methods do not get trapped in local minima. However, they are very inefficient at searching the parameter space, since they search it randomly without taking into account the geometry of the objective function. A successful strategy for non-linear problems containing many local minima will combine both random and guided elements. Recently, genetic algorithms (GAs) have attracted wide-spread interest for nonlinear optimization problems as they overcome many of the problems associated with the more traditional optimization strategies mentioned above.

In this section, we apply the differential evolution (DE) algorithm (Storn and Price 1995, 1997) to the analysis of XRR data using a data-fitting method. We were attracted to the DE algorithm over other GAs for several reasons: the algorithm was straightforward, it used real numbers (rather than binary strings or integers) to encode the problem parameters, and it was reported to be both versatile and reasonably efficient. For a more detailed discussion of our application, the interested reader is referred to the original publication (Wormington et al. 1999) on which this section is largely based. Additional information regarding the characterization of advanced materials using X-ray techniques can be found in Bowen and Tanner (1998) and Holý et al. (1999), and the references contained therein.

7.10.2 The Data-Fitting Procedure

The Differential Evolution (DE) Algorithm

Let us assume that the experimental data contains N measured points (θ_j, I_j) where θ_j is the incidence angle, I_j is the intensity measured at θ_j and $j = 1, 2, ..., N$. Simulated data $I(\theta_j, \mathbf{p})$ is computed assuming a structural model with n continuous, adjustable parameters represented by the vector $\mathbf{p} = [p_1, p_2, ... p_n]$ and is compared to the experimental data using some objective function $f(\mathbf{p})$. Guided by $f(\mathbf{p})$, the DE algorithm attempts to optimize the parameter vector $\mathbf{p}$ starting with an initial population of randomly generated parameter vectors, by a repeated cycle of mutation, recombination and selection.

As discussed in previous chapters, several versions of the basic DE algorithm have been proposed. The DE/Best/1/Bin version of the algorithm was used in this work. The population was initialized by assigning the parameter vector $\mathbf{p}_0$ the user's initial guess at the structure, while the remaining vectors were initialized by assigning each parameter with a randomly chosen value from within its allowed range. The control parameters of the DE algorithm, namely the mutation constant and crossover constants, respectively designated F and C_r, must be empirically determined to give fast convergence (minutes). Values of $F = 0.7$ and $C_r = 0.5$ were used in this work. The population size was also determined empirically and a population size of about 25 to 50 seemed to work well for most cases.

The recombination operation of the DE algorithm is able to produce parameter values outside of the range specified by the user (or physically implausible solutions such as a thickness being less than zero). In order to prevent this from happening we modified the basic DE algorithm such that any temporary (trial) parameter p'_j that falls outside the specified constraints is replaced by a random value selected according to the expression

$$p'_j = p_j^{min} + \text{rand}(p_j^{max} - p_j^{min}) \tag{7.108}$$

where p_j^{min} and p_j^{max} are the minimum and maximum permissible values of parameter j, respectively. The function rand(x) designates a real uniform random number drawn from the range $[0, x]$. We found that setting the out-of-range trial parameter to a random value within the specified constraints improved the data-fitting performance.

Finally, we added several stopping criteria to the DE algorithm such that the algorithm will stop if:

1. the user cancels the data-fitting procedure;
2. a specified number of simulations is reached;
3. a specified elapsed time is exceeded;
4. a specified value of the objective function is attained.

The Objective Function

The choice of an appropriate objective function is crucial for any data-fitting procedure regardless of the optimization method used. The DE algorithm gives us a great deal of flexibility in this choice since we need only choose a continuous function and do not require the function to have continuous derivatives. When fitting XRR data, the objective function should have the following additional properties:

1. a single deep global minimum;
2. local minima that are much less deep than the global minimum;
3. fast and simple to calculate;
4. relative insensitivity to the absolute magnitude of the data, since XRR data often spans many orders of magnitude;
5. does not overemphasize outlying points in the experimental data, since we expect a Poisson distribution of statistical noise.

Point 4 suggests that a logarithmic function could be appropriate since it linearizes data spanning several orders of magnitude. Point 5 suggests that a robust objective function (Press et al. 1989) will be one that is more suitable than the mean-squared objective function commonly encountered in least squares fitting. To confirm these conjectures, we have investigated a number of objective functions that have been applied to fitting problems, namely:

Mean-square difference of the data:

$$f(\mathbf{p}) = \frac{1}{N-1} \sum_{j=1}^{N} [I_j - I(\theta_j; \mathbf{p})]^2 .$$

(7.109)

Mean-absolute difference of the data:

$$f(\mathbf{p}) = \frac{1}{N-1} \sum_{j=1}^{N} |I_j - I(\theta_j; \mathbf{p})| .$$

(7.110)

Mean-square difference of the log transformed data:

$$f(\mathbf{p}) = \frac{1}{N-1} \sum_{j=1}^{N} [\log I_j - \log I(\theta_j; \mathbf{p})]^2 .$$

(7.111)

Mean-absolute difference of the log transformed data:

$$f(\mathbf{p}) = \frac{1}{N-1} \sum_{j=1}^{N} |\log I_j - \log I(\theta_j; \mathbf{p})| .$$

(7.112)

Tests of these functions with XRR data showed that the non-logarithmic objective functions (Eqs. 7.109 and 7.110), as expected, did not effectively fit the data at low intensities. These occur at large scattering angles and contain information on the smallest length scales present in the structure, which are often those at which the X-ray characterization is aimed. The objective functions in Eqs. 7.111 and 7.112 both could cope adequately with such data, but Eq. 7.112 is preferred because of its lower sensitivity to outlying data points (due mainly to statistical noise in the experimental data). Clearly we cannot assert that Eq. 7.112 is the best possible objective function but it is very effective, and this is sufficient.

Performance

The performance of the data-fitting procedure is primarily affected by the following four factors:

1. The quality and size of the experimental data. If the experimental data is noisy or contains a very large number of points it will take longer to determine the best-fit parameter values.
2. The quality of the initial estimates for the parameters. If the initial values for the parameters are grossly different from the optimized values the fitting procedure will take longer to converge.

3. The search range of the parameter values. If a large search range is specified the fitting procedure may take longer to converge to the global minimum of the objective function. However, because the DE algorithm is rather good at finding the global minimum, without becoming trapped in local minima, it tends not to falsely converge to incorrect values for the parameters.

4. The number of adjustable parameters. The ability of the procedure to determine the optimum parameter values decreases as the number of parameters increases. In practice we find that up to ten parameters can be optimized in a matter of minutes, and that several tens of parameters can be optimized during an overnight run.

We have developed an efficient program for analyzing XRR data based on the fitting procedure presented in this section. The program, which was developed using Borland Delphi, runs on personal computers (PCs) under the Microsoft Windows operating system. All benchmarks reported in this work were obtained using a 1 GHz Pentium III-based notebook computer fitted with 512 Mbytes of memory.

7.10.3 The Model and Simulation

The simulation method used is taken from the Bede REFS program (Wormington et al. 1992). We will consider a multi-layer on a thick substrate in which the refractive index of each layer is assumed constant. For X-rays, the refractive index of a material is slightly less than unity and can be written as

$$n = 1 - r_e \frac{\lambda^2}{2\pi} \sum_a (f_a + f_a' + i f_a'') N_a, \tag{7.113}$$

where r_e is the classical electron radius and λ is the X-ray wavelength. The atomic scattering factor is denoted by f_a and the real and imaginary parts of the dispersion correction are f_a' and f_a'', respectively. Values for the scattering factor and its corrections are tabulated in the International Tables for Crystallography (Ibers and Hamilton 1974). The summation is taken over all constituent atoms, a, of number density N_a.

The amplitude ratio $X_j = E_j^r / E_j^t$ of the reflected and transmitted waves at the bottom of layer j within the multi-layer is obtained by solving Maxwell's equations and the appropriate boundary conditions. According to Parratt (1954), we may write

$$X_j = \frac{r_j + X_{j+1}\varphi_{j+1}^2}{1 + r_j X_{j+1}\varphi_{j+1}^2}$$

(7.114)

where r_j is the Fresnel coefficient for reflection from the interface between layers j and $j + 1$. For a sharp interface, r_j is given by the expression

$$r_j = \frac{k_{j,z} - k_{j+1,z}}{k_{j,z} + k_{j+1,z}}$$

(7.115)

where $k_{j,z} = 2\pi / \lambda \times (n_j^2 - \cos^2 \theta)^{1/2}$ is the component of the wavevector in layer j perpendicular to the surface of the multi-layer (i.e., along the z axis), n_j is the refractive index of the layer and θ is the grazing angle of the incident plane wave. The complex phase factor for wave propagation through the layer thickness t_j is denoted by $\varphi_j = \exp(ik_{j,z}t_j)$. To include the effects of grading (interdiffusion) and roughness within this formalism we need only modify the form of the Fresnel coefficient. From the work of Névot and Croce (1980) an appropriate modification is given by

$$r_j = \frac{k_{j,z} - k_{j+1,z}}{k_{j,z} + k_{j+1,z}} \exp[-2(k_{j,z}k_{j+1,z})^{1/2}\sigma_{j+1}]$$

(7.116)

where σ_{j+1} denotes the width of the interface between layers j and $j + 1$ due to both grading and roughness.

To calculate the amplitude ratio at the top of the multi-layer, X_0, Eqs. 7.113 and 7.115 are applied recursively for all interfaces starting at the substrate (layer $N + 1$), where $X_N = r_N$. The plane-wave reflectivity is then given by $R = |X_0|^2$ and is related to the reflected intensity through the correlation function

$$I(\theta;\mathbf{p}) = I_0 \int F(\theta')R(\theta' - \theta)\,d\theta' + I_b$$

(7.117)

where I_0 and I_b denote the incident and background intensity, respectively. Here $F(\theta)$ denotes an instrument function and takes into account the finite divergence of the incident X-ray beam. We have included both this incident angle θ and the adjustable parameters $\mathbf{p}$ in our notation. Specifically $\mathbf{p}$ contains the following:

1. the incident intensity I_0;
2. the background intensity I_b;
3. the densities ρ_j of the layers $j = 1, 2, \dots N$;
4. the thicknesses d_j of the layers $j = 1, 2, \dots N$;
5. the widths σ_{j+1} of the interfaces between layers j and $j + 1$ due to grading and/or roughness.

Finally, we note that X-ray reflectivity measurements cannot usually distinguish between layers of high atomic number Z and low mass density, and those with low Z and high density. We have therefore chosen to fit the density of layers in the structural model and assume their chemical composition.

7.10.4 Examples

Example I – Ta Layer on Al$_2$O$_3$

In our first example, we consider a Ta (10 nm) layer deposited on an Al$_2$O$_3$ substrate. Figure 7.54a shows the measured and simulated XRR curves before fitting. At very small angles of incidence, $\theta \leq 0.5°$, the intensity is very high as a result of total external reflection from the Ta layer. As the incidence angle is increased, the reflected intensity decreases rapidly and prominent oscillations (Kiessig fringes) are clearly visible due to the interference of the waves partially reflected from the Ta layer and the underlying Al$_2$O$_3$ substrate. The period, $\Delta\theta$, of the Kiessig fringes is related to the thickness, t, of the Ta layer according to the relation $\Delta\theta \approx \lambda / 2t$. The amplitude of the Kiessig fringes depends on the difference in the refractive index of the layer and the substrate, which is quite large in this example. Figure 7.54b shows the measured curve together with its best-fit simulation.

The time for the fitting procedure to converge was about 30 seconds, fitting a total of nine adjustable parameters. The best-fit parameter values and their uncertainties are given in Table 7.15. It should be noted that a surface oxide layer had to be included in the structural model to obtain close agreement of the measured and simulated curves.

Table 7.15. Best-fit parameter values for the Ta layer on Al$_2$O$_3$

Layer	Material	t (nm)	σ (nm)	ρ (g/cm^3)
2	Ta$_2$O$_5$	2.70 ± 0.05	0.71 ± 0.03	8.6 ± 0.2
1	Ta	10.49 ± 0.02	0.45 ± 0.02	16.1 ± 0.2
Substrate	Al$_2$O$_3$	∞	0.38 ± 0.02	3.99

a)

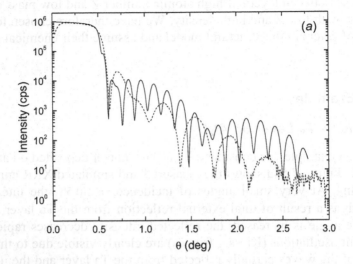

b)

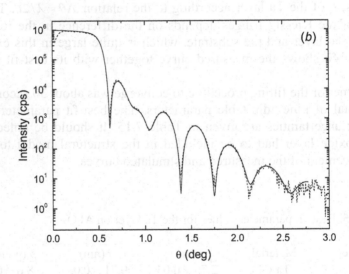

Fig. 7.54. Comparison of experimental and simulated X-ray reflectivity curves for a Ta layer on Al_2O_3, **a** before and **b** after the fitting procedure has converged. The dashed lines represent the measurements and the solid lines are the simulations.

The parameter values for the simulated curve shown in Fig. 7.54a were chosen to be far from the anticipated best-fit parameter values. This was a deliberate choice in order to demonstrate that the fitting procedure rapidly converges to the global minimum in the objective function without getting trapped in local minima. The progress of the fitting procedure is illustrated in Fig. 7.55, which shows the value of the objective function versus the number of generations (iterations of the DE algorithm). Horizontal sections are times during which the fitting procedure is temporarily in local minima. The fitting procedure is seen to have converged to the global minimum after only 1000 generations.

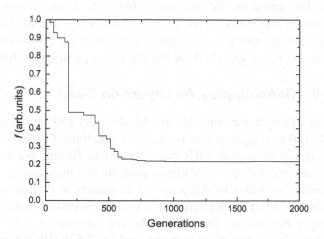

Fig. 7.55. Variation of the objective function, f, with the number of DE generations. The fitting procedure has converged after approximately 1000 generations.

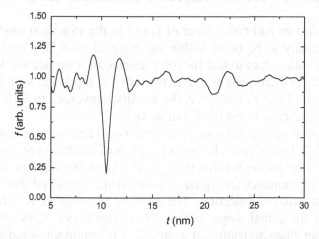

Fig. 7.56. Variation of the objective function, f, with the Ta layer thickness. All other adjustable parameters in the model are held constant at their best-fit values.

Figure 7.56 shows the value of the objective function as a function of the thickness of the Ta layer, with all other parameters held at their best-fit values. We note that the objective function has a single, deep global minimum and many local minima. Harmonic minima, which occur at half and twice the best-fit Ta layer thickness, are the deepest of the local minima. The global minimum is "shielded" by fairly large maxima on either side. This is a typical feature in such curves when the thickness is varied, and turns out to be caused by the beating of two sets of oscillations (Kiessig fringes) in which one period is fixed and the other is variable. This characteristic shape is very useful for recognizing whether the global minimum is in fact within the range specified for the thickness parameters in question.

Example II – GaAs/Al$_{0.3}$Ga$_{0.7}$As Layers on GaAs

For our next example we consider an Al$_{0.3}$Ga$_{0.7}$As (50 nm) layer capped with a GaAs (50 nm) layer grown on a GaAs substrate. Figure 7.57a shows the measured and simulated XRR curves prior to fitting. We see that the Kiessig fringes are far less prominent than in the previous example because the refractive index of Al$_{0.3}$Ga$_{0.7}$As is similar to that of GaAs. Furthermore, the period of the Kiessig fringes is much smaller than in the previous example because of the thicker layers considered. The measured curve and best-fit simulation are shown in Fig. 7.57b. Whilst more iterations are required than in the previous example, the fitting procedure still took about 45 seconds to converge and fit ten adjustable parameters. The best-fit parameters and their respective uncertainties are given in Table 7.16.

We included an additional layer of GaAs in the structural model and allowed its density to be fitted within the range 2.66–5.32 g/cm^3 (i.e., 50–100% of its bulk value) to test for the presence of a surface oxide layer. If no surface layer were present, the density of the top layer would naturally converge to 5.32 g/cm^3. However, the density converged to 3.19 g/cm^3, indicating the presence (most likely) of an oxide.

Finally, we note that detector saturation is evident in the measured data (see Fig. 7.57), but this has clearly not prevented the fitting procedure from converging. The reason for this is that we treated the incident intensity as an adjustable parameter and ignored most of the measured data in the region of total external reflection. The incident intensity is, in effect, determined from the initial slope of the reflectivity curve. However, for the most accurate characterization of a sample it is important to reduce this effect experimentally, e.g., by using an Al absorber to attenuate the reflected beam at very low incident angles, or a high dynamic-range detector.

a)

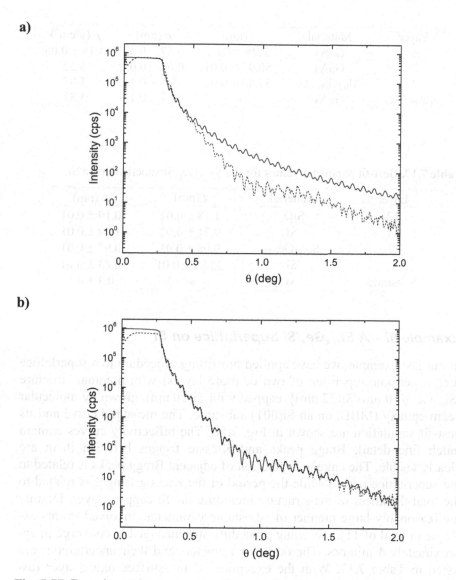

b)

Fig. 7.57. Experimental and simulated X-ray reflectivity curves for an $Al_xGa_{1-x}As$ layer capped with GaAs on a GaAs substrate, **a** before and **b** after fitting. The dashed lines are the measurements and the solid lines represent the simulations. Experimental data courtesy of Prof. B. K. Tanner (University of Durham).

Table 7.16. Best-fit parameter values for the GaAs/Al$_{0.3}$Ga$_{0.7}$As layers on GaAs

Layer	Material	t (nm)	σ (nm)	ρ (g/cm^3)
3	GaAs	2.28 ± 0.02	0.57 ± 0.02	3.19 ± 0.05
2	GaAs	50.97 ± 0.01	0.64 ± 0.03	5.32
1	Al$_{0.3}$Ga$_{0.7}$As	50.40 ± 0.02	0.5 ± 0.1	4.87
Substrate	GaAs	∞	0.7 ± 0.1	5.32

Table 7.17. Best-fit parameter values for the Si$_{1-x}$Ge$_x$/Si superlattice on Si

Layer	Material	t (nm)	σ (nm)
12	SiO$_2$	1.18 ± 0.02	0.19 ± 0.01
11	Si	9.31 ± 0.02	1.19 ± 0.01
...10	Si$_{0.43}$Ge$_{0.57}$	9.08 ± 0.01	0.97 ± 0.01
1...	Si	22.89 ± 0.01	0.43 ± 0.01
Substrate	Si	∞	0.3 ± 0.1

Example III – A Si$_{1-x}$Ge$_x$/Si Superlattice on Si

In our last example, we have applied our fitting procedure to a superlattice (i.e., a periodic repetition of two or more layers) with nominal structure [Si$_{0.5}$Ge$_{0.5}$(10 nm)/Si(22 nm)]$_5$ capped with Si(10 nm), grown by molecular beam epitaxy (MBE) on an Si(001) substrate. The measured curve and its best-fit simulation are shown in Fig. 7.58. The reflectivity curves contain much fine detail; Bragg peaks and Kiessig fringes between them are clearly visible. The angular separation of adjacent Bragg peaks is related to the superlattice period while the period of the Kiessig fringes is related to the total thickness of the structure including the Si capping layer. Despite the reasonably large number of adjustable parameters involved in this example (a total of 11), the fitting procedure still managed to converge in approximately 4 minutes. The best-fit parameters and their uncertainties are listed in Table 7.17. With the exception of the surface oxide layer (assumed to be SiO$_2$), which had a fitted density of 1.5 ± 0.1 g/cm^3, all density values were fixed at their bulk values during the fitting. The Ge concentration was determined precisely by high-resolution X-ray diffraction to be $x = 57 \pm 5\%$ and also remained fixed during the fitting.

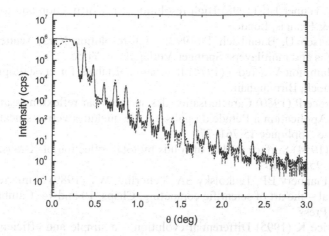

Fig. 7.58. X-ray reflectivity curves for an $Si_{1-x}Ge_x/Si$ superlattice on an Si substrate. The dashed line represents the measurements and the solid line is the best-fit simulation.

Note that the fitting procedure automatically found an asymmetry between the SiGe-on-Si and Si-on-SiGe interfaces. The width of the latter interfaces was almost twice that of the former. This asymmetry is due to "waviness" at the Si-on-SiGe interfaces and has been observed by imaging the interface using transmission electron microscopy (TEM).

7.10.5 Conclusions

We have developed an effective data-fitting and parameter optimization method using a combination of the differential evolution algorithm and a thoughtful consideration of the objective function. The procedure is robust against nonlinearity, local minima in the objective function, data that spans many orders of magnitude, and the choice of initial parameter values. The method is conceptually simple, easy to implement and rapid in execution. We have specifically applied the data-fitting method to the analysis of X-ray reflectivity data; however, the method is general and could be applied to many problems in science and engineering.

References

Bevington PR (1969) Data reduction and error analysis for the physical sciences. McGraw-Hill, New York

Bowen DK, Tanner BK (1998) High-resolution x-ray diffraction and topography. Taylor & Francis, London

Holý V, Pietsch U, Baumbach T (1999) High-resolution X-ray scattering from thin-films and multilayers. Springer-Verlag, New York

Ibers JA, Hamilton WC (eds) (1974) International tables for crystallography, vol IV. Kynoch, Birmingham

Névot L, Croce P (1980) Caractérisation des surfaces par reflexion rasante de rayons X. Application à l'étude du polissage de quelques verres silicates. Revue Physique Appliquée 15:761–779

Parratt LG (1954) Surface studies of solids by total reflection of X-rays. Phys.ical Review 95:359–369

Press WH, Flannery BP, Teukolsky SA, Vetterling WT (1989) Numerical recipes in Pascal – the art of scientific computing, Chaps. 10 and 14. Cambridge University Press

Storn R, Price K (1995) Differential evolution – a simple and efficient adaptive scheme for global optimization over continuous spaces. Technical report TR-95-012, ICSI, March 1995. PostScript file is downloadable from: http://www.icsi.berkeley.edu/techreports/1995.abstracts/tr-95-012.html

Storn R, Price K (1997) Differential evolution – a simple evolution strategy for fast optimization. Dr. Dobb's Journal 22(4):18–24 and 78

Wormington M, Bowen DK, Tanner BK (1992) Principles and performance of a PC-based program for simulation of grazing incidence X-ray reflectivity profiles. Materials Research Society Symposium Proceedings 238:119–124

Wormington M, Panaccione C, Matney KM, Bowen DK (1999) Characterization of structures from X-ray scattering data using genetic algorithms. Philosophical Transactions of the Royal Society: Mathematics, Physics and Engineering Science 357(1761):2827–2848

7.11 Inverse Fractal Problem

Ivan Zelinka

Abstract. This contribution focuses on the so-called inverse fractal problem and its solution by means of a new evolutionary algorithm – the differential evolution algorithm. The principles behind the inverse fractal problem are briefly explained here. The contribution then discusses the use of differential evolution for the solution of the inverse fractal problem and selected results.

7.11.1 General Introduction

The inverse fractal problem (IFP) arises where some appropriate mathematical methods are searching for so-called "coefficients of affine transformations". These coefficients are used in fractal geometry to generate "special" objects called fractals. The main aim of IFP is such that the difference between an observed object and an object generated by means of identified coefficients of affine transformations is minimal. A good description of this problem can be found in Barnsley (1993) and a description of fractal geometry in Peitgen et al. (1992), Barnsley (1993), Hastings and Sugihara (1993) and Bunde and Shlomo (1996). The solution of IFP can be based on two algorithms, i.e., IFS (Iterated Function System, Barnsley 1993) or TEA (Time Escape Algorithm, Barnsley 1993). Because there is a clear relation between the IFS and TEA (Barnsley 1993), the IFS solved in this contribution focuses on use of the TEA for ease of cost function construction.

Other research articles on IFP from various points of view and by means of various methods can be found in the literature (e.g., Arneodo et al. 1994, 1995; Muzy et al. 1994; Arul and Kanmani 1995; Struzik 1995, 1996; Deliu et al. 1997; Gutierrez et al. 2000).

Fractal Geometry – An Introduction

The basic idea of fractal geometry is that the geometric structure of the main object body repeats itself at smaller scales inside the main body of the observed object (Peitgen et al. 1992; Barnsley 1993; Hastings and Sugihara 1993; Bunde and Shlomo 1996). This geometrical repetition can be extended to infinity but only in a mathematical sense. The term "infinity" has to be taken into account only in such a sense, because in the real world physical borders appear. Fractal structures behind these borders usually disappear, such as, for example, in a bush–root system.

In fractal geometry fractals are divided into two main groups, i.e.,

- self-similar
- self-affine.

Self-similar fractals can usually be observed in the artificial–mathematical world. Their characteristic attribute is that the structure of the main body of the observed object repeats itself everywhere in the main body at different levels of magnification (Fig. 7.59a). Any subset of such an object is an exact copy of the main object.

Self-affine fractals are objects which can be observed everywhere at any time. Examples are trees (Fig. 7.59b), clouds, water surface, etc. Their characteristic attribute is that the structure of the main body of the observed object does not repeat itself in the main body as an exact copy of the original object. Any subset of such an object is an affine copy of the main object.

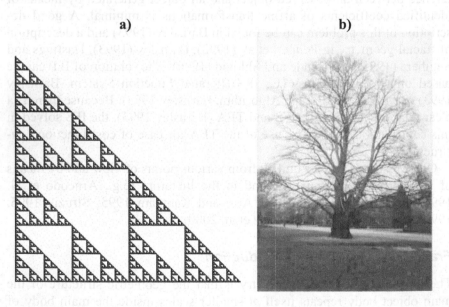

Fig. 7.59. Self-similar **a** and self-affine **b** fractals

Fractal construction can be performed by means of so-called *affine transformations*. Transformations of this kind involve three geometrical operations with the object under consideration – rotation, scaling and shift. The mathematical description of the affine transformation is given by Eq. 7.118:

$$w(x) = w\begin{pmatrix} x_1 \\ x_1 \end{pmatrix} = \begin{pmatrix} r_1 \cos\theta & -r_2 \sin\vartheta \\ r_1 \sin\theta & r_2 \cos\vartheta \end{pmatrix}\begin{pmatrix} x_1 \\ x_2 \end{pmatrix} + \begin{pmatrix} e \\ f \end{pmatrix}. \qquad (7.118)$$

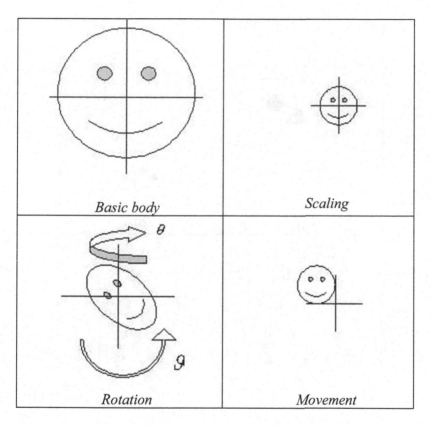

Basic body

Scaling

Rotation

Movement

Fig. 7.60. Affine transformations

The parameters in Eq. 7.118 have the following sense: r_1 and r_2 are so-called scaling parameters which define the change in size – the reduction along the x and y axes. The angles define rotation of the object around the x and y axes, and parameters e and f define the shift along the x and y axes. The action of all three types of transformations is represented artificially in

Fig. 7.60. By means of their composition very complicated structures can be generated (Figs. 7.61–7.65).

Usually more than one affine transformation is needed to build up fractal structure. For example, the fractal called the "Sierpinski triangle" (Fig. 7.59a) needs three affine transformations to be properly generated. By means of repetition of a whole set of affine transformations, also called iteration, the fractal structure begins to emerge from the main object body (Figs. 7.61–7.65).

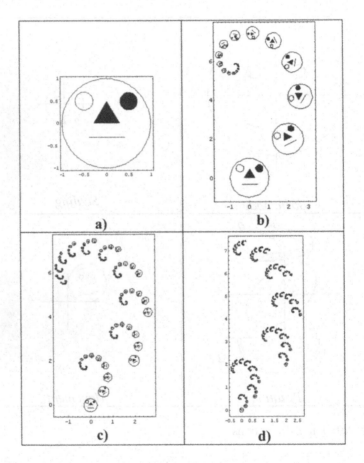

a)   b)

c)   d)

Fig. 7.61. Affine transformations in action

As examples of the principles mentioned above, some fractals are generated here. One concerns three transformations applied on a basic object called "Mr. Head", namely rotation, shrinking and shifting (see second head from bottom in Fig. 7.61b. They were repeated on the second head and the third head was obtained. This was done again and again till 20 iterations had been completed and b) was finished. The same approach and set of iterations as used on b) were used on one graphical object and c) and then d) were created. A similar philosophy but with different transformations was used in Fig. 7.62. From both figures it is clearly visible that the shape of the initial object is not important – which means that any structure can be generated from any other structure.

Fractal construction by means of these described affine transformations is also referred to as the iterated function system (IFS) algorithm (Barnsley 1993); it is able to generate black and white fractals.

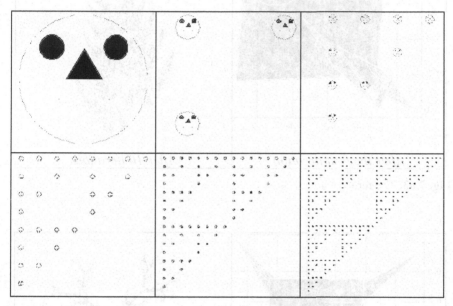

Fig. 7.62. The same basic object and different final shape

The second algorithm which can be used for fractal creation is the so-called time escape algorithm (TEA), also well described in Barnsley (1993). This algorithm also has an iterative nature like the IFS and is based on complex numbers using the philosophy of a user-defined "area" inside which the trajectory starts. This trajectory is calculated in an iterative way (the next point from the previous one) and after each calculation it is checked whether the border of the user-defined area has been overstepped. If so, then the start point is marked by a color which is proportional to the number of iterations needed for overstepping, otherwise the point is black. This means that black points are points which "show" or "are sources" of "stable" behavior (the trajectory does not run away from the user-defined area). Thus if coordinates of recalculated points are control parameters of some dynamical system, then the TEA can be used for example for checking the stability of a system etc.

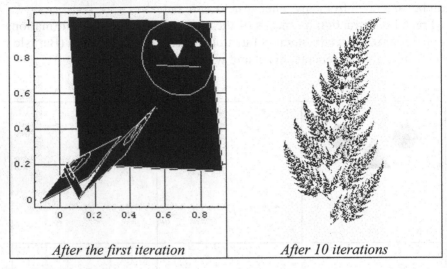

After the first iteration | After 10 iterations

Fig. 7.63. The fern

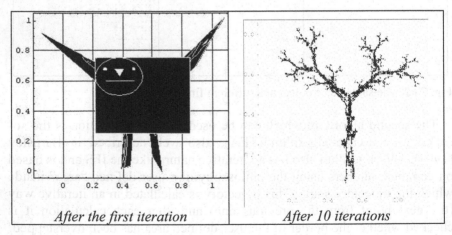

After the first iteration | After 10 iterations

Fig. 7.64. The tree

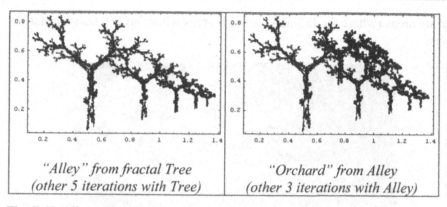

| "Alley" from fractal Tree (other 5 iterations with Tree) | "Orchard" from Alley (other 3 iterations with Alley) |

Fig. 7.65. Alley and orchard

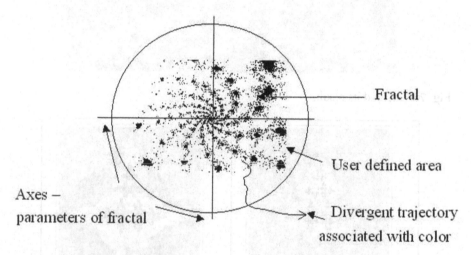

Fig. 7.66. Principle of TEA...

The whole principle of the TEA is demonstrated in Fig. 7.66 and Fig. 7.67, which are in principle based on Mandelbrot's simple fractal given by the equation

$$z_{n+1} = z_n^2 + c \quad \text{where } "z" \text{ and } "c" \in C \tag{7.119}$$

Transformation into the colored version reveals a few "reefs". The first one is that fractal conversion from a black and white version into a colored one is easy only for those fractals that do not contain the transformation of rotation, and sets (black points) are not mutually overlapping. The second "reef" is that the color composition of the final colored version of black and white fractals depends on what color "rule" is associated with appropriate areas (where the IFS acts) or with areas where no IFS transformation acts.

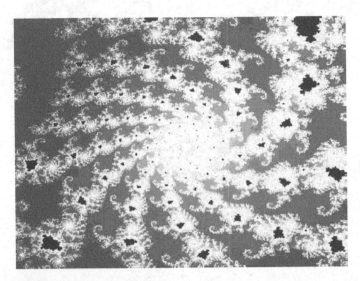

Fig. 7.67. . . . and its result.

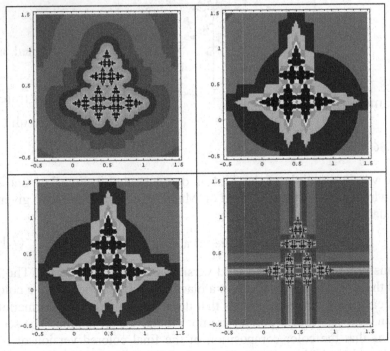

Fig. 7.68. Various versions of a Christmas tree by TEA

Different ways of how to paint sectors are shown in Fig. 7.68 for a fractal Christmas tree. Shown in Fig. 7.69 is a mix of both, i.e., the painting of areas where no IFS transformation acts (left upper sector of the Sierpinski triangle) and different color rules for whole fractal repainting (all four pictures). All this means that there are different ways of how to generate colored fractals. Only one part of the colored fractals is "stable": a set of black points which represents the original fractal from the IFS algorithm. This is the main reason why only black and white versions of TEA fractals were used for cost functions, as is described in the next subsection.

The third "reef" concerns the size of the grid which is used for TEA calculation. From Fig. 7.66 and Fig. 7.67 it is clear that in the ideal case each point should be recalculated, but practically this is impossible. Thus some grid has to be defined (see Fig. 7.71). The size of the grid used then determines the time of calculations and precision of the depicted (or identified) fractals.

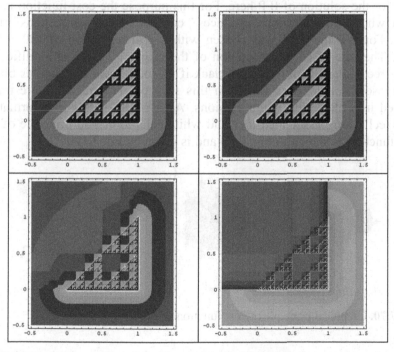

Fig. 7.69. Various versions of the Sierpinski triangle by TEA

Inverse Fractal Problem

IFP is a process during which so-called coefficients of affine transformations are identified, or the coefficients of the TEA, which is complementary to the IFS algorithm (Barnsley 1993) as mentioned above. Fractal reconstruction is one of the well-known problems of fractal geometry and was solved not just by evolutionary algorithms. IFP might be an artificial problem, but its solution can help in time series processing (see fractal interpolation of time series in Barnsley 1993) or as part of artificial intelligence in computer vision (Hlaváč and Šonka 1992). In computer vision fractal geometry can be used in object description. Such a description is not large in terms of data size and is exact – each pixel position is given by the coefficients of affine transformations (Peitgen et al. 1992; Barnsley 1993; Hastings and Sugihara 1993; Bunde and Shlomo 1996; Zelinka 1999a).

Algorithms like the differential evolution algorithm (Lampinen 1999; Lampinen and Zelinka 1999; Zelinka 2002a, 2002b; Storn 2002) can be used for the solution of IFP here. For this purpose the cost function is defined whose minimization "produces" coefficients of affine transformations. For IFP the TEA was chosen with a grid resolution of 100×100 cells. The rule of color association of the original TEA was modified so that a recalculated cell remained black if its color was calculated as black, otherwise the color was white. In this way colored fractals were transformed into black and white versions. As a result of this transformation black cells were associated to 1 and white cells to 0. The principle of the cost function is shown in Fig. 7.70 and is given by Eq. 7.120:

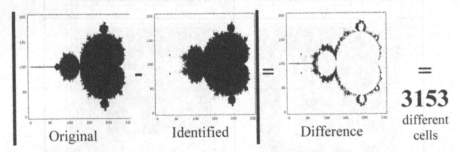

Original   Identified   Difference   **3153** different cells

Fig. 7.70. Graphical principle of cost function of IFP

$$f_{uc} = \sum_{x,y} \left| P_{org_{x,y}} - P_{ident_{x,y}} \right|$$

(7.120)

In Eq. 7.120 the terms P_{org} and P_{ident} represent respectively the original and actually identified elements of the matrix of cells of the fractal picture. This formula calculates the sum of cells in which both fractals differ. Minimization of this cost function (zero in the ideal case) leads to optimal coefficients, which should be the same as the coefficients of the original fractal. For each combination of coefficients the cost function returns the total sum – a scalar. For simple and maximally identified two-coefficient fractals such a cost function can be depicted like the geometrical surface shown in Fig. 7.71. This picture represents the surface of the cost function of the fractal "Spider" (for the equation, see Eq. 7.123). The global extreme (the black spot set on position {41,11}) is surrounded by a flat landscape which makes finding the global extreme harder especially for gradient-based methods. This is not a problem for the class of evolutionary algorithms. IFP has been solved many times by different evolutionary algorithms such as the hybrid evolutive–genetic strategy (Gutierrez et al. 2000), genetic algorithms, etc. Also, hybrid or different approaches like wavelet analysis have been used for the solution of IFP (Arneodo et al. 1994; Muzy et al. 1994). Differential evolution was chosen for the solution of IFP here for two reasons, as follows.

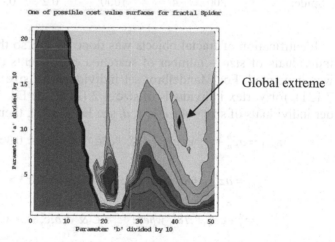

Fig. 7.71. Landscape of the "Spider" cost function. The global extreme is approximately located near coordinates $b = 4.15$ and $a = 1.11$ (axes divided by 10). The global extreme is in fact a point – here a black "spot" due to quite a rough grid that was used to increase the speed of calculations.

The first reason is that DE's population is floating-point encoded. Individuals of the population, which is binary encoded, have limited precision (in recalculating from the binary to the decimal domain), which can lead evolutionary process into a local extreme instead of global one, while floating-point encoded population doesn't suffer by such "property". The second reason is the fact that DE is very capable of finding global extremes on surfaces which are surrounded by a "flat landscape"; it does not matter what size of dimension is determined for the solved problem. This ability of DE was demonstrated for example in a "pathological function", see Zelinka (2001, 2002b). For all simulations here the DE version "DE/Rand/1/Bin" was chosen, which in the author's opinion is one of the best versions of DE. It was used for many problems with good success (see for example Lampinen 1999; Storn 2002; Zelinka 2001, 2002a, 2002b). The parameter settings of DE in the frame of IFP were similar for all simulations. Table 7.18 describes the parameter settings for the basic simulations of IFP.

Table 7.18. Parameter settings for basic simulations of IFP (Eqs. 7.121–7.123)

Fractal	NP	D	Generations	F	CR	Parameters (a,b,c,d)
Mandelbrot	20	2	100	0.82	0.3	<0, 3>
Vortex	20	2	100	0.82	0.3	<0, 3>
Spider	200	4	1000	0.2	0.73	<0, 5>

Identification of fractal objects was done by DE so that a population of individuals of size = number of searched coefficients (Eqs. 7.121–7.123) was generated. For Mandelbrot set individuals of size = 2 (a, b, see Eq. 7.121), for vortex individuals of size = 2 (a, b, see Eq. 7.122) and for spider individuals of size = 4 (a, b, c, d, see Eq. 7.123), then

$$z_{n+1} = az_n^b + c \quad where \text{ "}z\text{" \& "}c\text{"} \in C \ \& \ a,b \in <0,3>$$ (7.121)

$$z_{n+1} = az_n^b + .68i \quad where \text{ "}z\text{"} \in C \ \& \ a,b \in <0,3>$$ (7.122)

$$z_{n+1} = az_n^b + cz_n^d + .78 \quad where \text{ "}z\text{"} \in C \ \& \ a,b,c,d \in <0,3>$$ (7.123)

In this basic IFP three fractals were identified in total, i.e., the Mandelbrot set (Eq. 7.121), the vortex (Eq. 7.122) and the spider (Eq. 7.123) (see Figs. 7.73–7.75). Each IFP was repeated 100 times to check the validity and robustness of the results obtained. Shown in Fig. 7.72 for demonstration purposes are the histories of all 100 simulations done by DE for the Mandelbrot set. Original fractals for which IFP was done are not depicted here because the results of IFP (Figs. 7.73–7.75) are almost the same.

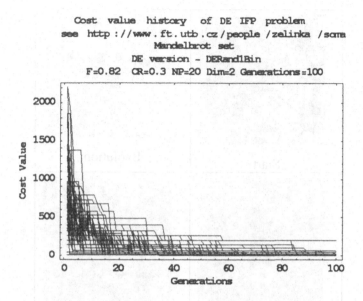

Fig. 7.72. Simulation of IFP repeated 100 times for the Mandelbrot set by DE

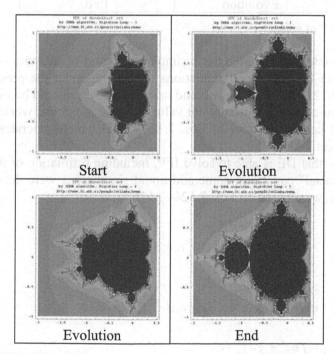

Fig. 7.73. IFP of Mandelbrot set

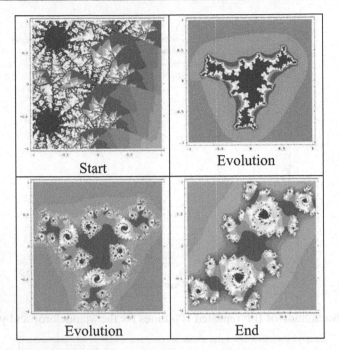

Fig. 7.74. IFP of vortex

In the context of all three basic simulations good results were obtained by means of DE. All three simulations were done under the presumption that Eqs. 7.121–7.123 are known and only parameters should be estimated. These presumptions are usually not valid, because in the real world there is no a priori known structure of Eqs. 7.121–7.123 which generates the observed object.

To show that DE is able to solve IFP, including estimation of the structure of "fractal" equations, some additional simulations were done. These simulations were based on Eqs. 7.124–7.126:

$$z_{n+1} = az_n^b + dz_n^e + c \qquad (7.124)$$
$$\textit{where } "z" \& "c" \in C \ \& \ a,b,d,e \in <0,3>$$

$$z_{n+1} = az_n^b + dz_n^e + fz_n^g + c \qquad (7.125)$$
$$\textit{where } "z" \& "c" \in C \ \& \ a,b,d,e,f,g \in <0,3>$$

$$z_{n+1} = faz_n^b + gdz_n^e + c \qquad (7.126)$$
$$\textit{where } "z" \& "c" \in C \ \& \ a,b,d,e \in <0,3> \ \& \ f,g \in [0,1]$$

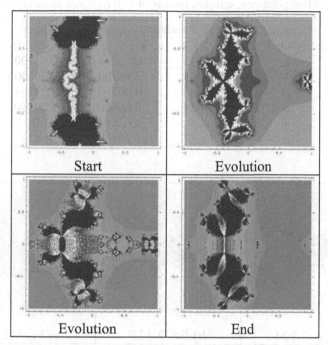

Fig. 7.75. IFP of spider

The first equation, Eq. 7.124, is basically Eq. 7.121 enriched with an additional term as well as Eq. 7.125 and Eq. 7.126. IFP simulations based on Eqs. 7.124–7.125 demonstrate by their numerical and graphical outputs (Table 7.19, Fig. 7.76) that the same fractal can be generated by different equations, which can be called "redundant" for the sake of additional terms, which are not needed for generation of the final fractal structure.

If the different equations can generate the same fractal, then the question arises: "how can one select the optimal equation (not redundant) of all possible equations?" There is probably more than one approach to solving this problem. Here the simplest one is discussed and demonstrated. The main idea is that redundant terms can be "disabled" from the equation during evolutionary searching for IFP. This disabling is satisfied by converting Eq. 7.124 to Eq. 7.126 by means of additional parameters f and g which belong to the discrete set [0,1]. Their values in multiplication are responsible for disabling or enabling the appropriate term in the relevant equation.

Table 7.19. IFP simulations, selected results of 100 repeated simulations.

Estimated parameters	IFP simulation according to equation		
	Eq. 7.124	Eq. 7.125	Eq. 7.126
a	0.53743	0.01714	0.45900
b	1.94713	2.96700	2.26800
d	0.50508	0.06780	1.03154
e	2.08631	0.00600	2.00110
f	–	1.02654	0
g	–	1.99800	1

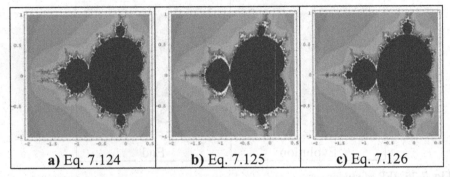

| a) Eq. 7.124 | b) Eq. 7.125 | c) Eq. 7.126 |

Fig. 7.76. Graphical "output" of simulations based on Eqs. 7.124–7.126, selected results of 100 repeated simulations.

The results in Table 7.19, selected from 100 repeated simulations, show that this approach is viable because parameter $f = 0$, i.e., the first term in Eq. 7.126 is disabled, and because $g = 1$ the second term in Eq. 7.126 is enabled. It can be seen that the enabled second term has exactly the same parameters as the Mandelbrot set which was identified originally. The graphical representation of the finally identified and consequently generated structure appears in Fig. 7.76c.

Software Support

All the above-mentioned examples are supported in the C language, a Mathematica environment notebook and are accessible on the Internet (Zelinka 2002b). In C++ for running IFP special initial files have to be used. The syntax for their use in a DOS environment is "Der1b IFP". For DE, version DE/Rand1//Bin is used (remaining versions are in development now). Term IFP represents one of the three names of initializing files, i.e., "Mandelbrot", "Vortex" or "Spider". After completion two files will be created with extension *.HST and *.OUT (where * represents the appropriate name of the IFP initial file). In the HST file the history of the evolution (migration loops or generations) is shown. In the OUT file is a report with all the details about the simulation including estimated parameters. The content of initializing files can be changed, but it is important to remember that results can differ from simulation to simulation because the parameters of the algorithm may be modified. For more details see Zelinka (2002b).

The cost function of, for example, the Mandelbrot set is shown in Fig. 7.77. Code for other fractals is similar to this one.

For the Mathematica environment Zelinka (2002b) provides an accessible notebook which contains the DE, some examples and selected simulations of basic IFP depicted in this section. All-important information on how to handle this notebook is given on the web site.

7.11.2 Conclusion

In this contribution the inverse fractal problem was solved by means of a differential evolution algorithm. Two kinds of inverse fractal problem simulations were undertaken using differential evolution. Both sets of simulations were based on the time escape algorithm, which is the source of colored fractals. This algorithm was chosen because of its "complementarity" with the algorithm called the iterated function system (Barnsley 1993) or multiple reduction copy machine (Peitgen et al. 1992). The term "complementarity" here means that the iterated function system (or multiple reduction copy machine) can be converted into the time escape algorithm and vice versa (Barnsley 1993). In the case of the time escape algorithm it is easier to build up the cost function – the main reason why this algorithm was chosen for the inverse fractal problem.

The first set of simulations focused on identification of appropriate parameters with a priori knowledge of the equation structure. Three well-known fractals, i.e., the Mandelbrot set (Eq. 7.121), vortex (Eq. 7.122) and spider (Eq. 7.123), were selected for these simulations. According to results from the simulations in Figs. 7.72–7.75 it can be seen that the inverse fractal problem is solvable by means of differential evolution as well as by other evolutionary algorithms. In the second set of simulations attention was focused on parameter estimation as well as on optimal equation structure. Results were again satisfactory. Despite the fact that the simulations here can look like easy ones, results show that the inverse fractal problem is solvable quite well if the appropriate technique is used.

```
//Inverse Fractal Problem - Mandelbrot
//See http://www.ft.utb.cz/people/zelinka/soma.htm
//**  Mandelbrot for constant k and power n -> kz^2-c   **

#ifdef __cplusplus
  #include <complex.h>
#endif
double absval,Im,Re,incIm,incRe,m,n;
int       TEA,UserArea,UserNunmIts,Org,Ident,Difference;
#ifdef __cplusplus /* if C++, use class complex */
complex Mand(0,0);
complex MandOrg(0,0);
complex C(0,0);
complex COrg(0,0);

m=getPopulation(0,Individual);
n=getPopulation(1,Individual);
UserArea=10;
UserNunmIts=50;
Difference=0;
for(Im=-2;Im<0.5+incIm;Im=Im+incIm)
  for(Re=-1;Re<1+incRe;Re=Re+incRe)
    {
    Org=Ident=0;
    Mand = complex(0,0);
    C = complex(Re,Im);
    MandOrg = complex(0,0);
    COrg = complex(Re,Im);
    for(TEA=0;TEA<UserNunmIts;TEA++)
      {
      if(Ident=0)
        {
        if(abs(Mand)>UserArea)
        Mand = m * pow(Mand,n) + C;
        Ident=1;
        }
      if(Org=0)
        {
        if(abs(MandOrg)>UserArea)
        MandOrg = 1 * pow(MandOrg,2) + COrg;
        Org=1;
        }
      }
    Difference = Difference + abs(Org-Ident);
    }
CostValue=Difference;
#endif
```

Fig. 7.77. Cost function for "Mandelbrot set" in the C language, see Zelinka (2002b).

From a general point of view the inverse fractal problem is not just an "academic toy" because its use can be very practical. Future possible uses lie for example in computer vision (Zelinka 1999b, 1999c) where the selected object in the digital picture can be described by fractal geometry.

This type of description consists in high compression of the object described (see the fern, 24 numbers in the fractal approach or megabytes in, for example, the bitmap description) without loss of any information about its structure. In standard computer vision methods this is a problem. The object is usually described roughly (low amount of data, no information about structure) or described in high detail (huge amount of data) and the object classification based on such a large data set is thus problematic. For more details see Hlaváč and Šonka (1992) and Zelinka (1999b, 1999c). Another possible use of the inverse fractal problem is in fitting missing data if a data set has fractal properties. For more details see fractal interpolation in Arneodo et al. (1995).

The simulations done here can be taken as basic simulations, which show not only how the inverse fractal problem can be solved and to what quality, but also how robust differential evolution is in finding the global extreme.

References

Arneodo A, Bacry E, Muzy J (1994) Solving the Inverse Fractal Problem from Wavelet Analysis. Europhysics Letters 25(7):479–484

Arneodo A, Bacry E, Muzy J (1995) The Thermodynamics of Fractals Revisited with Wavelets. Physica A 213(1–2):232–275

Arul A, Kanmani S (1995) A Conjugate Transform Solution to the Inverse Fractal Problem. Europhysics Letters 32(1):1–5

Barnsley MF (1993) Fractals everywhere. Academic Press, New York

Bunde A, Shlomo H (1996) Fractals and disordered systems. Springer, Berlin

Deliu A, Geronimo J, Shonkwiler R (1997) On the inverse fractal problem for two-dimensional attractors. Philosophical Transactions of The Royal Society of London, Series A: Mathematical, Physical and Engineering Sciences 355(1726):1017–1062

Gutierrez J, Cofino A, Ivanissevich M (2000) An hybrid evolutive-genetic strategy for the inverse fractal problem of IFS models. In: Advances in Artificial Intelligence. Lecture notes in artificial intelligence, vol 1952. Springer, Berlin, pp 467–476

Hastings HM, Sugihara G (1993) Fractals: a user's guide for the natural sciences. Oxford University Press

Hlaváč V, Šonka M (1992) Počítačové vidění (Computer Vision, Czech edn). Grada, Prague

Lampinen J (1999) A bibliography of differential evolution algorithms. Technical report, Lappeenranta University of Technology, Department of Information Technology, Laboratory of Information Processing, October 16. Available at: http://www.lut.fi/~jlampine/debiblio.htm

Lampinen J, Zelinka I (1999) Mechanical engineering design optimization by differential evolution. In: New ideas in optimization. McGraw-Hill, London

Muzy J, Bacry E, Arneodo A (1994) The multifractal formalism revisited with wavelets. International Journal of Bifurcation and Chaos 4(2):245–302

Peitgen HO, Jürgens H, Saupe D (1992) Chaos and fractals: new frontiers of science. Springer, Berlin

Storn R (2002) Homepage of DE. Available at:
http://www.icsi.berkeley.edu/~storn/code.html

Struzik Z (1995) The wavelet transform in the solution to the inverse fractal problem. Fractals – An Interdisciplinary Journal on the Complex Geometry of Nature 3(2):329–350

Struzik Z (1996) Solving the two-dimensional inverse fractal problem with the wavelet transform. Fractals – An Interdisciplinary Journal on the Complex Geometry of Nature 4(4):469–475

Zelinka I (1999a) Aplikovaná informatika, vol 1. Ediční středisko, Fakulty technologické, VUT, Zlín

Zelinka I (1999b) Inverse fractal problem by means of evolutionary algorithms. In: Mendel'99, 5th international conference on soft computing, PC-DIR, Brno, pp 430–435

Zelinka I (1999c) Fraktální vidění pomocí neuronových sítí. In: Process control '99, vol 1. Slovak University of Technology, Vydavatelstvo STU Bratislava, pp 318–322

Zelinka I (2001) Prediction and analysis of behavior of dynamical systems by means of artificial intelligence and synergetics, Ph.D. thesis, Tomas Bata University, Zlín, Czech Republic

Zelinka I (2002a) Umělá inteligence v problémech globální optimalizace (Artificial intelligence in problems of global optimization, Czech ed). BEN, Prague. ISBN 80-7300-069-5

Zelinka I (2002b) Information on DE algorithm. Available at: http://www.ft.utb.cz/people/zelinka/soma

7.12 Active Compensation in RF-Driven Plasmas by Means of Differential Evolution

Ivan Zelinka and Lars Nolle

Abstract. In this section two different stochastic optimization methods are discussed and compared. They were applied to the deduction of 14 Fourier terms in a radio-frequency (RF) waveform to tune a Langmuir probe. Langmuir probes are diagnostic tools used to analyze the electron energy distribution in plasma processes. RF plasmas are inherently nonlinear, and many harmonics of the driving fundamental are generated in the plasma. RF components across the probe sheath distort the measurements made by the probes. To improve the quality of the measurements, these RF components must be removed. In this research, this was achieved by applying an RF signal to the probe tip that matches both the phase and amplitude of the driving RF signal. It also had to match the waveform of the plasma, which is determined by the nonlinearity of the plasma. Here, seven harmonics are used to generate the waveform. Therefore, 14 mutually interacting parameters (seven phases and seven amplitudes) had to be tuned on-line. In this work, two stochastic optimization algorithms were used for automated tuning of the probe – simulated annealing (SA) and differential evolution (DE). SA was previously used for this problem, whereas DE was chosen and compared with SA because of its reported global optimization performance.

7.12.1 Introduction

Radio-frequency (RF) driven discharge plasmas, which are partially ionized gases that are not in thermal equilibrium with their surroundings, are widely used in the material processing industry for etching, deposition and surface treatment, e.g., in the semiconductor industry. In order to achieve best-quality products, it is essential for users of such plasmas to have tight control over the plasma and hence they need appropriate diagnostic tools in order to close the control loop. Better diagnostics lead to better control of the plasma and hence to better-quality products. In this work, simulated annealing (SA) and differential evolution (DE) are used and compared to control a Langmuir probe, which is a diagnostic measurement system for industrial RF-driven plasma systems.

7.12.2 RF-Driven Plasmas

Under normal conditions gases do not conduct electrically. Almost all electrons are bound to atoms or molecules. If, however, electrons are introduced and given enough energy by an external power source (e.g., electromagnetic fields, light, heat, etc.), then they have the potential on colliding with gas atoms or surfaces to release more electrons, which themselves may release other electrons. This resulting electrical breakdown is known as an avalanche effect. The ionized gas or plasma so formed is now conducting.

In the case of industrial RF-powered plasmas, an RF generator is used as an external power source, usually operating at 13.56 MHz. The use of RF rather than DC has developed for a number of reasons including efficiency and compatibility with systems in which direct electrical contact with the plasma is not feasible. This frequency is assigned for industrial, non-telecommunications use. The RF is inductively or capacitively coupled into a constant gas flow through a vacuum vessel using electrodes, see Fig. 7.78.

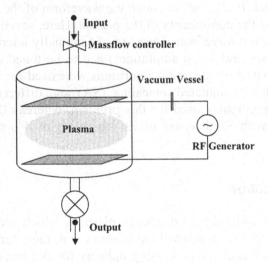

Fig. 7.78. Principle of RF-driven plasmas

The main application of RF-powered plasmas is to produce a flux of energetic ions, which can be applied continuously to a large area of a workpiece, e.g., for etching or deposition. This flux is generated by the RF plasma because the mass of the electrons is only a fraction of the mass of the atoms, and hence they can follow the electric field, while the ions respond only to slower variations in electrical structure. Electrons near the electrodes can thus escape, which results in electric fields, pointing from the plasma to the electrodes. These fields then generate a flux of energetic ions.

7.12.3 Langmuir Probes

Langmuir probes, developed in 1924 by Irving Langmuir, are one of the oldest probes used to obtain information about low-pressure plasma properties. They are metallic electrodes, which are inserted into a plasma. By applying a positive or negative DC potential to the probe, either an ion or an electron current can be drawn from the plasma, returning via a large conducting surface such as the walls of the vacuum vessel or an electrode. This current is used to analyze the plasma properties, e.g., for the determination of the energy of electrons, electron particle density, etc.

The region of space charge (or sheath) that forms around a probe immersed in a plasma has a highly nonlinear electrical characteristic. As a result, harmonic components of potential across this layer give rise to serious distortion of the probe's signal. In RF-generated plasmas this is a major issue as the excitation process necessarily leads to the space potential in the plasma having RF components. As a consequence of this fact a serious distortion of the probe's signal can be observed. It is caused by harmonic components of potential across this layer.

7.12.4 Active Compensation in RF-Driven Plasmas

To eliminate the time variation of RF potential difference, which is between the probe and plasma, the probe potential has to follow that of the exciting RF signal (Benjamin et al. 1988). This can be achieved by superimposing a synchronous signal of appropriate amplitude and phase onto the probe tip. Because plasmas are inherently nonlinear, they generate many harmonics of the exciting fundamental. As a consequence, the RF signal necessary for satisfactory compensation has not only to match the amplitude and the phase of the exciting RF, but also to match the waveform of the harmonics generated in the plasma.

Conveniently, the electrostatic probe and the plasma spontaneously generate a useful control signal. In the presence of plasma, an isolated electrostatic probe adopts a "floating potential", at which it draws zero current. The effect of inadequate compensation on a probe in an RF plasma is to drive the DC potential of the probe less positive (or less negative). Thus, optimal tuning is identical with the probe adopting the most positive (or least negative) potential. The "floating potential" is also referred to here as a DC bias.

7.12.5 Automated Control System Structure and Fitness Function

During previous work, an additive synthesizer (harmonic box) with seven harmonics was developed (Nolle et al. 2002) to generate the appropriate waveform for a Langmuir probe system attached to a Gaseous Electronics Conference (GEC) reference reactor (Sobolewski 1992). Figure 7.79 shows the schematics of the control system for waveform tuning.

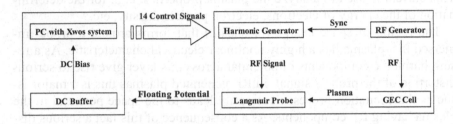

Fig. 7.79. Closed control loop for waveform tuning

The control software selects set points for the harmonic generator and sets the parameters using 14 D/A converter channels. The harmonic box, which is synchronized with the main RF power generator, outputs the required waveform to the Langmuir probe. The Langmuir probe's floating potential (DC bias) is used as a fitness measure. It is read on-line via a DC buffer and an A/D converter by the computer system. Depending on the optimization algorithm used in the system, the software then calculates a new set point based on the actual measure of the fitness (DC bias). It can be seen that all the fitness evaluations are actually measurements rather than simulation results. This implies time restrictions on the search process.

The 14 input parameters (seven amplitudes and seven phases) are strongly interacting due to the technical realization of the synthesizer and the nature of the problem. For example, the slightest departure from an ideal sinusoidal shape in one of the channels introduces harmonics. In practice, even after careful electronic design, it is found that there is a weak but significant coupling between amplitude control and phase and vice versa. As a consequence, the number of points in the discrete search space has to be calculated as follows:

$$n = \left(2^b\right)^p \tag{7.127}$$

where :

n = number of points in search space

b = resolution per channel in bits

p = number of parameters to be optimized.

The D/A and A/D converters used in this project had a resolution of 12 bits and the dimensionality of the search space was 14. Hence, the search space consisted of $n \approx 3.7 \times 10^{50}$ search points. In this case, mapping out the entire search space would take approximately 10^{41} years with the plasma system used. This was clearly not an option!

In previous work, SA was used successfully to tune the Langmuir probe (Nolle et al. 2002). The results could even be improved further by introducing step width adaptation to SA (Nolle et al. 2001). However, small variations in fitness indicated that the SA algorithm had not always found the precise global solution, even if it always came quite close to it. Therefore, in this research DE (Price 1999) was used for the optimization and compared with SA.

For this application domain, only simplified physical models are available. For the tuning of the Langmuir probe, these models were not accurate enough to simulate the complex plasma system used. Hence, the fitness function f could not be calculated, but was an actual measurement of the DC bias produced by the real process during the experiments (Fig. 7.79):

$$f_{fitness} = DCBias = f_{reactor}\left(a_1, \cdots, a_7, p_1, \ldots, p_7\right) \tag{7.128}$$

where

$a_1, \cdots, a_7$ seven amplitudes

$p_1, \ldots, p_7$ seven phases.

Therefore, the fitness value was obtained on-line from the real process. This means that from the observer's point of view, the fitness function was basically a "black box".

a)

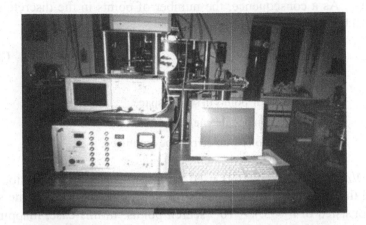

b)

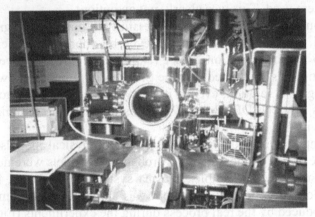

Fig. 7.80. Laboratory equipment: **a** computer with control software (right), wave synthesizer (left bottom) and oscilloscope (left top); **b** plasma reactor with Langmuir probe.

7.12.6 Experimental Setup

All experiments were carried out at the Oxford Research Unit, The Open University, UK. Figure 7.80 shows the experimental setup. Apart from the control system described above, a digital oscilloscope was used to measure the actual waveforms found by the two optimization algorithms. The control software was run on a PC under the Linux operating system. The algorithms used for this experiments were written in C++ and integrated in the existing Langmuir probe control software. The plasma system used was a standard GEC cell.

7.12.7 Parameters and Experimental Design

Table 7.20 shows the plasma parameters used for the experiments, whereas Table 7.21 states the parameter settings for the optimization algorithms.

The parameter settings for SA were chosen to be the same as in previous experiments (Nolle et al. 2001). The DE parameters were then selected empirically (Lampinen and Zelinka 1999; Price 1999; Zelinka 2002). Finding the optimum parameter settings is not an easy task, because the plasma drifts over time, i.e., its behavior is not constant. The plasma can change its behavior constantly over time as well as spontaneously.

Table 7.20. Plasma parameters

Plasma parameters	
Gas	Argon
Power	50 W
Pressure	100 mTorr
Flow rate	95 sccm

Table 7.21. The best parameter settings used in experiments

Simulated annealing		Differential evolution	
T_{start}	25,000	CR	0.5
Temperature coeff.	0.8	F	0.8
Iterations per temperature	50	NP	50
S_{max}	4000	Generations	250
Number of particles	3		
Iterations	4000		

The experiments were designed so that for both algorithms one optimization cycle took no longer than 4 minutes, which was a requirement made by users of such plasma diagnostic systems. Despite the fact that the search time was limited to 4 minutes, approximately 12,000 cost function evaluations, i.e., DC bias measurements, were achieved in one optimization run.

a) DE

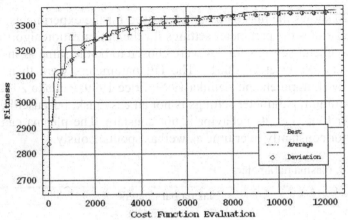

b) SA

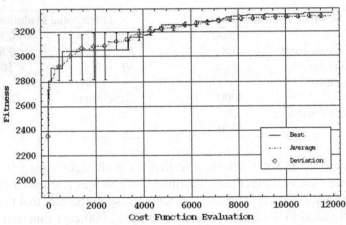

Fig. 7.81. Fitness value history

a) DE

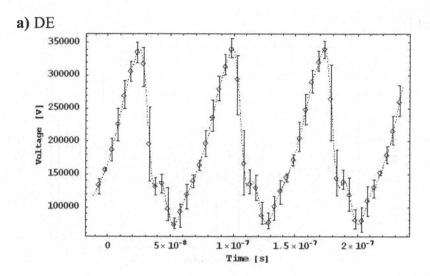

b) SA

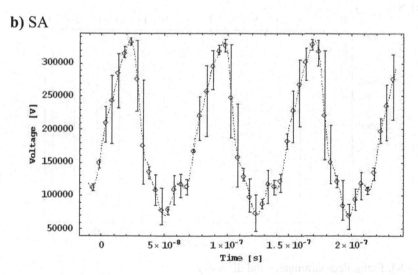

Fig. 7.82. Waveform

a) DE

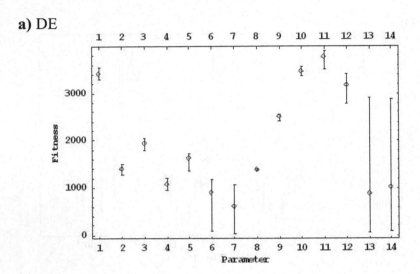

b) SA

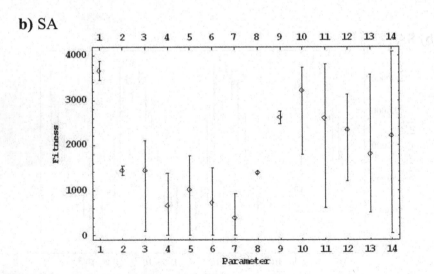

Fig. 7.83. Estimated parameters and diversity

7.12.8 Results

The experimental results can be seen in Figs. 7.81–7.84 for both algorithms. Figure 7.81 shows a typical search run over time. The average fitness of the population, the best individual in the current generation and the standard deviation are given. In Fig. 7.82 the average waveform found by the algorithms is depicted and Fig. 7.83 shows the average values and the deviation for all the 14 parameters found by the algorithm.

In Fig. 7.84 it can be seen that DE had outperformed SA not only by finding a greater average fitness, but also by having a smaller standard deviation than SA.

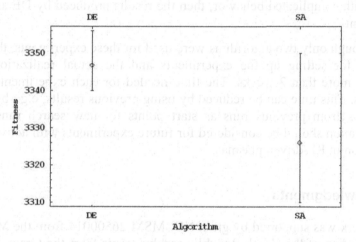

Fig. 7.84. Diversity and average value of fitness value (highest DC bias).

7.12.9 Conclusion

Two stochastic optimization algorithms, SA and DE, were used for on-line optimization to tune an actively compensated Langmuir probe system. These algorithms were selected because of the complexity of the problem. Based on the experimental results, which are depicted in Figs. 7.81–7.84, one can draw the following conclusions:

Ability to be used. Both algorithms can be used for active compensation in RF-driven plasmas. However, based on the results in Figs. 7.81–7.83 it is clear that DE has greater potential for this task.

Preciseness and reproducibility. One of the crucial points in science is reproducibility, i.e., the ability to achieve the same results for two identical experiments. In practical applications like this one, a high degree of repro-ducibility is needed. From Fig. 7.83 it can be seen that DE has a greater reproducibility than SA. It is also more precise than SA.

Speed. The speed of the optimization process was not determined by the computer power available, but by the time constants of the analogue equipment, e.g., the harmonic box. Therefore, both algorithms have shown similar speed performance in this specific application.

Diversity. This is closely connected with preciseness and reproducibil-ity. From this point of view, DE performed almost three times better than SA. If one remembers that plasmas are highly nonlinear dynamical sys-tems with complicated behavior, then the results produced by DE are very sufficient.

Although only two algorithms were used for these experiments, the time needed for setting up the experiments and the actual realization took slightly more than 2 weeks. The time needed for each experiment was 4 minutes. This time can be reduced by using previous results, e.g., by using solutions from previous runs as start points for new search runs. This modification should be considered for future experiments with active com-pensation in RF-driven plasmas.

Acknowledgments

This work was supported by grant NO. MSM 26500014 from the Ministry of Education of the Czech Republic and by grants from the Grant Agency of the Czech Republic (GACR 102/00/0526 and GACR 102/02/0204). The authors wish to express their thanks to Professor Nicholas St. J. Braithwaite, Director of the Oxford Research Unit, and Professor Adrian Hopgood, Head of School of Computing and Mathematics, The Notting-ham Trent University, for supplying resources and advice.

References

Benjamin NMP, Braithwaite NSJ, Allen JE (1988) Self bias of an r.f. driven probe in an r.f. plasma. Materials Research Society Symposium Proceedings 117:275–280
Lampinen J, Zelinka I (1999) Mechanical engineering design optimization by dif-ferential evolution. In: Corne D, Dorigo M, Glover F (eds) New ideas in op-timization. McGraw-Hill, London, pp 127–146
Nolle L, Goodyear A, Hopgood AA, Picton PD, Braithwaite NSJ (2001) On step width adaptation in simulated annealing for continuous parameter optimisa-

tion. In: Reusch B (ed) Computational intelligence – theory and applications. Lecture notes in computer science, vol 2206. Springer, Berlin Heidelberg New York, pp 589–598

Nolle L, Goodyear A, Hopgood AA, Piction PD, Braithwaite NSJ (2002) Automated control of an actively compensated Langmuir probe system using simulated annealing. Knowledge-Based Systems 15(5–6):349–354

Price K (1999) Differential evolution. In: Corne D, Dorigo M, Glover F (eds) New ideas in optimization, McGraw-Hill, London

Sobolewski MA (1992) Electrical characterization of radio frequency discharges in the Gaseous Electronics Conference reference cell. Journal of Vacuum Science & Technology A – Vacuum Surfaces and Films 10(6):3550–3562

Zelinka I (2002) Umělá inteligence v problémech globální optimalizace (Artificial intelligence in problems of global optimization, Czech ed). BEN, Prague. ISBN 80-7300-069-5

Appendix

This appendix contains a total of twenty test functions that have been divided into three categories: unconstrained uni-modal (5), unconstrained multi-modal (12) and bound-constrained multi-modal (3). Functions range from trivially simple to very challenging. Most test functions can be evaluated at more than one dimension. Each function definition includes a set of upper and lower initial parameter bounds. In this test bed, a function's initial parameter bounds apply to all of its parameters. For bound-constrained functions, initial parameter bounds also constrain the search.

Each test problem description also lists the minimum objective function value. In most cases, the minimum is independent of the function's dimension. If, however, the objective function's minimum depends on its dimension, then the optimal function values for selected dimensions are given. Problem descriptions also include a value for ε to indicate how close to the minimum objective value an optimizer's best point must be before the optimization can be considered a success. In all cases, this *value-to-reach*, or "VTR", is VTR $= f(\mathbf{x}^*) + \varepsilon$, where $\mathbf{x}^*$ is the globally optimal vector.

In most cases, a picture is included that illustrates the function's two-dimensional landscape. Usually, these two-dimensional figures provide insight into the characteristics of the higher-dimensional functions, but in a couple of cases (e.g., Storn's Chebyshev function) the restriction to two dimensions does not permit the difficulties posed by the function to be faithfully rendered. In some cases, figures are not possible due to technical reasons. Not all functions listed here are used in the text of this book, but all appear on the accompanying CD-ROM.

Some functions in this appendix are taken from the test bed developed for the Second International Contest on Evolutionary Optimization (2nd ICEO). The actual contest was not held due to a lack of participation, but code and data for functions described as 2nd ICEO functions can be found on both the CD-ROM that accompanies this book and on the Internet (Second ICEO 1997). For a moderately challenging test bed of constrained functions, the interested reader is referred to Michalewicz and Shoenauer (1996).

A.1 Unconstrained Uni-Modal Test Functions

This section includes five unconstrained, uni-modal test functions, none of which should pose a problem for a robust optimizer. Not all sources agree on the initial parameter bounds for these functions, but in practice these variations do not dramatically affect run times or the probability of success. For many EAs, the most difficult function to optimize in this uni-modal test bed is the generalized Rosenbrock function. In addition, some GAs may have problems solving Schwefel's ridge function because it is a highly eccentric, rotated hyper-ellipsoid with dependent parameters.

A.1.1 Sphere

This simple function tests a search method's local optimization speed and its response to changing dimension. To accommodate bit-encoded GAs, early test beds usually specified the initial parameter bounds as [−5.12, 5.12], but Yao and Liu's more recent and widely referenced test bed (Yao and Liu 1997) initializes parameters with values chosen from the interval [−100, 100].

$$f(\mathbf{x}) = \sum_{j=0}^{D-1} x_j^2,$$ (A.1)

$$-100 \le x_j \le 100, \quad j = 0,2,...,D-1,$$

$$f(\mathbf{x}^*) = 0, \quad x_j^* = 0, \quad \varepsilon = 1.0 \times 10^{-6}.$$

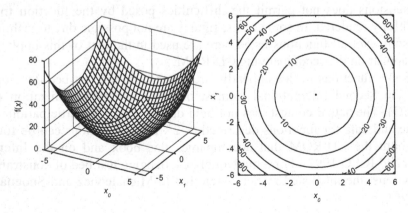

Fig. A.1. The sphere function

A.1.2 Hyper-Ellipsoid

Some literature specifies [−5.12, 5.12] as the bounds for initializing this function, but this book adopts the limits given in Yao and Liu (1997). To decrease the eccentricity of the hyper-ellipsoid, some versions of this function use a term like $(j + 1)^2$ as the pre-factor to x_j instead of putting the parameter index in the exponent. This function can take a long time to solve if an optimizer cannot adapt step sizes to suit each dimension.

$$f(\mathbf{x}) = \sum_{j=0}^{D-1} 2^j \cdot x_j^2, \tag{A.2}$$

$$-100 \leq x_j \leq 100, \quad j = 0,1,...,D-1,$$

$$f(\mathbf{x}^*) = 0, \quad x_j^* = 0, \quad \varepsilon = 1.0 \times 10^{-6}.$$

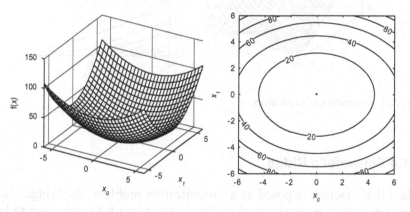

Fig. A.2. The unrotated hyper-ellipsoid

A.1.3 Generalized Rosenbrock

The original Rosenbrock function was just two-dimensional, but it was later generalized to this higher-dimensional version. The ridge in Fig. A.3 shows that this uni-modal function is non-convex. This function exhibits limited parameter dependence that poses a problem for many optimizers. Some studies use [−2.048, 2.048], while others use [−5.12, 5.12] for initial parameter bounds. Yao and Liu initialized parameters with values chosen from [−32, 32], but initial parameter bounds were [−30, 30] for studies in this book.

$$f(\mathbf{x}) = \sum_{j=0}^{D-2}\left(100\cdot\left(x_{j+1}-x_j^2\right)^2+\left(x_j-1\right)^2\right),$$

(A.3)

$$-30\le x_j\le 30,\quad j=0,1,...,D-1,\quad D>1,$$

$$f\left(\mathbf{x}^*\right)=0,\quad x_j^*=1,\quad \varepsilon=1.0\times10^{-6}.$$

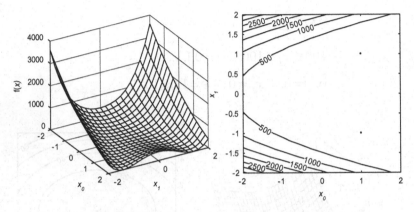

Fig. A.3. Rosenbrock's function

A.1.4 Schwefel's Ridge

When this function is posed as a minimization problem, the "ridge" in its landscape becomes an elliptical "valley". For some EAs, adapting to both the orientation and high eccentricity of the ellipse can be a significant challenge. Some studies have used [–65.536, 65.536] as initial parameter bounds, but this book adopts the bounds published in Yao and Liu (1997).

$$f(\mathbf{x}) = \sum_{k=0}^{D-1}\left(\sum_{j=0}^{k}x_j\right)^2,$$

(A.4)

$$-100\le x_j\le 100,\quad j=0,1,...,D-1,$$

$$f\left(\mathbf{x}^*\right)=0,\quad x_j^*=0,\quad \varepsilon=1.0\times10^{-6}.$$

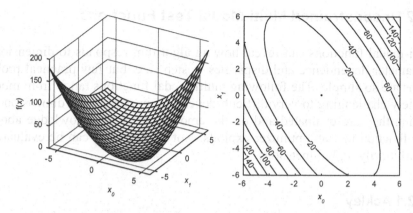

Fig. A.4. Schwefel's ridge function

A.1.5 Neumaier #3

This function also displays elliptical contours that are aligned with coordinate diagonals, but the optimum is not centered in the initial bounding box.

$$f(\mathbf{x}) = \sum_{j=0}^{D-1} (x_j - 1)^2 - \sum_{j=1}^{D-1} x_j \cdot x_{j-1}, \quad D > 1, \tag{A.5}$$

$$-D^2 \le x_j \le D^2, \quad j = 0,1,\dots,D-1, \quad \varepsilon = 1.0 \times 10^{-6},$$

$$f(\mathbf{x}^*) = -D \cdot (D+4) \cdot (D-1)/6, \quad x_j^* = (j+1) \cdot (D-j).$$

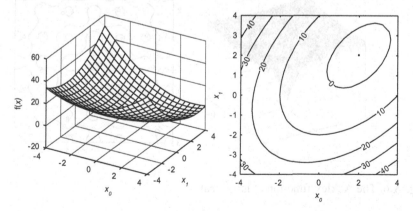

Fig. A.5. Neumaier's function number three

A.2 Unconstrained Multi-Modal Test Functions

Uni-modal functions can reveal how an algorithm responds to dimension, parameter dependence and disparities in step size, but few practical problems are so simple. The following multi-modal functions range from moderately challenging to very difficult depending in part on the dimension at which they are evaluated and on the amount of special knowledge about the function that an optimizer exploits. Not all functions can be evaluated at arbitrarily high dimensions.

A.2.1 Ackley

One of the most commonly cited multi-modal test functions is Ackley's function. At high dimension (e.g., $D \geq 30$), care must be taken with computer code to ensure a precise result. For example, the constant $e = 2.71828...$ in Eq. A.6 is best implemented as $e = \exp(1)$. Bounds are usually given as [−32, 32], but this book uses [−30, 30].

$$f(\mathbf{x}) = -20 \cdot \exp\left(-0.2 \cdot \sqrt{\frac{1}{D} \cdot \sum_{j=0}^{D-1} x_j^2}\right) - \exp\left(\frac{1}{D} \cdot \sum_{j=0}^{D-1} \cos(2\pi \cdot x_j)\right) + 20 + e, \tag{A.6}$$

$$-30 \leq x_j \leq 30, \quad j = 0,1,...,D-1,$$

$$f(\mathbf{x}^*) = 0, \quad x_j^* = 0, \quad \varepsilon = 1.0 \times 10^{-6}.$$

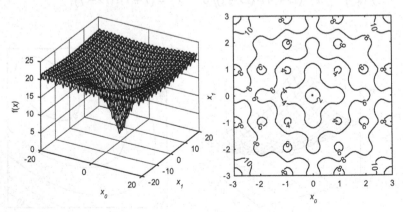

Fig. A.6. The Ackley function at large scale

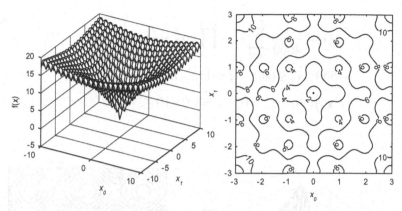

Fig. A.7. The Ackley function at small scale

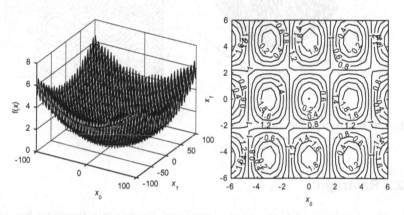

Fig. A.8. Griewangk's function

A.2.2 Griewangk

This mildly parameter-dependent function becomes relatively easier to solve as D increases. The summation term creates a parabolic bowl while the product of cosine terms generates the local optima. As D increases, the contribution from the cosine terms becomes less significant and the local basins of attraction become shallower. At the same time, the relative size of the optimal basin of attraction increases. See Whitley et al. (1996) for details. It is not uncommon that this function will require a relatively large population to forestall premature convergence.

$$f(\mathbf{x}) = \frac{1}{4000} \cdot \sum_{j=0}^{D-1} x_j^2 - \prod_{j=0}^{D-1} \cos\left(\frac{x_i}{\sqrt{j+1}}\right) + 1, \qquad \text{(A.7)}$$

$$-600 \le x_j \le 600, \quad j = 0,1,...,D-1,$$

$$f(\mathbf{x}^*) = 0, \quad x_j^* = 0, \quad \varepsilon = 1.0 \times 10^{-6}.$$

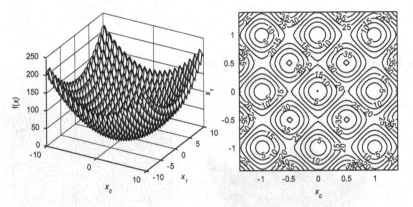

Fig. A.9. Rastrigin's function

A.2.3 Rastrigin

Like the Ackley and Griewangk functions, Rastrigin's function has many local optima arrayed on the side of a larger bowl-shaped depression. This function is separable as written and easily solved by methods that can exploit decomposable functions. It is *much* harder to solve when rotated. Like Rosenbrock's function, Rastrigin's function is a generalization of a two-dimensional function. Like the Ackley and Griewangk's functions, Rastrigin's function is symmetric about its solution. Optimizers that search the vicinity of the mean population vector will do well on these symmetric functions because, like the local minima, the population will also be symmetrically distributed.

$$f(\mathbf{x}) = \sum_{j=0}^{D-1} \left(x_j^2 - 10 \cdot \cos\left(2\pi \cdot x_j\right) + 10 \right), \tag{A.8}$$

$$-5.12 \le x_j \le 5.12, \quad j = 0,1,...,D-1,$$

$$f(\mathbf{x}^*) = 0, \quad x_j^* = 0, \quad \varepsilon = 1.0 \times 10^{-6}.$$

A.2.4 Salomon

The landscape for this parameter-dependent function resembles a pond with ripples. Because this function is symmetric, methods that search the vicinity of the population's mean vector will likely perform well.

$$f(\mathbf{x}) = -\cos\left(2\pi \cdot \|\mathbf{x}\|\right) + 0.1 \cdot \|\mathbf{x}\| + 1, \tag{A.9}$$

$$\|\mathbf{x}\| = \sqrt{\sum_{j=0}^{D-1} x_j^2},$$

$$-100 \le x_j \le 100, \quad j = 0,1,...,D-1,$$

$$f(\mathbf{x}^*) = 0, \quad x_j^* = 0, \quad \varepsilon = 1.0 \times 10^{-6}.$$

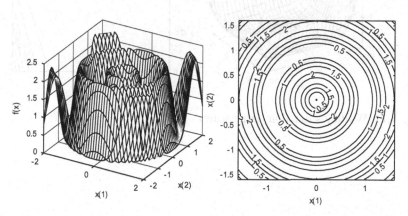

Fig. A.10. Salomon's function

A.2.5 Whitley

Whitley's function is a composition of the Griewangk and Rosenbrock functions. This implementation uses the unweighted full matrix expansion detailed in Whitley et al. (1996). This function's landscape resembles Rosenbrock's function at large scale and Griewangk's function at small scale.

$$f(\mathbf{x}) = \sum_{k=0}^{D-1}\sum_{j=0}^{D-1}\left(\frac{y_{j,k}^2}{4000} - \cos(y_{j,k}) + 1\right),$$ (A.10)

$$y_{j,k} = 100 \cdot \left(x_k - x_j^2\right)^2 + \left(1 - x_j\right)^2,$$

$$-100 \le x_j \le 100, \quad j = 0,1,...,D-1,$$

$$f(\mathbf{x}^*) = 0, \quad x_j^* = 1, \quad \varepsilon = 1.0 \times 10^{-6}.$$

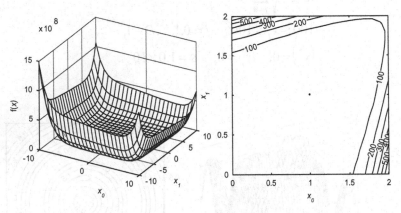

Fig. A.11. Whitley's function (large scale)

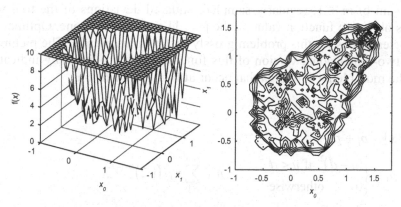

Fig. A.12. Whitley's function with values above 10 clipped

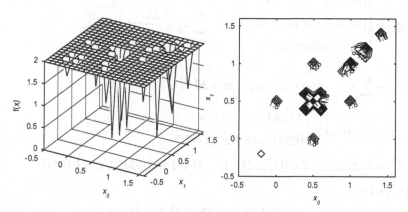

Fig. A.13. Whitley's function with values above 2 clipped

A.2.6 Storn's Chebyshev

The goal of this 2nd ICEO problem (Second ICEO 1997) is to find the coefficients of a polynomial such that the value of the polynomial oscillates between 1 and −1 as its argument, z, varies in the same range. In addition, the polynomial's value is also constrained when $z = 1.2$ and $z = -1.2$. The solution gives the coefficients of a Chebyshev polynomial. The coefficients for a Chebyshev polynomial of degree $D - 1$ can be expressed recursively as $T_{D+1}(z) = 2z \cdot T_D(z) - T_{D-1}(z)$, $D > 0$ and odd, $T_0(z) = 1$, $T_1(z) = z$. The objective function is designed as a three-term error function. The term, p_k, is the sum of $m + 1$, regularly sampled, squared deviations of the trial vector's objective function value in the $[-1, 1]$ containment zone. Optimal parameter values for this problem grossly differ in magnitude. The picture of the two-dimensional version of this function does not give any indication of the multiple local optima that occur at higher dimensions.

$$f(\mathbf{x}) = p_1 + p_2 + p_3, \qquad (A.11)$$

$$p_1 = \begin{cases} (u-d)^2 & \text{if } u < d \\ 0 & \text{otherwise} \end{cases}, \qquad u = \sum_{j=0}^{D-1} x_j \cdot (1.2)^{D-1-j},$$

$$p_2 = \begin{cases} (v-d)^2 & \text{if } v < d \\ 0 & \text{otherwise} \end{cases}, \qquad v = \sum_{j=0}^{D-1} x_j \cdot (-1.2)^{D-1-j},$$

$$p_k = \begin{cases} (w_k - 1)^2 & \text{if } w_k > 1 \\ (w_k + 1)^2 & \text{if } w_k < -1, \quad w_k = \sum_{j=0}^{D-1} x_j \cdot \left(\frac{2k}{m} - 1\right)^{D-1-j}, \\ 0 & \text{otherwise} \end{cases}$$

$$p_3 = \sum_{k=0}^{m} p_k, \quad k = 0,1,\dots,m, \quad m = 32 \cdot D,$$

$$d = T_{D-1}(1.2) \approx \begin{cases} 72.661 & \text{for } D = 9 \\ 10558.145 & \text{for } D = 17 \end{cases}$$

$$-2^D \le x_j \le 2^D, \quad j = 0,1,\dots,D-1, \quad D > 1 \text{ and odd}, \quad \varepsilon = 1.0 \times 10^{-8}$$

$$f(\mathbf{x}^*) = 0,$$

$$\mathbf{x}^* = \begin{cases} [128, 0, -256, 0, 160, 0, -32, 0, 1] & \text{for } D = 9 \\ [32768, 0, -131072, 0, 212992, 0, -180224, \\ \quad 0, 84480, 0, -21504, 0, 2688, 0, -128, 0, 1] & \text{for } D = 17. \end{cases}$$

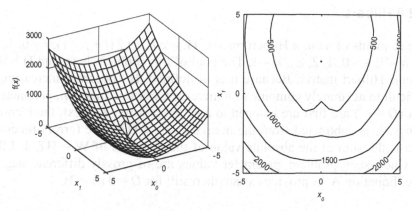

Fig. A.14. Storn's Chebyshev polynomial fitting problem

A.2.7 Lennard-Jones

This problem is based on the Lennard-Jones atomic potential energy function. The goal is to position n atoms in three-dimensional space to minimize their total potential energy. Since neither the cluster's position nor its orientation is specified, optimal parameter values are not unique.

$$f(\mathbf{x}) = \sum_{i=0}^{n-2} \sum_{j=i+1}^{n-1} \left(\frac{1}{d_{i,j}^2} - \frac{2}{d_{i,j}} \right), \quad d_{i,j} = \left(\sum_{k=0}^{2} \left(x_{3i+k} - x_{3j+k} \right)^2 \right)^3, \quad \text{(A.12)}$$

$$-2 \le x_j \le 2, \quad j = 0,1,...,D-1, \quad D = 3 \cdot n, \quad n = 2,3,... \quad \varepsilon = 0.01.$$

Table A.1. Optimal function values for $n=2$ to $n=19$ "atoms"

n	$f(\mathbf{x}^*)$	n	$f(\mathbf{x}^*)$
2	−1.0	11	−37.967600
3	−3.0	12	−44.326801
4	−6.0	13	−47.845157
5	−12.712062	14	−52.322627
6	−16.505384	15	−56.815742
7	−19.821489	16	−61.317995
8	−24.113360	17	−66.530949
9	−28.422532	18	−72.659782
10	−32.765970	19	−77.177704

A.2.8 Hilbert

The elements of an $n \times n$ Hilbert matrix, $\mathbf{H}$, are $h_{i,j}=1 / (i + j + 1)$, $i = 0, 1, 2,$ ..., $n - 1, j = 0, 1, 2, ..., n - 1$. The goal of this problem is to find $\mathbf{H}^{-1}$, the inverse Hilbert matrix. Because it is ill defined, $\mathbf{H}^{-1}$ becomes increasingly difficult to accurately compute as n increases. For this function, parameters in $\mathbf{x}$ $(D = n^2)$ are first are mapped to a square matrix, $\mathbf{Z}$. Next, the identity matrix, $\mathbf{I}$, is subtracted from the matrix product $\mathbf{HZ}$. The (error) function returns the sum of the absolute value of the elements of $\mathbf{W} = \mathbf{HZ}–\mathbf{I}$. Like the Chebyshev problem, parameter values are of grossly different magnitude. Equation A.13 provides a sample result for $D = 9$ ($n = 3$).

$$f(\mathbf{x}) = \sum_{i=0}^{n-1}\sum_{k=0}^{n-1}|w_{i,k}|, \tag{A.13}$$

$$\mathbf{HZ} - \mathbf{I} = \mathbf{W} = (w_{i,k}), \quad \mathbf{I} = \begin{bmatrix} 1 & 0 & ... & 0 \\ 0 & 1 & ... & 0 \\ & \vdots & & \\ 0 & 0 & ... & 1 \end{bmatrix}$$

$$\mathbf{H} = (h_{i,k}), \quad h_{i,k} = \frac{1}{i+k+1}, \quad i,k = 0,1,...,n-1, \quad D = n^2,$$

$$\mathbf{Z} = (z_{i,k}), \quad z_{i,k} = x_{i+nk},$$

$$-2^D \le x_j \le 2^D, \quad j = 0,1,...,D-1, \quad \varepsilon = 1.0 \times 10^{-8},$$

$$f(\mathbf{x}^*) = 0,$$

$$\mathbf{Z}^* = \begin{bmatrix} 9 & -36 & 30 \\ -36 & 192 & -180 \\ 30 & -180 & 180 \end{bmatrix}, \quad \text{for } n = 3.$$

A.2.9 Modified Langerman

This 2nd ICEO function (Second ICEO 1997) function relies on a vector ($\mathbf{c}$ in Table A.2) and a matrix ($\mathbf{A}$ in Table A.3) of real-valued constants. The vector, $\mathbf{c}$, contains thirty constants, while $\mathbf{A}$ is a matrix that contains the coordinates of thirty points in ten dimensions. Points are indexed by rows and coordinates are indexed by columns, e.g., numbers in the kth row are the coordinates of the point $\mathbf{A}_k$, $k = 0, 1, 2, ..., 29$. The optimum is the point in **A** that has the lowest corresponding value of **c**. Although origi-nally designed to use all thirty points in **A**, this implementation, like the code posted for the 2nd ICEO, uses only the first five. Data for both **c** and **A** are available on the CD-ROM that accompanies this book.

$$f(\mathbf{x}) = \sum_{k=0}^{m-1} c_k \cdot \left(\exp\left(\frac{-\|\mathbf{x} - \mathbf{A}_k\|^2}{\pi} \right) \cdot \cos\left(\pi \cdot \|\mathbf{x} - \mathbf{A}_k\|^2 \right) \right), \tag{A.14}$$

$$\|\mathbf{x} - \mathbf{A}_k\|^2 = \sum_{j=0}^{D-1} (x_j - a_{k,j})^2, \quad k = 0,1,\dots m - 1 \le 29,$$

$$\mathbf{c} = (c_k), \quad \mathbf{A} = (a_{k,j}),$$

$$0 \le x_j \le 10, \quad j = 0,1,\dots,D-1, \quad D \le 10, \quad \varepsilon = 0.001,$$

$$\text{for } m = 5 : f(\mathbf{x}^*) = \mathbf{c}_5 = -0.96500, \quad \mathbf{x}^* = \mathbf{A}_5.$$

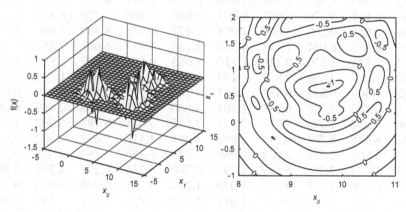

Fig. A.15. The Modified Langerman function

Table A.2. Values for $\mathbf{c} = (c_k)$

k	c_k	k	c_k	k	c_k	k	c_k	k	c_k
0	0.806	6	0.524	12	0.463	18	0.828	24	0.332
1	0.517	7	0.902	13	0.714	49	0.964	25	0.817
2	0.100	8	0.531	14	0.352	20	0.789	26	0.632
3	0.908	9	0.876	15	0.869	21	0.360	27	0.883
4	0.965	10	0.462	16	0.813	22	0.369	28	0.608
5	0.669	11	0.491	17	0.811	23	0.992	29	0.326

Table A.3. Values for $\mathbf{A}=(a_{j,k})$. The columns are counted by j (parameter index) while the points, $\mathbf{A}_k$, are numbered by row and are counted by k.

9.681	0.667	4.783	9.095	3.517	9.325	6.544	0.211	5.122	2.020
9.400	2.041	3.788	7.931	2.882	2.672	3.568	1.284	7.033	7.374
8.025	9.152	5.114	7.621	4.564	4.711	2.996	6.126	0.734	4.982
2.196	0.415	5.649	6.979	9.510	9.166	6.304	6.054	9.377	1.426
8.074	8.777	3.467	1.863	6.708	6.349	4.534	0.276	7.633	1.567
7.650	5.658	0.720	2.764	3.278	5.283	7.474	6.274	1.409	8.208
1.256	3.605	8.623	6.905	4.584	8.133	6.071	6.888	4.187	5.448
8.314	2.261	4.224	1.781	4.124	0.932	8.129	8.658	1.208	5.762
0.226	8.858	1.420	0.945	1.622	4.698	6.228	9.096	0.972	7.637
7.305	2.228	1.242	5.928	9.133	1.826	4.060	5.204	8.713	8.247
0.652	7.027	0.508	4.876	8.807	4.632	5.808	6.937	3.291	7.016
2.699	3.516	5.874	4.119	4.461	7.496	8.817	0.690	6.593	9.789
8.327	3.897	2.017	9.570	9.825	1.150	1.395	3.885	6.354	0.109
2.132	7.006	7.136	2.641	1.882	5.943	7.273	7.691	2.880	0.564
4.707	5.579	4.080	0.581	9.698	8.542	8.077	8.515	9.231	4.670
8.304	7.559	8.567	0.322	7.128	8.392	1.472	8.524	2.277	7.826
8.632	4.409	4.832	5.768	7.050	6.715	1.711	4.323	4.405	4.591
4.887	9.112	0.170	8.967	9.693	9.867	7.508	7.770	8.382	6.740
2.440	6.686	4.299	1.007	7.008	1.427	9.398	8.480	9.950	1.675
6.306	8.583	6.084	1.138	4.350	3.134	7.853	6.061	7.457	2.258
0.652	2.343	1.370	0.821	1.310	1.063	0.689	8.819	8.833	9.070
5.558	1.272	5.756	9.857	2.279	2.764	1.284	1.677	1.244	1.234
3.352	7.549	9.817	9.437	8.687	4.167	2.570	6.540	0.228	0.027
8.798	0.880	2.370	0.168	1.701	3.680	1.231	2.390	2.499	0.064
1.460	8.057	1.336	7.217	7.914	3.615	9.981	9.198	5.292	1.224
0.432	8.645	8.774	0.249	8.081	7.461	4.416	0.652	4.002	4.644
0.679	2.800	5.523	3.049	2.968	7.225	6.730	4.199	9.614	9.229
4.263	1.074	7.286	5.599	8.291	5.200	9.214	8.272	4.398	4.506
9.496	4.830	3.150	8.270	5.079	1.231	5.731	9.494	1.883	9.732
4.138	2.562	2.532	9.661	5.611	5.500	6.886	2.341	9.699	6.500

A.2.10 Shekel's Foxholes

This 2nd ICEO version of the Shekel's foxholes function (Second ICEO 1997) also relies on the set of points listed in $\mathbf{A}$ and on the constants in $\mathbf{c}$, but unlike the Modified Langerman function, this function uses all thirty points. Minima for both $D = 5$ and $D = 10$ are provided below. This function is hard for optimizers that tend to prematurely converge.

$$f(\mathbf{x}) = -\sum_{k=0}^{m-1} \frac{1}{\|\mathbf{x} - \mathbf{A}_k\|^2 - c_k}, \tag{A.15}$$

$$\mathbf{c} = (c_k), \quad \mathbf{A} = (a_{j,k}), \quad k = 0,1,\dots,m-1, \quad m = 30,$$

$$0 \le x_j \le 10, \quad j = 0,1,\dots,D-1, \quad D \le 10, \quad \varepsilon = 0.01,$$

$$f(\mathbf{x}^*) = \begin{cases} -10.4056 & \text{for } D = 5 \\ -10.2088 & \text{for } D = 10, \end{cases}$$

$$\mathbf{x}^* = \mathbf{A}_3 \quad \text{for } D = 5, 10.$$

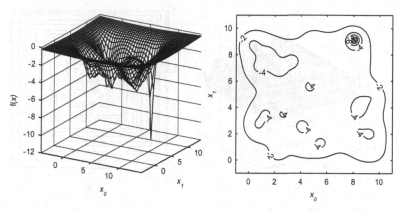

Fig. A.16. The Shekel's foxholes function

A.2.11 Odd Square

This 2nd ICEO function (Second ICEO 1997) resembles Salomon's function except that the ripples are rectangular, not circular. Because the Odd Square is symmetric about the solution, methods that search the vicinity of the population's mean vector will likely do well on this problem. In Eq. A.16, d is D times the square of the single, largest coordinate difference between the trial vector and the center point, **b**.

$$f(\mathbf{x}) = -\exp\left(\frac{-d}{2\pi}\right) \cdot \cos(\pi \cdot d) \cdot \left(1 + \frac{0.02 \cdot h}{d + 0.01}\right), \qquad (A.16)$$

$$d = D \cdot \max\left((x_j - b_j)^2\right), \quad h = \sum_{j=0}^{D-1}(x_j - b_j)^2,$$

$$-5 \cdot \pi \le x_j \le 5 \cdot \pi, \quad j = 0,1,...,D-1, \quad D \le 20, \quad \varepsilon = 0.01,$$

$$f(\mathbf{x}^*) = -1.14383, \quad \mathbf{x}^* = \text{many solutions near } \mathbf{b},$$

$$\mathbf{b} = [1, 1.3, 0.8, -0.4, -1.3, 1.6, -0.2, -0.6, 0.5, 1.4,$$
$$1, 1.3, 0.8, -0.4, -1.3, 1.6, -0.2, -0.6, 0.5, 1.4].$$

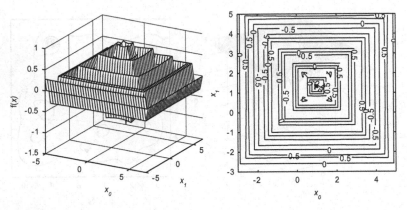

Fig. A.17. The Odd Square function

A.2.12 Katsuura

To be computed accurately, this function needs a floating-point format that supports more than 32 bits of precision when $m \ge 32$. The function "nint()" returns the nearest integer to the argument.

$$f(\mathbf{x}) = \prod_{j=0}^{D-1}\left(1 + (j+1) \cdot \sum_{k=1}^{m} \text{nint}(2^k \cdot x_j) \cdot 2^{-k}\right), \quad k = 0,1,...,m = 32, \qquad (A.17)$$

$$-1000 \le x_j \le 1000, \quad j = 0,1,...,D-1,$$

$$f(\mathbf{x}^*) = 1, \quad x_j = 0, \quad \varepsilon = 1.0 \times 10^{-6}.$$

A.3 Bound-Constrained Test Functions

A.3.1 Schwefel

This classic test function has a solution that lies on a coordinate system diagonal. In this version, the objective function is normalized by D so that $f(\mathbf{x}^*)$ is the same regardless of dimension. Success here can depend heavily on how bound constraints are handled. This function is separable.

$$f(\mathbf{x}) = -\frac{1}{D}\sum_{j=0}^{D-1} x_j \cdot \sin\left(\sqrt{|x_j|}\right),$$

(A.18)

$$-500 \le x_j \le 500, \quad j = 0,1,...,D\text{-}1, \quad \varepsilon = 0.01,$$

$$f(\mathbf{x}^*) = -418.983, \quad x_j^* = 420.968746.$$

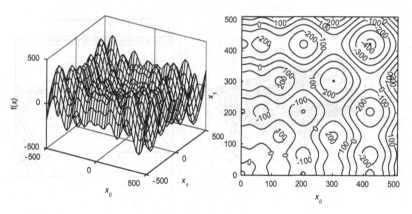

Fig. A.18. Schwefel's function

A.3.2 Epistatic Michalewicz

This 2nd ICEO function (Second ICEO 1997) also has a solution that lies near the limits of the allowed search space.

$$f(x) = \sum_{j=0}^{D-1} \sin(y_j) \cdot \left(\sin\left(\frac{(j+1) \cdot y_j^2}{\pi} \right) \right)^{2m}, \quad m = 10, \tag{A.19}$$

$$y_j = \begin{cases} x_j \cdot \cos\left(\dfrac{\pi}{6}\right) - x_{j+1} \cdot \sin\left(\dfrac{\pi}{6}\right) & \text{if } (j+1)\bmod(2) = 1 \\[2ex] x_{j-1} \cdot \sin\left(\dfrac{\pi}{6}\right) + x_j \cdot \cos\left(\dfrac{\pi}{6}\right) & \text{if } (j+1)\bmod(2) = 0, \ j \neq D-1 \\[2ex] y_{D-1} = x_{D-1} & \text{if } j = D-1 \end{cases}$$

$$0 \leq x_j \leq \pi, \quad j = 0,1,\dots,D-1, \quad D > 1,$$

$$f(\mathbf{x}^*) = \begin{cases} -4.68766 & \text{for } D = 5 \\ -9.66015 & \text{for } D = 10 \end{cases}$$

$$\mathbf{x}^* = \begin{cases} [2.693170, 0.258897, 2.074365, 1.022922, 1.720470] & \text{for } D = 5 \\ [2.693170, 0.258897, 2.074365, 1.022922, 2.275369, \\ \quad 0.500115, 2.137603, 0.793609, 2.818757, 1.570796] & \text{for } D = 10 \end{cases}$$

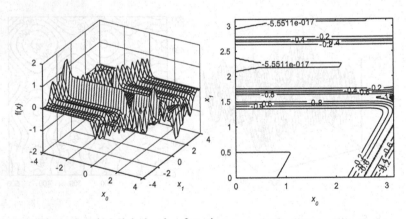

Fig. A.19. The epistatic Michalewicz function

A.3.3 Rana

This is one of the extended functions described in Whitley et al. (1996) in which a two-dimensional primitive function is evaluated with consecutive pairs of parameters, e.g., (0, 1), (1, 2), …, (D – 1, 0), so that the last term pairs the trial vector's first and last parameters (a "full-wrap" evaluation).

$$f(\mathbf{x}) = \sum_{j=0}^{D-1} x_j \cdot \sin(\alpha) \cdot \cos(\beta) + x_{(j+1)\bmod D} \cdot \cos(\alpha) \cdot \sin(\beta), \tag{A.20}$$

$$\alpha = \sqrt{|x_{j+1} + 1 - x_j|}, \quad \beta = \sqrt{|x_{j+1} + 1 + x_j|},$$

$$-512 \le x_j \le 512, \quad j = 0,1,...,D-1, \quad D > 1,$$

$$f(\mathbf{x}^*) = -511.708, \quad x_j^* = -512, \quad \varepsilon = 0.01.$$

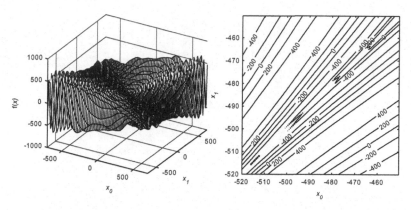

Fig. A.20. Rana's function

References

Michalewicz Z, Shoenauer M (1996) Evolutionary algorithms for constrained parameter optimization problems. Evolutionary Computation 4(1):1–32; the test problems are also available via the Internet at:
http://www.lut.fi/~jlampine/testset.pdf

Second ICEO (1997) Code for 2nd ICEO test functions is available via the Internet at: http://iridia.ulb.ac.be/~aroli/ICEO/Functions/Functions.html

Yao X, Liu Y (1997) Fast evolution strategies. In: Angeline PJ, Reynolds RG, McDonnell JR, Eberhart R (eds) Evolutionary programming VI, Lecture notes in computer science 1213, Springer, Berlin, pp 151–161

Whitley D, Mathias K, Rana S, Dzubera J (1996) Evaluating evolutionary algorithms. Artificial Intelligence 85:1–32

Index

Natural Computing Series

W.M. Spears: **Evolutionary Algorithms. The Role of Mutation and Recombination.**
XIV, 222 pages, 55 figs., 23 tables. 2000

H.-G. Beyer: **The Theory of Evolution Strategies.** XIX, 380 pages, 52 figs., 9 tables. 2001

L. Kallel, B. Naudts, A. Rogers (Eds.): **Theoretical Aspects of Evolutionary Computing.**
X, 497 pages. 2001

G. Păun: **Membrane Computing. An Introduction.** XI, 429 pages, 37 figs., 5 tables. 2002

A.A. Freitas: **Data Mining and Knowledge Discovery with Evolutionary Algorithms.**
XIV, 264 pages, 74 figs., 10 tables. 2002

H.-P. Schwefel, I. Wegener, K. Weinert (Eds.): **Advances in Computational Intelligence.**
Theory and Practice. VIII, 325 pages. 2003

A. Ghosh, S. Tsutsui (Eds.): **Advances in Evolutionary Computing. Theory and**
Applications. XVI, 1006 pages. 2003

L.F. Landweber, E. Winfree (Eds.): **Evolution as Computation.** DIMACS Workshop,
Princeton, January 1999. XV, 332 pages. 2002

M. Hirvensalo: **Quantum Computing.** 2nd ed., XI, 214 pages. 2004 (first edition
published in the series)

A.E. Eiben, J.E. Smith: **Introduction to Evolutionary Computing.** XV, 299 pages. 2003

A. Ehrenfeucht, T. Harju, I. Petre, D.M. Prescott, G. Rozenberg: **Computation in Living**
Cells. Gene Assembly in Ciliates. XIV, 202 pages. 2004

L. Sekanina: **Evolvable Components. From Theory to Hardware Implementations.**
XVI, 194 pages. 2004

G. Ciobanu, G. Rozenberg (Eds.): **Modelling in Molecular Biology.** X, 310 pages. 2004

R.W. Morrison: **Designing Evolutionary Algorithms for Dynamic Environments.**
XII, 148 pages, 78 figs. 2004

R. Paton[†], H. Bolouri, M. Holcombe, J.H. Parish, R. Tateson (Eds.): **Computation in Cells**
and Tissues. Perspectives and Tools of Thought. XIV, 358 pages, 134 figs. 2004

M. Amos: **Theoretical and Experimental DNA Computation.** XIV, 170 pages, 78 figs. 2005

M. Tomassini: **Spatially Structured Evolutionary Algorithms.** XIV, 192 pages, 91 figs.,
21 tables. 2005

G. Ciobanu, G. Păun, M.J. Pérez-Jiménez (Eds.): **Applications of Membrane Computing.**
X, 441 pages, 99 figs., 24 tables. 2006

K.V. Price, R.M. Storn, J.A. Lampinen: **Differential Evolution.** XX, 538 pages,
292 figs., 48 tables and CD-ROM. 2006

A. Brabazon, M. O'Neill: **Biologically Inspired Algorithms for Financial Modelling.**
XVI, 275 pages, 92 figs., 39 tables. 2006